AF361838

Wastewater Treatment: Current and Future Techniques

Wastewater Treatment: Current and Future Techniques

Editors

Amin Mojiri
Mohammed J.K. Bashir

MDPI • Basel • Beijing • Wuhan • Barcelona • Belgrade • Manchester • Tokyo • Cluj • Tianjin

Editors
Amin Mojiri
Hiroshima University
Japan

Mohammed J.K. Bashir
Universiti Tunku Abdul Rahman
Malaysia

Editorial Office
MDPI
St. Alban-Anlage 66
4052 Basel, Switzerland

This is a reprint of articles from the Special Issue published online in the open access journal *Water* (ISSN 2073-4441) (available at: https://www.mdpi.com/journal/water/special_issues/Wastewater_Techniques).

For citation purposes, cite each article independently as indicated on the article page online and as indicated below:

LastName, A.A.; LastName, B.B.; LastName, C.C. Article Title. *Journal Name* **Year**, *Volume Number*, Page Range.

ISBN 978-3-0365-3274-5 (Hbk)
ISBN 978-3-0365-3275-2 (PDF)

Contents

About the Editors

Amin Mojiri is currently working at Hiroshima University (Japan) as an Assistant Professor. He received his PhD from Universiti Sains Malaysia in 2015. His research interests include water and wastewater treatment, the microalgae-based system, anammox, water quality, emerging contaminants, and the application of biochar. With more than 8 years of experience in academia, he has published more than 70 papers in peer-reviewed journals. Moreover, he received several international awards. In 2017, he received three silver medals and one gold medal in the International Invention and Innovation Competitions for his projects related to the treatment of water and wastewater. Additionally, he was selected as an Excellent Young Researcher candidate by the Ministry of Education, Culture, Sport, Science and Technology (MEXT, Japan) in 2020.

Mohammed J.K. Bashir received his Masters and PhD degrees in Environmental Engineering from Universiti Sains Malaysia. He is currently serving as an Associate Professor and leads the Department of Environmental Engineering at Universiti Tunku Abdul Rahman, Malaysia. He is actively engaged in research and international collaborations and is serving as a Visiting Professor at the University of Indonesia, as well as an external expert at the National Centre of Scientific and Technical Evaluation (NCSTE) in Kazakhstan. He also serves as an editorial member and guest editor in various prestigious journals. Dr. Bashir has published more than 200 articles in high impact factor journals (WOS), books, and international conferences. Today his h-index is 34, with a total of 4350 citations. He has received more than 20 awards, including the best PhD research award from the School of Civil Engineering (Universiti Sains Malaysia, 2011), excellent teaching performance awards from UTAR between 2015 to 2021, top cited paper, a medal from the international exhibition in science and technology, and many others. He has secured various research funding grants in the area of wastewater treatment, waste management, resource recovery, and emerging contaminants. Dr. Bashir's expertise covers the treatment of water and wastewater, the identification and treatment of emerging contaminates, waste reutilization, and waste management technologies.

Preface to "Wastewater Treatment: Current and Future Techniques"

With the rapid growth in industrialization and urbanization, water contamination has worsened as a result of the incessant discharge of toxic substances into water bodies, which has become a worldwide problem. Therefore, one of the most vital challenges for sustainable development is to supply clean water. Additionally, one of the sustainable development goals of the United Nations emphasizes the access to water and sanitation for all. Consequently, the treatment and reuse of wastewater have become vital concepts in the attempt to improve water availability.

The main goal of this Special Issue is to address the existing knowledge gaps concerning the new techniques in water and wastewater treatment with maximal efficiency and minimal energy consumption. Due to the demands of clean water, it is expected that the new techniques in the removal of wide ranges and types of contaminants from water bodies will play a more prominent role in the future global water supply. Finally, we would like to express our appreciation to the staff at MDPI, the editorial team of Water, the assistant editor of this Special Issue, the talented authors, and professional reviewers.

Amin Mojiri, Mohammed J.K. Bashir
Editors

water

Editorial

Wastewater Treatment: Current and Future Techniques

Amin Mojiri [1,*] and Mohammed J. K. Bashir [2]

[1] Department of Civil and Environmental Engineering, Graduate School of Advanced Science and Engineering, Hiroshima University, Higashihiroshima 739-8527, Japan
[2] Department of Environmental Engineering, Engineering and Green Technology Faculty, Universiti Tunku Abdul Rahman, Kampar 31900, Malaysia; jkbashir@utar.edu.my
* Correspondence: amin.mojiri@gmail.com

1. Introduction

With the rapid growth in urbanization and industrialization, environmental contamination has worsened due to the incessant discharge of toxic substances into water bodies, which has become a worldwide problem [1]. Furthermore, the demand for water in domestic and industrial activities has significantly increased, which has accordingly increased the amount of wastewater that is released into sewage systems. Thus, the reuse and treatment of wastewater have become important concepts in the attempt to increase water availability [2]. The wastewater industry is in a state of transition [3] due to the recent wastewater effluent standards and emerging contaminants such as pharmaceutical and personal care products, and dyes in water bodies [4]. At present, several physicochemical methods (e.g., advanced oxidation process, adsorption, and membrane technologies), biological methods (e.g., activated sludge process, phytoremediation, bioremediation, and anammox), and hybrid methods have been developed to treat polluted water [1]. However, a treatment method with maximum efficiency in the removal of all kinds of contaminants is still far being realized. Moreover, the United Nations' sustainable development goal (https://www.un.org/sustainabledevelopment/water-and-sanitation/ (accessed on 25 January 2022)) emphasizes access to water and sanitation for all. All these issues led to the proposal of a Special Issue (SI) entitled, "Wastewater Treatment: Current and Future Techniques". This SI discusses state-of-the-art wastewater and water treatment technologies that could be used to develop a sustainable treatment method in the future. On this topic, studies have focused on measurements, modeling, and experiments under laboratory and field conditions.

2. Summary of the SI

Original research and review papers (12 papers in total) on advanced technologies applied to the treatment of industrial wastewater, domestic wastewater, and sludge were published after the peer-review process. The studies presented in this SI include the following themes.

One of the current main concerns is the emerging contaminants in water bodies. For instance, widespread water contamination with perfluoroalkyl and polyfluoroalkyl substances (PFASs) has become a great concern [5]. In this SI, Abunada et al. [5] monitored the concentrations of PFASs worldwide. Moreover, previous studies [6,7] have reported that conventional wastewater treatments have failed to remove emerging contaminants from water bodies. Therefore, researchers have tried to propose new systems with maximum performance in removing emerging pollutants. In a study published in this SI, 89.73% of the amount of polycyclic aromatic hydrocarbons was removed from water at optimal conditions using the ferrate (VI) oxidation process [8]. In another study, UV light and oxidizing disinfectants removed from 0% to 99.9% of the amounts of cetirizine, furosemide, diclofenac, losartan, venlafaxine, benzotriazole, and lamotrigine [9]. In addition, Alazaiza et al. [10]

Citation: Mojiri, A.; Bashir, M.J.K. Wastewater Treatment: Current and Future Techniques. *Water* 2022, 14, 448. https://doi.org/10.3390/w14030448

Received: 27 January 2022
Accepted: 31 January 2022
Published: 1 February 2022

Publisher's Note: MDPI stays neutral with regard to jurisdictional claims in published maps and institutional affiliations.

discussed the performance of several natural coagulants in eliminating pharmaceuticals and personal care products from water bodies. The use of these natural coagulants has been described as an efficient method of removing emerging micropollutants.

Furthermore, in recent years, water contamination with dyes, which are harmful organic pollutants, has become a serious issue. Consequently, the elimination of these contaminants from water is a global demand to ensure human and environmental safety. In one study, more than 80% of anionic dye reactive black 5 (RB5) was removed with hybrid hexadecylamine-impregnated chitosan powder-activated carbon beads [11]. In this study, adsorption data were fitted to the Freundlich and pseudo-second-order models.

Another toxic compound found in water bodies is ammonia, and ammonia contamination in wastewater and water bodies has become a major environmental problem [12]. Several techniques have been established for the treatment of nitrogen. Among these techniques, anaerobic ammonium oxidation (anammox) has received researchers' attention for nitrogen removal purposes. Anammox is a microbial procedure in which ammonia is oxidized to nitrogen gas, with nitrite as the electron acceptor [13]. Hosokawa et al. [13] studied the cometabolism of *Patescibacteria* with anammox in an anammox reactor. On the basis of their study, *Patescibacteria* might play an ecological role in supplying lactate and formate to other coexisting bacteria, supporting growth in the anammox reactor.

The discharge of heavy metals into the environment has significantly increased. The main source of heavy metal ions is the industrial effluents of various processing industries [14]. The toxicity of heavy metals has already been proven to be a major threat to humans and the environment [15]. Almost 75% of hexavalent chromium (Cr VI) was removed using rice husk. In the study by Bhattacharjee et al. [15], the adsorption data were more fitted to the Dubinin–Radushkevich and Langmuir models.

Wastewater and landfill leachate contain different organic and inorganic contaminants. Among the several techniques for removing a wide range of pollutants, membrane filtration could provide a suitable purification process [16]. By using a new polyvinylidene fluoride membrane synthesized by integrating powdered activated carbon, 35.3% of chemical oxygen demand, 48.7% of color, and 22% of ammonia were removed from landfill leachate [16]. In addition, different types of membrane techniques were discussed in terms of their performance in the treatment of poultry slaughterhouse wastewater [17]. Moreover, several treatment methods, such as membrane and biological methods, were discussed and compared by Gutu et al. in terms of their performance in the removal of organics and nutrients [18].

Finally, green and sustainable wastewater technologies (GSWTs) have recently attracted researchers' attention. GSWT represents a term that denotes sustainable and environmentally friendly approaches to wastewater treatment [19]. Nanoremediation and microalgae-based systems can be considered important GSWTs. Alazaiza et al. [20] mentioned the advantages of using nanoremediation technologies for remediation. In another study, microalgae harvesting with biopolymers was described by Ang et al. [21] as a sustainable algae-based system.

Author Contributions: Writing—original draft preparation, A.M.; writing—review and editing, M.J.K.B. All authors have read and agreed to the published version of the manuscript.

Funding: This research received no external funding.

Conflicts of Interest: The authors declare no conflict of interest.

References

1. Lin, R.; Li, Y.; Yong, T.; Cao, W.; Wu, J.; Shen, Y. Synergistic effects of oxidation, coagulation and adsorption in the integrated fenton-based process for wastewater treatment: A review. *J. Environ. Manag.* **2022**, *306*, 114460. [CrossRef] [PubMed]
2. Sakr, M.; Mohamed, M.M.; Maraqa, M.A.; Hamouda, M.A.; Aly Hassan, A.; Ali, J.; Jung, J. A critical review of the recent developments in micro–nano bubbles applications for domestic and industrial wastewater treatment. *Alex. Eng. J.* 2021, *in press*. [CrossRef]

3.   Soares, A. Wastewater treatment in 2050: Challenges ahead and future vision in a European context. *Environ. Sci. Ecotechnol.* **2020**, *2*, 100030. [CrossRef]

4.   Li, S.; Show, P.L.; Ngo, H.H.; Ho, S.-H. Algae-mediated antibiotic wastewater treatment: A critical review. *Environ. Sci. Ecotechnol.* **2022**, *in press*. [CrossRef]

5.   Abunada, Z.; Alazaiza, M.Y.D.; Bashir, M.J.K. An Overview of Per- and Polyfluoroalkyl Substances (PFAS) in the Environment: Source, Fate, Risk and Regulations. *Water* **2020**, *12*, 3590. [CrossRef]

6.   Mojiri, A.; Zhou, J.; Vakili, M.; Van Le, H. Removal performance and optimisation of pharmaceutical micropollutants from synthetic domestic wastewater by hybrid treatment. *J. Contam. Hydrol.* **2020**, *235*, 103736. [CrossRef]

7.   Ahmed, M.B.; Zhou, J.L.; Ngo, H.H.; Guo, W. Adsorptive removal of antibiotics from water and wastewater: Progress and challenges. *Sci. Total Environ.* **2015**, *532*, 112–126. [CrossRef]

8.   Haneef, T.; Mustafa, M.R.U.; Wan Yusof, K.; Isa, M.H.; Bashir, M.J.K.; Ahmad, M.; Zafar, M. Removal of Polycyclic Aromatic Hydrocarbons (PAHs) from Produced Water by Ferrate (VI) Oxidation. *Water* **2020**, *12*, 3132. [CrossRef]

9.   Ikonen, J.; Nuutinen, I.; Niittynen, M.; Hokajärvi, A.-M.; Pitkänen, T.; Antikainen, E.; Miettinen, I.T. Presence and Reduction of Anthropogenic Substances with UV Light and Oxidizing Disinfectants in Wastewater—A Case Study at Kuopio, Finland. *Water* **2021**, *13*, 360. [CrossRef]

10. Alazaiza, M.; Albahnasawi, A.; Ali, G.; Bashir, M.; Nassani, D.; Al Maskari, T.; Amr, S.; Abujazar, M. Application of Natural Coagulants for Pharmaceutical Removal from Water and Wastewater: A Review. *Water* **2022**, *14*, 140. [CrossRef]

11. Vakili, M.; Zwain, H.M.; Mojiri, A.; Wang, W.; Gholami, F.; Gholami, Z.; Giwa, A.S.; Wang, B.; Cagnetta, G.; Salamatinia, B. Effective Adsorption of Reactive Black 5 onto Hybrid Hexadecylamine Impregnated Chitosan-Powdered Activated Carbon Beads. *Water* **2020**, *12*, 2242. [CrossRef]

12. Azreen, I.; Lija, Y.; Zahrim, A.Y. Ammonia nitrogen removal from aqueous solution by local agricultural wastes. *IOP Conf. Ser. Mater. Sci. Eng.* **2017**, *206*, 012077. [CrossRef]

13. Hosokawa, S.; Kuroda, K.; Narihiro, T.; Aoi, Y.; Ozaki, N.; Ohashi, A.; Kindaichi, T. Cometabolism of the Superphylum Patescibacteria with Anammox Bacteria in a Long-Term Freshwater Anammox Column Reactor. *Water* **2021**, *13*, 208. [CrossRef]

14. Francis Xavier, L.; Money, B.K.; John, A.; Rohit, B. Removal of cadmium heavy metal ion using recycled black toner powder. *Mater. Today Proc.* **2021**, *in press*. [CrossRef]

15. Bhattacharjee, T.; Islam, M.; Chowdhury, D.; Majumdar, G. In-situ generated carbon dot modified filter paper for heavy metals removal in water. *Environ. Nanotechnol. Monit. Manag.* **2021**, *16*, 100582. [CrossRef]

16. Abuabdou, S.M.A.; Jaffari, Z.H.; Ng, C.-A.; Ho, Y.-C.; Bashir, M.J.K. A New Polyvinylidene Fluoride Membrane Synthesized by Integrating of Powdered Activated Carbon for Treatment of Stabilized Leachate. *Water* **2021**, *13*, 2282. [CrossRef]

17. Fatima, F.; Du, H.; Kommalapati, R.R. Treatment of Poultry Slaughterhouse Wastewater with Membrane Technologies: A Review. *Water* **2021**, *13*, 1905. [CrossRef]

18. Gutu, L.; Basitere, M.; Harding, T.; Ikumi, D.; Njoya, M.; Gaszynski, C. Multi-Integrated Systems for Treatment of Abattoir Wastewater: A Review. *Water* **2021**, *13*, 2462. [CrossRef]

19. Paruch, A.M.; Mæhlum, T.; Eltun, R.; Tapu, E.; Spinu, O. Green wastewater treatment technology for agritourism business in Romania. *Ecol. Eng.* **2019**, *138*, 133–137. [CrossRef]

20. Alazaiza, M.Y.D.; Albahnasawi, A.; Ali, G.A.M.; Bashir, M.J.K.; Copty, N.K.; Amr, S.S.A.; Abushammala, M.F.M.; Al Maskari, T. Recent Advances of Nanoremediation Technologies for Soil and Groundwater Remediation: A Review. *Water* **2021**, *13*, 2186. [CrossRef]

21. Ang, T.-H.; Kiatkittipong, K.; Kiatkittipong, W.; Chua, S.-C.; Lim, J.W.; Show, P.-L.; Bashir, M.J.K.; Ho, Y.-C. Insight on Extraction and Characterisation of Biopolymers as the Green Coagulants for Microalgae Harvesting. *Water* **2020**, *12*, 1388. [CrossRef]

Review

An Overview of Per- and Polyfluoroalkyl Substances (PFAS) in the Environment: Source, Fate, Risk and Regulations

Ziyad Abunada [1], Motasem Y. D. Alazaiza [2] and Mohammed J. K. Bashir [3,*

[1] School of Engineering & Technology, Central Queensland University, Coastal Marine Ecosystems Research Centre (CMERC), 120 Spencer St., Melbourne, QLD 3000, Australia; z.abunada@cqu.edu.au

[2] Department of Civil and Environmental Engineering, College of Engineering, A'Sharqiyah University (ASU), Ibra 400, Oman; my.azaiza@gmail.com

[3] Department of Environmental Engineering, Faculty of Engineering and Green Technology (FEGT), Universiti Tunku Abdul Rahman, Kampar 31900, Malaysia

* Correspondence: jkbashir@utar.edu.my

Received: 3 November 2020; Accepted: 17 December 2020; Published: 21 December 2020

Abstract: The current article reviews the state of art of the perfluoroalkyl and polyfluoroalkyl substances (PFASs) compounds and provides an overview of PFASs occurrence in the environment, wildlife, and humans. This study reviews the issues concerning PFASs exposure and potential risks generated with a focus on PFAS occurrence and transformation in various media, discusses their physicochemical characterization and treatment technologies, before discussing the potential human exposure routes. The various toxicological impacts to human health are also discussed. The article pays particular attention to the complexity and challenging issue of regulating PFAS compounds due to the arising uncertainty and lack of epidemiological evidence encountered. The variation in PFAS regulatory values across the globe can be easily addressed due to the influence of multiple scientific, technical, and social factors. The varied toxicology and the insufficient definition of PFAS exposure rate are among the main factors contributing to this discrepancy. The lack of proven standard approaches for examining PFAS in surface water, groundwater, wastewater, or solids adds more technical complexity. Although it is agreed that PFASs pose potential health risks in various media, the link between the extent of PFAS exposure and the significance of PFAS risk remain among the evolving research areas. There is a growing need to address the correlation between the frequency and the likelihood of human exposure to PFAS and the possible health risks encountered. Although USEPA (United States Environmental Protection Agency) recommends the 70 ng/L lifetime health advisory in drinking water for both perfluorooctane sulfonate (PFO) perfluorooctanoic acid (PFOA), which is similar to the Australian regulations, the German Ministry of Health proposed a health-based guidance of maximum of 300 ng/L for the combination of PFOA and PFOS. Moreover, there are significant discrepancies among the US states where the water guideline levels for the different states ranged from 13 to 1000 ng L^{-1} for PFOA and/or PFOS. The current review highlighted the significance of the future research required to fill in the knowledge gap in PFAS toxicology and to better understand this through real field data and long-term monitoring programs.

Keywords: poly-fluoroalkyl substances (PFASs); toxicology; PFAS health risk; regulatory values

---

1. Introduction

Widespread surface and groundwater contamination with perfluoroalkyl and polyfluoroalkyl substances (PFASs) has become of great concern in the last few years. PFAS was first realized in the globe through the identification of perfluorooctane sulfonic acid, $C_8F_{17}SO_3H$ (PFOS), in wildlife [1,2].

PFASs have recently received increasing global attention because of their persistence and toxicity in the environment, bioaccumulation potential, and possible adverse health impacts [3]. PFAS are commonly have an aliphatic carbon composition in which hydrogen molecules have been replaced by fluorine completely (prefix: per-) or partially (prefix: poly-) [4]. These compounds are characterized by their highly polar and strong carbon fluorine bonds [5]. They are considered as highly fluorinated surfactants that have been applied in numerous industrial applications and manufactured goods including food packaging, firefighting foams, clothes and protective coatings for fabrics and carpets, electronics and fluoropolymer manufacturing [1,2,5–9]. The most extensively produced and frequently detected PFASs in the environments are perfluorooctanoic acid, $C_7F_{15}COOH$ (PFOA) and perfluorooctane sulfonic acid, $C8F17SO3H$ (PFOS) [1]. PFASs have been discovered in different environmental compartments, including water, sediment organisms, and air [6,10–12].

PFAS has been a serious concern to industry, governments scientists, and even to the public worldwide [13]. It has been detected in various aquatic matrixes, including rain, snow, groundwater, tap water, lakes, and rivers with the C8-based substances PFOS and PFOA typically being the dominating compounds [4,9,14]. PFAS degradation products can be freely mobile in water, soil, and air, and can be extremely resistant to breakdown by different processes. The complexity of measuring PFAS in various media, and the associated unknown risks are among the challenges facing the current regulatory bodies [4]. Typical concentrations of PFASs in water are very low, however, higher concentrations of (mg/L) have been observed in surface and groundwater after firefighting activities closed to fluorochemical manufacturing facilities. PFASs spread worldwide has triggered the governmental concern towards regulating the exposure and spread of PFASs [15,16]. Although there is enough evidence about the negative impacts of PFAS on human and animal health, the scale of the risk imposed by PFAS compounds is not fully understood. The current regulations tend to address the potential risk limit for various wildlife where the PFASs persistence, bioaccumulation potential, and toxicity (PBT) raise a great concern [6]. Several studies have reviewed various aspects related to PFASs fate and behavior in different environments. They also reviewed the sources and occurrence of PFOA in drinking water, toxicokinetic, and health impacts [17–21]. Other reviews on PFASs have discussed different aspects such as environmental biodegradation of PFASs, PFASs removal from drinking water treatment plants, wastewater treatment plants and PFASs transformation in landfills [22–24]. The authors are aware of the developing research concerning PFAS and the many reviews investigating the PFAS human exposure, fate, transport, accumulation, health hazard and guidelines [2,5,16,25–31]. The current mini review investigates the PFAS occurrence in collective all geo-environmental compartments and is the first to collate the various international PFAS standards in one article. The current study reviews existing publications in the field of PFAS and aims to: (i) summarize the recent publication in the field of PFAS and ensure easy access of the research on the occurrence and behavior of PFASs in various environments, (ii) to identify knowledge gaps in the PFAS field, particularly the discrepancies in the current prevailing legislation and practices across various countries, and (iii) to present the key future research directions to better address the PFAS issue.

2. The Developing Trend in PFAS Research

PFAS was first detected in the early 1950s in the form of PFAO and PFOS as a part of the Teflon production process [4]. A few decades later, in the early 1990s, and due to the development of the analytical techniques and instrumentation advancement, PFAS was detected in environment at low concentrations. The investigation on PFASs was evolved in early 2000s when a voluntary phase- out in production of the parent chemical to PFOS was undertaken [32]. Due to the significant development of PFAS production in 2009, more attention was paid to limiting PFAS production whereby many researchers and institutions investigated the source, fate, and impact of PFAS compounds (Figure 1). The related articles published recently were extracted from Scopus based on the following keywords: perfluoroalkyl and polyfluoroalkyl substances, assessment of perfluoroalkyl substances, accumilation toxicity of PFASs, treatment of perfluoroalkyl and polyfluoroalkyl. In total, 122 articles were selected based on their relevancy, scope, and depth of discussion.

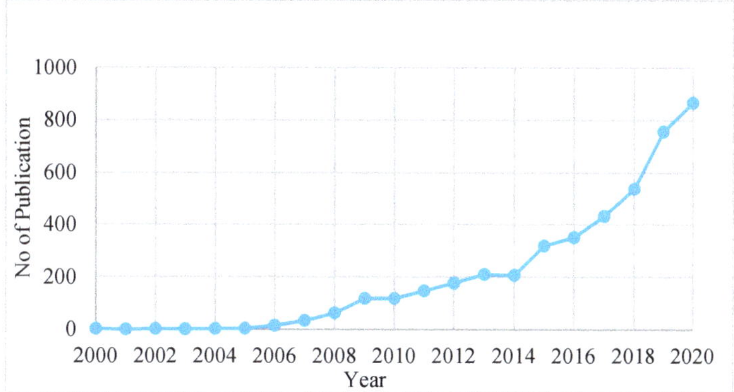

Figure 1. Total number of documents published, concerning PFAS from 2000 to 2020 (data extracted from Scopus; October 2020).

As a result, the research trend reflected by the number of publications concerning perfluoroalkyl and polyfluoroalkyl compounds have been augmented considerably in the last two decades and resulted in a much better understanding of the adverse health effects related to the exposure of PFOA and PFOS [33].

Moreover, it was observed that developed countries have invested much more than other countries on PFAS research, which was reflected by the research funding number of publications as shown in Figure 2. Nevertheless, not all countries share the same concerns and interests due to many various reasons. Some of these reasons are driven by economic factors, where the PFAS is still not on the priority list, as shown in Figure 2. Other factors are associated with industrial and sociopolitical factors, where PFAS form a significant part in industry.

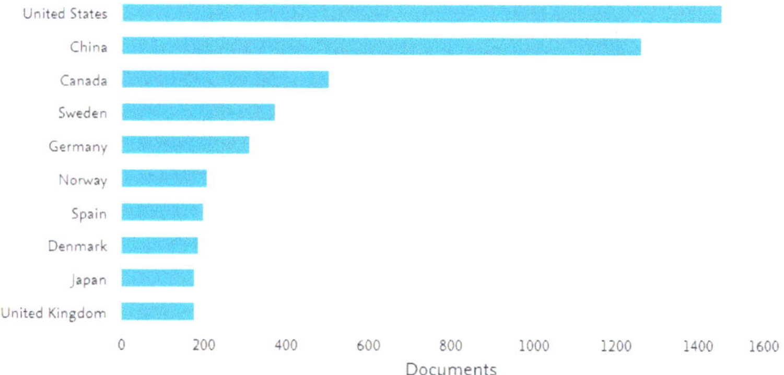

Figure 2. Total number of documents published by top 10 countries concerning PFAS from 2000 to 2020 (data extracted from Scopus; October 2020).

3. PFASs Occurrence and Transformation

Due to the strong C-F bonds in PFASs, they are highly stable and PFASs are unlikely to degrade easily in the environmental matrices [34]. PFASs in the environment has been resulted from several sources. The sources of PFASs in groundwater, drinking water, and surface water could be categorized into (i) point as well as (ii) diffuse sources. Wastewater treatment plants are considered as the most common point sources of PFASs to surface water. Other forms of point sources have been found to have a high impact on surface water in the USA including industrial pollution from PFAS production sites. In addition, it was observed that high concentrations of PFASs can be existed in surface water closed to commercials and military places due to the usage of aqueous film forming foams (AFFF) that contains PFASs. Landfills are considered as important point sources for PFASs in groundwater that comprise PFAS polluted waste in China, and that they could cause a hazard for tap water pollution [35]. Also, in Europe, landfills have not been supported to a significant degree in terms of their capacity for groundwater PFAS contamination. The following subsections illustrate the different occurrence of PFASs in the environments.

3.1. PFAS in Environment

3.1.1. PFASs in Water

The level of PFAS as well as their fate in water bodies have been investigated be many researchers as water represents one of the main pathways for human exposure to [36–41]. The consistent detection of PFAS compounds such as perfluoroalkane sulfonates (PFSAs) and perfluoroalkyl carboxylates (PFCAs) in tap water samples at various locations has raised the concern over their potential health risk [37]. Such detection has been reported in drinking water samples in Europe, China, Malaysia, Thailand, USA, Singapore, Vietnam, and Brazil [4,14,20,36,37,40–48]. Other compounds, including perfluoro hexanoic acid (PFHxA) [38], perfluorooctanesulfonamide (PFOSA) [32], and perfluorinated phosphonic acids (PFPAs) [49], are also among the commonly detected compounds in water samples. This relatively persistent level of PFAS exposure increases with the increased drinking water contamination events where it was estimated that the average human daily PFAS intake ranges from 0.17 to 0.21 ng/kg bodyweight/day for PFOS and PFOA, respectively [50]. The results by Gellrich et al. [50] revealed that short chain PFAS (<8 carbon atoms) were dominant in samples collected from tap water with a maximum level of 42.7 ng/L followed by mineral water and spring water [50]. One of the kay aspects on PFAS level in drinking water is the difference in PFAS level in both treated and source water. An interesting finding by Lu et al. [37] indicated that PFAS concentration was higher in treated water compared with the source water which could be accounted by the potential contamination that may occur whilst treated water moving through the water network and the treatment plant facilities [37]. Moreover, literature showed that various and inconsistent pattern of PFAS compounds was found on many occasions. While PFOA was dominant PFAS compound in tap water samples tested from Shanghai, Beijing, and Nanjing, PFOS was the key PFAS compound in water samples collected from Shenzhen and Hong Kong, accounting for more than 50% of the total PFAS compounds.

Moreover, researchers found mysterious temporal and spatial patterns of PFAO and PFOS across the various events where a comparison of PFAS levels in tap water in various Chinese cities revealed that the PFAS level varied significantly from one city to another with the highest was reported in Shanghai [37]. Moreover, the inconsistent and varied PFAS level was also significant across various countries. Quinete et al. [42] found that, unlike the pattern and the level of PFAO and PFAS level in the USA and Japan tap water samples, PFOS level was higher than PFOA in tap water samples collected in China [42]. An average level of total PFCs of 130 ng/L was measured in tap water samples from Shanghai (China), and a much lower maximum PFCs level was identified in water samples from Toyama (Japan) (0.62 ng/L) [44]. An average of 2000 ng/L was identified in in treated drinking water distribution system at the city of Oakdale, USA. This seems to be a very extreme level of PFAS where a health-based drinking water level of 0.04 g/L was assessed as a protective lifetime exposure through risk assessment exposure [12].

One of the main concerns about PFAS contamination is their persistence and bioaccumulation properties as well as the potential to travel with either water streams or sediments. Traditional treatment facilities seem unable to eliminate PFASs during normal treatment processes [51]. PFAS discharge into water bodies was also reported by Boiteux et al. [47] where river water proved to be impacted by the nearby fluorochemical manufacturing industry. This confirms that discharge of PFOA and PFOS are still detected in nearby industrial and manufacturing facilities. Results showed that river water and sediment samples as well as treated water samples at various stages from the main treatment plant have all showed various level of PFCA compounds coming from the manufacturing industry. Interestingly, PFCA was also detected at sediment samples at 62 km away from the source in almost 50% of the samples [47]

The occurrence of PFASs in surface water is frequently happening across many countries around the world [52]. Researchers have conducted several field studies for different types of surface water to investigate the occurrence and presence of PFASs [2,23,29,53,54]. A previous study was conducted to assess the level of PFASs from different locations in Gangs River, India. Results showed that around 15 types of PFASs were found in the water samples where the highest level detected was for PFHxA and PFBS. In addition, significant relationships were detected ($p < 0.05$) between the different PFASs substances such as PFCAs, PFSAs, PFBA, and PFHxS, indicating chemical binding and co-transport with dissolved oxygen carbon (DOC) in fresh and seawater. Consequently, assessed the pollutants concentration and spatial distribution of PFASs in Shuangtaizi Estuary, China. Results showed that the Shuangtaizi Estuary was in general polluted by PFASs. The total concentration of PFASs varied from 66.2 to 185 ng/L and from 44.8 to 209 ng/L in surface and bottom water of the Shuangtaizi Estuary, respectively, where the maximum concentration was reported for PFBS and PFBA. The level of PFASs in different environmental matrices was tested including surface runoff water rain, snow, and lake water in an urban area, to identify the sources of PFASs to urban water bodies [36,39,40]. Another research conducted by Yin et al. [55] discovered a significant temporal variation of PFASs compounds level over 12 months period due to the seasonal and climatic dry and wet conditions. Moroer, PFASs concentration was a function of the chain length where the level of short-chain compounds including PFBS, PFHxA and PFHpA tend to be highly influenced and decreased by the wet conditions. On contrary the level of long-chain PFASs compounds was more stable in both wet and dry conditions. These findings provide a good understanding to the leachate of PFAS compounds from point source pollution as landfills and treatment plant. The leaching of long chain is more likely controlled by the partitioning effect whilst the short chain leaching is influenced by the climatic conditions [55].

Another main finding in the field of PFAS in water is the variation of the PFAS where PFOA was the major compound with an average concentration of 35% of the total PFASs levels, in all environmental matrices investigated. In addition, the concentrations, and relative substances of PFASs in surface water were comparable to the concentrations found for urban lakes. Surface water leads to PFOA pollution in urban lakes. A sampling campaign was conducted in different seas in China in 2012. The results revealed that the higher concentration of PFAS was detected in the South Yellow Sea, where FTOH was the predominant substance, contributing 92–95% of the total PFAS [2].

3.1.2. PFASs in Soil

PFAS was detected in soil at various concentrations due to the reach out from various pollutions sources where PFAS compounds retain in soil due to sorption, partition and other complex reaction [56]. Table 1 shows the range of PFAS concentration in soil. The application and the reuse of sludge from wastewater treatment plants in farmlands is one of the main sources that contribute to soil contamination [57]. Other sources could be due to the degradation of fluorotelomer-based materials that lead to the release of PFCAs [43], precipitation, and water irrigation [58]. PFAS compound in soil in coastal areas can be emitted from direct sources which could level the PFAS concentration up to around 8–50 µg kg^{-1} soil as was reported in Chinese soil [45] which is somehow higher than the proposed PFCs in soil proposed by the USEPA (6 mg/kg for PFOS and 16 mg/kg for PFOA). One main concern about PFAS in soil is the potential PFAS release and carryover by plants as well as the possible PFAS leaching to the underneath soil layers and the groundwater. This carryover of PFOA and PFOS to the plant was evident where the PFAS level in plants was proportionally related to the PFOA/PFOS in the soil [59]. PFCs uptake from contaminated soil by crops was reported [60], where samples from rye grass, grain, and potatoes showed high potential of PFCs transfer from soil to crops [60]. This resulted in proposing a preventative PFCs limit of 100 ng/g dry soil in sludge to be reuse for farming purposes as to limit the potential transfer of PFCs from soil to plants and crops [59]. Moreover, the potential leaching of PFAS from soil through vadose zone is another threat that requires more attention and understanding where insufficient data about in-situ soil remediation and contaminants leaching to the groundwater are available [60]. The development of PFAS compounds in the soil system is complex since PFAS compounds can attain both hydrophilic and hydrophobic characteristics [56]. While the transfer of PFAS from soil to plant roots undergo through diffusion and sorption onto roots, there are still insufficient details about the PFC transfer rates in various crops and vegetables [59]. This result was confirmed [59] where they found the straw and grains of maize plants have the same carboxylic and sulfonic functional groups as in the contaminated soil referring to a direct correlation between soil and crops PFASs contamination.

PFASs toxicity their impact on soil microorganisms is among the other factors that can deteriorate the soil quality. Research found that the PFCs can negatively affect the soil functionality where it may disturb soil enzyme activity as well as change the microbial availability and damage the cellular structure [61–63]. The same result was confirmed by Sun et al. [64] as the soil contaminated with PFASs compounds had less bacterial diversity [64]. PFOA and PFOS are the dominating compounds reported in soil where their concentration ranged from <1 to around 13,000 ng/g in soil [56]. The fate of PFASs in the soil is a function of many parameters including soil pH, soil structure, clay content, organic matter content (OM), PFAS characteristics (long versus short chain), and climatic conditions [61]. OM seems to be the most significant controlling factor determining the PFAS toxicity level where the PFAS toxicity is inversely proportional with the soil OM content [34,65]. Additional research on PFAS adsorption and migration from soil to the groundwater and how this can migrate with the groundwater is still a research gap needs more investigation and modelling to account for the various PFAS concentrations in various groundwater conditions [60]. Contaminated soil with PFAS is a challenge since there is no definite remediation strategy to address the in situ PFAS remediation. Although soil stabilization using various reagents such as clay and Portland cement seems to be a promising technique for soil remediation, it does not provide an elimination for PFAS where it does not remove PFAS permanently [60]. Finally, the PFAS uptake by plant poses a direct human risk where the food chain represents a main risk pathway. Therefore, a toxicological risk assessment addressing the maximum allowed levels of 1.5 and 0.15 µg/kg body weight as TDI µg/kg for PFOS and PFOA, respectively, were identified by the European Food Safety Agency (EFSA) as a function of the respective tolerable daily intakes (TDI) of the compounds [59,66].

PFASs compounds are soluble in water and have the potential to leach down to the groundwater particularly in areas with potential source pollution like landfills and treatment plants. PFASs occurrence and leaching was reported by many researchers around the world [22,55,67–69]. The potential PFAS leaching could be alarming in many cases where PFASs were detected at large depths (15 m) below ground [70]. Yet, the leaching speed and behavior vary from one PFASs to another which depends on the soil binding, retardation and adsorption capacity [56]. The leaching characteristics of PFASs compound is a function of the chain length where short chain is more mobile than long ones. An analysis of the landfill leachate from 27 landfills in Australia was investigated by Gallen et al. [22]. Interesting findings presented in their study showed that the landfill leachate was significantly different from one landfill to another with an average PFASs of 1700 ng/L and a maximum PFAS level of 25,000 ng/L [22]. In contrast, the reported PFAO range in USA landfills was ranging between (7280–290,000) ng/L compared with 214,000 ng/L in China [36]. Nonetheless, these PFAS concentration are highly likely to vary due to the heterogeneous nature of waste dumped in landfills as well as the varied PFAS content in the generated landfilled materials. Operating landfills receiving municipal waste had much more PFAS level than closed ones and the leachate from landfills with construction and demolished materials seems to leach more PFAS than municipal landfills. Another study investigated the leachate from 11 landfills in USA and found that PFAO was detected in all samples [70]. Table 2 presents the level of various PFCS and PFAS compounds in leachate and compare the PFASs in water and solid. It can be seen that PFASs levels vary from one compound to another as a function of chain length and climatic conditions as illustrated in the previous sections. The risk associated from the landfill leachate is the potential volumes leachate generated particularly in wet climates, which contributes to the groundwater contamination. The total leachate volume in the USA was estimated to be around 61 million m^3 with around 80% coming from landfills [70]. Meanwhile, the leachate mass of $\sum$PFA in China was estimated by around 3 ton per year with the landfill leachate contribute to around 35% of this quantity [35]. Interestingly, analysis of leachate from young landfills showed much higher PFAS concentration in many occasions confirming the fact that the complexity and persistence of PFAS compound has been developed in the recent years where more frequent PFAS containing materials are in use [67,70]. The uniqueness of the landfills associated with its design capacity, climate, age, engineering, dumbed materials and frequency and other factors made it hard to predict the amount of PFAS leachate in various landfills where ad-hoc studies to be conducted. The results from various areas across the globe showed significant variation of PFAS leachate from one country to another where a maximum was reported in Australia (25,000 ng/L). This was evident while the leachate was significantly lower in Norway (590 to 757 ng/L), Germany (<0.37 to 2509 ng/L), and China (146 to 4430 ng/L) [35]. In conclusion, although the phasing out of PFAS materials and the ongoing effort to eliminate the PFAS release in the environment, yet there seems to be a need to consider more adaptation strategies dealing with PFAS risk. The increasing evidence of PFAS in newly designed and operated landfills indicates the potential exposure to higher leaching risk with greater PFAS concentrations is leaching to the environment is growing [70].

Table 1. Ranges of PFAS concentration in soil.

Header	Å. Høisæter, et al. [71] ng/g PFASS in Soil at Various Depths				Cai et al. [63] ng/g Dry Weight in	Chen et al. [69] ng/g Dry Weight	Cai et al. [61] ng/g Dry Weight	Gao et al. [72] ng/g Dry Weigh from 32 Samples	Wang et al. [73] The Mean Values ng/g Dry Weight	Liu et al. [74] ng/g Dry Weight	Chen et al. [75] ng/g Dry Weight	Dalahmeh et al. [76] ng/g Dry Weight	Armstrong et al. [77]
	0–1.0 m	1–2 m	2–3 m	3–4 m	Dry soil in China	Tot.PFC 0.34–65.8			PFAS in soil at varying distances				
PFOS	500–3000	1000–6500	1000–3500	1000–1200	130	70.5	8.6–10.4	0.06	2583,	87	0–2	0.6–3	23
PFOA	NA	NA	NA	NA	NA	93	3.3–47.5	0.32	50	0.3–8	63	0.5–0.9	24
PFHxS	NA	NA	NA	-	NA	61	NA	0.19	36		65		
∑PFCs	NA	NA	NA	NA	NA	-	NA	-	NA	99	NA	8	126–809
PFHxA	NA	NA	NA	NA	NA	NA	NA	0.09	NA	NA	NA	0.2–0.5	8
PFAA	NA	NA	NA	NA	NA	NA	NA	1.30–913	NA	NA	NA	NA	NA
PFBS	NA	NA	NA	NA	NA	NA	NA	0.05	NA	NA	NA	NA	NA

Table 2. Ranges and mean concentration of individual PFCs in landfill leachate.

Header	Gallen et al. [22] (ng L⁻¹)		Busch et al. [68] (ng L⁻¹)	Herzke et al. [78] (ng L⁻¹)	Clarke et al. [79] (ng L⁻¹)	Yin et al. [55] (ng L⁻¹)	Benskin et al. [80] (ng L⁻¹)	Eggen et al. [81] (ng L⁻¹)	Robey et al. [69] (ng L⁻¹)	Fuertes et al. [82] (ng L⁻¹)		Eggen et al. [81] (ng L⁻¹)		Huset et al. [83] (ng L⁻¹)	Garg et al. [84] (ng L⁻¹)
	landfill (>50% SW)	landfill (>50% C&D)	Compounds in landfill leachates	Coated textiles, Teflon waste, fire-fighting foam, papers, and furniture		Leachate from CW outlet system (Max. level)	Municipal landfill leachate	Municipal land fill leachates	Foam produced via the bubble aeration of landfill leachate	Raw Leachate in MSW landfill	Treated Leachate in MSW landfill	PFCs analysis-untreated leachate Water	PFCs analysis -untreated leachate Particles	Leachates from six landfill	Manufacture and disposal of electric and electronic products
PFOS	300	1100	235	570	187	439	4400	2920	104	25	NA	2920	34	56–160	128,670
PFOA	510	1200	926	9500	516	3457	1500	767	951	590	520	767	4	380–1100	118.3
PFHxS	940	3700	178	-	143	308	190	281	2058	630	870	281	ND*	120–700	133,330
PFDS	-		-	-	NA	0.72	63	<14	ND			-	-	0–16	-
PFHxA	1300	5000	2509	-	697	868	2500	757	2178	65	77	757	ND	270–2200	76
PFHpA	360	760	280		NA	486	690	277	454	-	-	277	ND	100–2800	9
PFNA	29	98	80	-	62	100	450	539	64	-	-	539	ND	19–140	8
PFDA	22	46	51	-	NA*	27	1100	75	87	-	-	70	ND	0.3–64	8
PFAA	NA	NA	NA	-	-	55	-	-	-	-	-	-	-	-	-
PFBS			1350	NA	112	1916	190	-	-	-	-	<5	ND	280–2300	-
PFPeA										11	325	-	-	-	

PFDA perfluorodecanoic acid, PFBS nonafluorobutane-1-sulfonic acid, PFHpA perfluoroheptanoic acid, PFHxS perfluorohexane sulfonate, PFNA perfluorononanoic acid. ND*: not detected, NA*: not analysed.

4. PFAS Treatment and Clean Up: Challenges and Achievement

Due to the persistence nature of the PFAS compounds, landfills and sewage treatment plants are highly likely to be a potential point source of PFAS emissions. Although the rapid advancement in PFAS testing and detection, yet the available standard analytical methods still short and there is very little experimental data detailing the physicochemical properties and partitioning constants of PFAS [19]. This place the treatment process of PFAS under the stress of producing precise and consistent outcomes. The literature presents various treatment methods of PFAS in various environment. Immobilization and plasma arc destruction are among the recommended methods to irreversibly transform PFAS waste. Some studies support the utilization of high temperature incineration as well as the usage of plasma destruction of PFAS in waste. It may be possible for certain types of waste. However, in the absence of regulation, there is no specific method to guarantee the universal adoption of such safer methods of disposing of PFOS wastes. In European wastewaters, it was found that PFOA is the most commonly found compound [84]. Conventional processes of wastewater treatment were found to be ineffective in removing of PFOA [85]. Other studies have recorded higher concentrations of PFOA in wastewater effluents than influences, possibly due to the transformation of the compounds of its precursor [21]. It was revealed that the degradation of precursor compound substances is a major supplier to environmental PFAS pollution. Thus, this part focuses on the treatability of PFAS compounds via conventional and modern water treatment processes. A comprehensive revision is judgmentally required to have a clear understanding on the transformation, migration and treatment of these substances in ecosystem and their potential influence on the secondary formation of PFOS and PFOA [86]. In order to eliminate and/or degrade PFAS, different pre-treatment methods have been tested in terms of their efficacy [19,20]. Some of these elements can theoretically be implemented as either a post-treatment or pre-treatment method with controlled aquifer recharge as summarized in Table 3.

Table 3. Potential Treatment technologies of PFAS.

Mechanism	Treatment Process
Destructive Treatment	Advance oxidation processes Electrochemical oxidation Incinerations Sono-chemical Biodegradation Photolysis
Non-Destructive treatment	Adsorption Ion exchange Fractionation

A detailed review summarizing the sorption mechanism along the sorption coefficients and capacity of PFAS on sediments is available [11]. Adsorption via activated carbon and ion exchange resins have been widely employed especially for pump-and-treat remediation to extract PFAS from polluted groundwater [52]. Compared to others, the use of GAC for PFAS removal has been recommended as a cheaper process, and it is the most recognized treatment technology for the groundwater contaminated by PFOS and PFOA [19]. Although the removal efficiency of polyfluoroalkyl substances by granular activated carbon was >90%, yet the sorption kinetics are normally faster for longer-chained PFAS [87]. For example, Kucharzyk et al. [19] reported that GC, which is optimized and applied effectively for the removal of PFOS, may not be appropriate for the removal of shorter-chained PFAS.

There is a risk that shorted-chained PFASs are more likely than their longer chain counterparts to split through a GAC medium. Otherwise, given the highly persistent existence of PFAS, stockpiling of spent GAC would turn out to be a serious hazardous waste management concern. Storage space is often restricted in MAR systems, and when leaching to groundwater, the disposal of PFAS polluted GAC at landfills can present a secondary contamination source to the ecosystem near the landfill site. Methods that can effectively kill PFAS are therefore highly desirable in both polluted water and stockpiled GAC. Other researchers have shown that nanofiltration can successfully extract PFOA from a spiked sample of groundwater [47]. In their study, three different levels of PFOA, including 5, 50, and 100 µg/L, were examined, and it was noticed that the remediation effectiveness was higher at a high PFOA level. At higher PFAS level, RO process could contribute to about 99% removal of PFOS with an initial concentration of 500 to 150,000 µg/L and a combination of nanofiltration TOGOTHER WITH reverse osmosis (RO) can achieve 99% PFOS removal and 90–99% PFOA (10,000 µg/L) removal throughout four days of treatment [88]. Nevertheless, the technique was not capable to ensure that the treated effluent was less than the recommended guideline values, even with high removal acacias (99%). Using RO and nanofiltration membranes showed that accumulation (fouling) of PFAS cause a substantial reduction in flux in the filtration process [88]. Unfortunately, the main drawbacks of nano-filtration processes are the low water recovery (75% to 80%) and the existence of high concentrations of inorganic substances comprising magnesium calcium, and silica in groundwater [89]. This provides an indication on the volume of brine water produced which also needs additional remediation before its final discharge. Table 4 illustrates the performance of numerous selected treatment technologies for PFAS at the laboratory-scale.

Table 4. Performance of numerous treatment technologies for PFAS (laboratory-scale).

Process	Treatment Mechanism	Operation Conditions	Performance	References
UV-Fenton	Oxidation	30.0 mM of H_2O_2, 2.0 mM of Fe^{2+}, pH 3.0. and 9 W UV lamp (max = 254 nm)	>95% PFOA destruction from 8.2 mg/L and defluorination efficiency of 53.2%	[88]
Oxidation	30.0 mM of H_2O_2, 2.0 mM of Fe^{2+}, pH 3.0. and 9 W UV lamp (max = 254 nm)	PFOA treatment >95% from 8.2 mg/L while defluorination effectiveness = 53%	Removal efficiency 100% (PFOA 559 mg/L)	[90]
Oxidation	Light-activated persulfate at 50 mM & radiation of 4 h of	Removal efficiency 100% (PFOA 559 mg/L)	73% removal efficiency of PFOS throughout 120 min	[91]
Sonolys.s	PFOS level from 65 µg/L to 13,100 µg/L) were treated through ultrasonic at frequency 505 kHz and power density 187.5 W/L).	73% removal of PFOS within 120 min	55–98% removals for different analyzed PFASs. Ozonation can create potentially toxic transformation products	[92]
Oxidaticn	Tested for 18 analyzed PFASs3 h of ozonation	55–98% removals for different analyzed PFASs. Ozonation can create potentially toxic transformation products which needs to be investigated in future research.	Adsorption capacity 41.3 mg/g of PFOA and 72.2 mg/g of PFOS	[93]
Adsorption	10 mg/L of PFOA; surface area: 534 m^2/g; time of equilibrium 24 h; pH 5	Adsorption capacity 41.3 mg/g of PFOA and 72.2 mg/g of PFOS	Adsorption capacity 510 mg/g of PFOA	[94]
Adsorption	700 mg/L of PFOA; surface area: 1539 m^2/g; time of equilibrium 24 h; pH 7	Adsorption capacity 510 mg/g of PFOA	Adsorption capacity 166 mg/g of PFHxA	[95]
Ion exchange using IRA 67	Particle size: 3–1.2 mm; 120 mg/L of PFHxA; time of equilibrium 12.5 h; pH 4	Adsorption capacity 166 mg/g of PFHxA	Adsorption capacity 2390 mg/g mg/g of PFOS	[96]
Ion exchange using IRA 67	Particle size: 3–1.2 mm; 200 mg/L of PFOS; time of equilibrium 20 h; pH 3	Adsorption capacity 2390 mg/g mg/g of PFOS		

In the degradation of PFOA and PFOS at ppm (mg/L) levels, methods such as advanced oxidation processes (AOPs) using ozone (O_3) and H_2O_2/Fe^{2+} were not effective. At 254 nm, direct UV irradiation was not capable of removing PFOA [84]. At relatively low temperatures (e.g., 40 °C), reaction rates were low and activation at higher temperatures was needed to speed up the reaction [19]. A functional and scalable approach for treating PFAS appears to be sonochemical therapy. For the removal of PFASs from water, AOPs based on heterogeneously catalyzed ozonation were used. Various combinations of ozone, a catalyst and persulfate were performed in laboratory-scale ozonation experiments. These combinations showed high removal efficiency, using all three parameters [97]. In the pilot-scale setup, within three hours of treatment, the concentrations of all 18 analyzed PFASs were reduced significantly. Given that the assessed ozonation treatment is already commercially available for large-scale applications today, it could easily be used in current water treatment trains, but ozonation will create potentially harmful conversion products that will need to be explored in future research. It promises to be used to decrease PFAS levels in PFAS-loaded sorbents as a destructive tool (e.g., spent GAC) [98]. Hydroxyls radicals are normally generated within the bubble from the cleavage of H_2O and O_2 to react with or abolish the pollutants. In another study, sonochemical treatment using a pilot-scale high-power sonicator was carried out for the treatment of 2,4,5-trichlorophenoxyacetic acid [99]. It can be concluded the sonochemical treatment process was effective in removing organic compounds (>90%) within a very short duration (a few minutes). Although the sonochemical technique appears promising for the large-scale treatment of PFAS contaminated products, incomplete PFAS destruction is currently viewed as a disadvantage. A recent study showed that 6:2 fluorotelomer sulfonate was less susceptible than perfluoroalkyl analogs (PFOA and PFOS) to sonochemical destruction and decreased the defluorination rate with a decreased degree of fluorination [100]. Given the recent focus on integrating deferential PFAS treatment processes in treatment trains in order to optimize the overall efficacy of PFAS destruction [98] In order to optimize the overall destruction of PFAS in the polluted media, it seems important to investigate the potential pairing of sonochemical treatment with alternative methods.

Table 5 summarizes the performance of various type of adsorbents in treating per- and polyfluoroalkyl substances. As a conclusion based on the treatment techniques performance investigated previously and partially summarized in Tables 3–5, traditional biological processes of wastewater treatments were found ineffective in removing of PFOA [85]. Due to the exceptional chemical futures of PFAS including high solubility, surfactant property, and thermal stability, various traditional and well-established treatment processes, including chemical oxidation, air stripping, and thermal treatment are invective in treating PFAS [19]. Similarly, Page et al. [21] indicated that numerous techniques like adsorption via granular/powdered activated carbon, ion exchange resins, reverse osmosis, membrane filtration, advance oxidation techniques, and sono-chemical decomposition have been investigated for the treatment of PFOA and PFOS from water and wastewater. Accordingly, among several physical and chemical treatment processes, adsorption process has been comprehensively tested and have shown to be effective methods for eliminating PFASs from water. It can be observed that the adsorption capacity and treatment performance using adsorption process were increased when the surface area increase [19]. Yet, in this adsorption process, the pollutant will be transferred from liquid phase to solid waste which will need to be managed as a hazardous waste. The main concern and limitation of RO and nanofiltration membranes is the fouling where PFAS cause a substantial reduction in flux in the filtration process [88]. Also, the low water recovery and the existence of high concentrations of inorganic substances comprising magnesium calcium, and silica in groundwater [89]. Yet, the brine water produced also needs additional remediation before its final discharge. Advanced oxidation processes are promising and have high potential in the removal of PFASs from water. AOPs were used recently at laboratory-scale and showed high removal efficiency, using all three parameters [97].

Table 5. Summary performance of various type of adsorbents in treating Per- and Polyfluoroalkyl Substances.

Sorbent	Adsorbate	Operation Conditions	Adsorption Capacity	References
Clay minerals (surface area: 67.52 m^2/g)	PFOS	Initial concentration of adsorbent 400 mg/L; pH7; concentration of adsorbate 0.2 mg/L	0.29–0.31 mg/g	[101]
Kaolinite (surface area: 11.9 m^2/g)	PFOS	Initial concentration of adsorbent 5000 mg/L; pH7; concentration of adsorbate 0.95 mg/L	0.08 mg/g	[102]
Alumina (surface area: 88.6 m^2/g)	PFOA	Initial concentration of adsorbent 10000 mg/L; pH4.3; concentration of adsorbate 0.1 mg/L	0.16×10^{-3} mg/g	[63]
Porous graphite (surface area: 2870 m^2/g)	PFOS	Initial concentration of adsorbent 100 mg/L; pH5; concentration of adsorbate 100 mg/L	1240 mg/g	[103]
Biochar from maize straw (surface area: 7.21 m^2/g)	PFOS	Initial concentration of adsorbent 200–1200 mg/L; pH7; concentration of adsorbate 100 mg/L	91.6 mg/g	[104]
Chitosan (surface area: 2870 m^2/g)	PFOS	Initial concentration of adsorbent 1350 mg/L; pH3; concentration of adsorbate 50 mg/L	645 mg/g	[105]
Zeolite) NaY80 surface area, 780 m^2/g	PFOS	Initial concentration of adsorbent 1000 mg/L; concentration of adsorbate 150 mg/L, Particle size: 3–1.2 mm	114.7 mg/g	[87]
Activated carbon from leaf biomass	PFOA PFOS	Modified activated carbons (AC-H$_3$PO$_4$) produced from leaf, uniform particle size of >64 µm	159.61 mg/g 208.64 mg/g	[106]
Modified silica	PFOS	Surface area 650 m^2/g, Particle size: 250–450 µm, Pore volume: 1.03 mL/g	55 mg/g	[107]
Boehmite	PFOS	Surface area 299 m^2/g, Average particle size 37.02 µm	0.1529 µg/m^2	[108]
Alumina nanoparticles	PFOS	Surface area: 83 m^2/g, Particle size: 13 nm	At 30 °C: 589 mg/g, at 40 °C: 485 mg/g, at 50 °C: 447 mg/g	[109]

5. The Key Knowledge Gaps and Future Research

PFOA and PFOS are the most well-known and well-studied PFASs, with an average removal half-life (t1/2) of 5.4 years and 3.8 years, respectively [7]. In order to ensure that the tested results represent real PFAS levels in the examined media, field sampling and laboratory hygiene procedures are important. With many sampling tools used in field and laboratory operations already containing PFAS, the process of sampling and PFAS testing remain uncertain and need lots of effort to alleviate the uncertainty involved. Figure 3 shows the molecular structure of PFOS and PFOA.

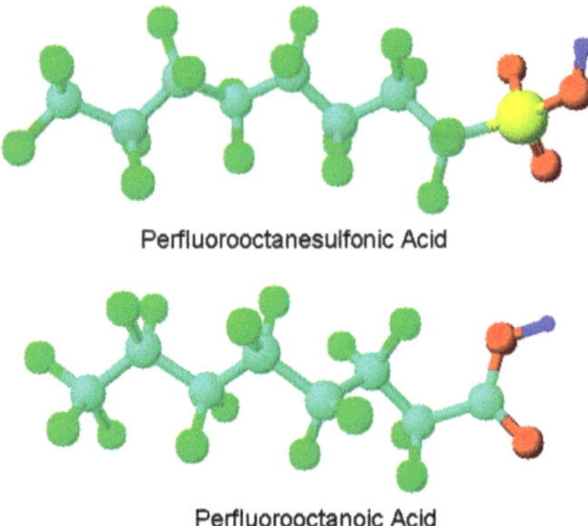

Figure 3. Molecular structures of two representative PFAAs: PFOS and PFOA [8].

Via their fact sheet collection, the Interstate Technology and Regulatory Council (ITRC) outlined site characterization, sampling safeguards, and analytical process concerns and choices. However, for the study of PFAS in surface water, wastewater, non-potable groundwater, and solids, there are currently no validated standard EPA methods [4]. Some US laboratories are applying adapted approaches based on EPA Method 537 for non-drinking water samples. These updated approaches do not have clear sample selection or analytical criteria and have not been checked or analyzed routinely for data quality [110]. As traditional water treatment techniques are unable to effectively remove PFASs, novel treatment methods are urgently needed to remove PFASs in water [92]. Although the intensive effort of phasing out many PFAS products with enforcement of alternative chemical production are in place in many areas across the globe, the risk of PFAS exposures due to the uptake and accumulation in the various media such as ocean and marine food chains as well as groundwater contamination represents a great challenge due to the complexity of the impact timescales [111]. The development and the propagation of the PFAS sites with increased possible exposures to newer PFASs have not been well defined yet.

Despite the agreed health impacts of PFAS particularly on aged or early aged groups, environment, no enforceable national drinking water limits and guidelines are in place in many parts across the globe [33]. There is still limited knowledge around the other PFAS substances. However, there is a growing evidence that these new detected compounds could have same potential on human health and may pose similar risks to human and the environment [33,112]. The use of engineered pre-treatment or post-treatment approaches must be based on a 'fit for purpose' definition and carefully combined with the planned water end use concept in order to make sure that both human and environmental health threats are properly managed and treated [21].

Another main challenge in PFAS potential health impact with considerable complexity was reported by Gebbink et al. [112]. They found that in samples collected from food, not only the PFOS and PFOA levels were overestimated by an order of magnitude, but also there is still a knowledge gap in identifying the precise percentage of these precursors can contribute to human PFCA exposure since the exposure pathway remains undefined [112].

In order to evaluate the development of degradation products and potentially undesirable by-products to track the occurrence of compounds in the gas process and to demonstrate the efficacy of treatment for other types of pollutants and to apply them to different types of water, further research is required. Whilst many studies demonstrate the link between PFAS exposure and the deteriorated immune system particularly in children, there is little evidence to map other health impacts, including cancer, as it is only noticeable in areas with exceptionally high exposures, with inadequate data to correlate these exposures to PFAS with neurodevelopment [111].

5.1. Risks Associated by PFAS

The developing use of the PFAS in various commercial and industrial sectors including aqueous fire-fighting foam, disposable food packaging, furniture, carpets, cookware, water treatment and many others poses a potential risk to the environment [15,16,49]. Like many other contaminants, PFAS can accumulate into the environment by either a direct or indirect pathway [113]. Direct sources and PFAS contamination released from different industries including wastewater treatment plants, sludge disposal, and landfill sites [4]. There are more than 4000 bioavailable PFAS compounds in the globe, however, the toxicity values for most of these compounds are still poorly understood with only few PFAS compounds have defined toxicity values and level [32]. Risk assessment aims at developing health-based guideline levels upon intensive review of PFAS toxicological level that cause harm to humans. PFOS was viewed as Persistent Organic Pollutant (POP) by the Conference of Parties, Stockholm Convention in 2009 where EPA has characterized PFOA as a "likely carcinogen" and its use was restricted [98]. Unfortunately, this PFAS toxicity level is still poorly understood and this has created the need to develop defined hazardous and risk registry for the PFAS toxicity values. This seems to be among the great challenges for the majority of PFAS compounds [53].

Human Exposure Pathways

The first ever published data about widespread PFOS in the environment was reported in 2001 by [6] where PFOS was found in fish and birds tissues as well as marine mammals. The finding revealed that the level of PFOS in animals is proportionally related with the population density and the industrialized activities. As such animals in these areas have much higher PFOS level than those live in remote marine areas [6,30]. Another pathway was found through the food chain where fish eating animals such as mink and bald eagles proved to have greater levels of PFOS than in their diets. Another toxicity realization of PFAS was introduced when PFAS compounds were reported in blood samples from various samples around the globe [32]. Potential health risks associated with PFAS exposure and the concern regarding their bioaccumulation indicated the tendency towards the potential exposure through various media and various sources. A study of consumer exposure using a scenario-based approach to PFOS and PFOA conducted by Trudel et al. [30] revealed an everyday exposure to PFAS is taking place in many countries resulting in the long-term uptake of PFOS and PFOA of 3–220 and 1–130 ng per kg body weight per day, respectively [30]. Many health impacts including cancer, liver damage, and immune system failure have been linked to the PFAS [113,114]. Moreover, whilst the routes of PFAS exposure remains somehow unclear, research agrees that diet is a potentially main source and well-established research suggests that PFOA is absorbed via inhalation and ingestion [30,115].

Human exposure pathways for PFOS, PFOA, and other PFAS related substances through various routes including drinking water, indoor polluted environment, food chain, long term contact with industries that produce PFAS compound including food packaging or cookware, breast milk, airborne dust, and air [1,116]. All of which result in cumulative uptake and PFAS build up [18,23,29]. This persistent exposure to PFAS can be poorly reversible due to slow elimination kinetics as well as the ongoing build up regardless of its magnitude and bioaccumulation potential [23]. China, for example, has witnessed a surge in PFAS release in the last decade due to the evolving industrial activities. The release of PFOS in China from industrial sources in 2010 was estimated by about 70 ton which is six times greater than the reported PFAS release in 2008 [20]. Table 6 shows the PFAS source- pathway and receptors.

Table 6. PFAS Source- Pathway and receptors.

Sources	Exposure Pathways	Receptors
Industrial and wastewater effluents Packaging Consumer products Landfills Fire-fighting foams	Soil Biosolids Dust Sediment Surface water Groundwater Drinking water Biota (including foods)	Ecological Aquatic Benthic Terrestrial Avian Human

Inappropriate treatment and disposal of waste and wastewater proved to pose significant PFAS risk and contamination [110]. Indirect PFAS generation and contamination take place through the transformation of perfluoroalkyl precursors and the breakdown of perfluoroalkyl-based products [116]. Both direct and the indirect PFAS generation can pose high contamination and risk to the surrounding environment including air and drinking water contamination, food poisonous where PFAS have been identified as a potential threat to public health [117]. The great effort made to counter the propagation of PFAS levels in various receptors resulted in PFOS and PFOA declines by 32% and 25%, respectively, in early 2000. Yet, other PFAS compounds continue to increase, suggesting various and more strict measures to quarantine the spread of the PFAS compounds [49]. This was evident through a long monitoring program of PFAS concentration in human blood in Norway (1977–2006) which resulted in the conclusion that PFAS levels increased by more than nine times in men aged between 40 and 50 years old over this period of time, pointing to the rapid development of PFAS contamination and therefore the increasing potential risks [49].

Due to their high solubility and high persistence, PFAS compounds can migrate in air and water bodies leading to concentrated level in environment and therefore, pose high risk and toxicity on public health [49]. This increasing aquatic bioaccumulation, soil/groundwater uptake, fish, seafood, meat, and vegetables were identified as the most PFAS sources that human can uptake at various levels [31]. On the other hand, soil contamination is one of the main environmental impacts due to PFAS spread. Research suggests that PFAS risk is associated with a range of impacts on ecosystem services. Although the impact and the risk of PFAS still not fully understood due to the poor evidence of the linkage between human public health and PFAS levels, previous research has found the exposure to PFAS including PFOA may delay bone development and accelerated male puberty in mice [5]. However, the adverse health effects of PFOA and some other compounds have been confirmed by several researchers [8,31,110].

PFAS proved to have high resistant to temperature and bio accumulative [118]. Recently, research has found that humans are normally have a long half-life of serum elimination of PFOS, PFHS, and PFOA with the recent realization of considerable PFAS levels in various media including fish, birds, mammals and human blood seems to be alarming and call for urgent intervention to alleviate any further degradation in the public health as well as the ecosystem [17,18,118,119]. Niu et al. [119] found that PFAS compounds, including PFOS and PFOA, pose a high risk to human and public health. The high linkage between the PFOS and PFOA and neuropsychological development in children was investigated and realized by Niu et al. [119]. They found that the PFOS and PFOA increased the risk of development problem and had significant impact on human health including the personal-social skills particularly among females [7,119]. The recent discovery proved that PFOA has been linked to the increased incidence of weight loss and even a disturbance in lipid system when the tests were conducted on laboratory animals [118]. Moreover, recent research on animals suggested that PFAO and PFOS are among the main causes of cancer where the two compounds were classified as carcinogenic substances. The same effect on humans is suggested by the World health Organization (WHO) who found that both PFOA and PFOS are potential carcinogenic materials to human bodies [111]. Other studies found high linkage between PFOA exposure and high cholesterol leading to liver enzymes and kidney cancer [110]. Other studies have found increasing level of PFOS and PFOA in the blood samples of human population and wildlife reflecting that severity of the exposure to the widespread of PFAS chemicals [8]. Another good evidence of PFAS exposure and potential risk was reported by the ATSDR [8] where blood serum concentrations with high PFOS and PFOA level were found in workers living near potential PFAS facilities and industries compared with normal population [8].

In summary, despite the many trials to limit the PFAS spread and the endorsement of phasing out the main PFAS substances (expressed in PFOS and PFOA), other compounds including PFAAs and related substances are still widely used in various industries including fire-fighting foams, photographic, semiconductor and others [20]. Therefore, the detection of PFAS in human bodies have not decreased [20,31,32,49]. Research has found the level of adverse health impact with the level of significance are depending on the extent of exposure, the duration, and the persistence [10].

A study by the French total diet [112] found that mothers are likely to provide a pathway of PFAO to their children through breast feeding where PFOA was noticed in 77% of the breast milk samples at an average level of 0.041 ng/mL and a maximum level of 0.308 ng/mL [119]. Also, because of the their immature developing immune system and fast body growth, children are probably much more sensitive to the impacts of PFAS [8]. A more valid link between PFAO and the adverse human health was realized when a sample of around 69,000 people in the Mid-Ohio Valley were tested for PFAS as the analysis of the water supply system there revealed a considerable level of PFAO (>50 ng/L of PFOA) [5].

6. PFAS Water Quality Guidelines

6.1. Current Llegislations and Practices in Various Countries

PFAS guideline threshold values are affected by several factors, including social, political, and economic influences [5]. The variation in the regulatory values of PFAS across different guidelines can be easily addressed. One of the main reasons is the differences in toxicology decisions and differences in exposure parameters [4]. Moreover, PFAS compounds encounter both temporal and spatial variation and as emerging contaminants, the regulations are rapidly changing to account for the developing knowledge. However, the protection of the human health remains the main focus of the PFAS regulations and guidance across all regulations and standards [4]. The Interstate Technology and Regulatory Council (ITRC) indicated the significant variation in PFAS regulations by identifying the states that have different guideline for PFOA and/or PFOS levels in drinking water and groundwater from EPA's health advisories (HAs) [4]. One main reason for that is due to the different bodies that regulate the PFAS. Whilst the environmental perspectives of PFAS is regulated by the (EPA). Toxic Substances Control Act (TSCA), their use in food is regulated by FDA which is normally associated by lack of certain scientific evidence on their hazardous impact and exposure rate [120–122]. Despite the drinking water contamination is an ongoing major issue, it is somehow puzzling that until now there seem to be no federal PFAS drinking water standards in the USA. The absence of such federal PFAS regulations has led multiple US states to develop specific water guidelines which can support the decisions regarding the cleaning of the contaminated site as well as drinking water surveillance and treatment [5]. Until recently, no MCLs were established for PFAS chemicals although great efforts are being made towards initiating MCL for PFOA and PFOS by the EPA and other agencies.

This lack of evidence between the public heath adverse impacts and the PFAS level has resulted in undefined epidemiological evidence which in turns created considerable variation among the different water guidelines due to the uncertainties involved [5]. There is still uncertainty around the potential

PFAS risk to the human health due to the limited data on how people are exposed to the PFAS and for how long. The exposure level and the consequences of this exposure are also poorly understood [53].

The need for extensive research to alleviate this uncertainty and how the PFAS impact the health risk is still among the top priorities in order to expect a safe exposure for humans. Until now, the authors are aware that there is still urgent need to conduct more research in the areas of PFAS toxicity and exposure as this remains one of the knowledge gaps in PFAS field. This is obvious since the first US EPA preliminary drinking water health advisory level towards the negative impacts of PFOS and PFOA was released before PFAS became a public issue in 2015 upon the growing public trend towards limiting PFAS release. This was upon the realization of the PFAs toxicity. More recent guidelines referring to the PFAS as a significant toxic substance were published by USEPA in 2019. Unfortunately, PFAS legislations and regulations are challenged by the significant differences across PFAS compounds associated with the limited information that can be utilized to establish uniform legislations across the globe [110].

Another main concern is the influence of the manufacturing companies and the bias attitude that some researcher may have depending on the interest of the funding agencies. Cordner et al. [5] indicated that economic factors play vital role in directing the guideline levels where a case of litigation was revealed by Minnesota Attorney General against 3 M when the company used a scientific researcher to manipulate others research findings and undermine the health impact of PFAS in what was considered as a violation of scientific norms [5].

6.2. EPA-US Guidelines

USEPA has released non-regulatory concentrations of PFAS that addresses the PFAS health impact in reference to the exposure time. Since 2006, EPA has reviewed and regulated around 191 PFAS compounds through a combination of orders [110]. According to the EPA, the lifetime health advisory (LHA) of 70 ng/L for PFOA and PFOS in drinking water is set as guideline. This LHA is applicable to PFOA and PFOS individually while it is applicable to the sum of both compounds in the case of accidental High concentration as in the case of Australian standards [4]. This value seems to be less conservative compared with other global regulations. The global focus on the PFAS was developed rapidly since 2002 where more conservative levels were in place due to the growing concerns. More restrictions were placed on the "long chain" (C8) molecule PFOS as it was withdrawn from markets. However, the current regulation seem to be driven by many factors including financial factors, detection limit and many others [5,121]. In 2016, the EPA suggested its lifetime limit for PFOA and PFOS of 70 ng/L, individually or combined. However, some US states argued that EPA's guidelines are insufficient and does not address the potential associated health risk and hence various PFAS threshold values were developed different from the EPA ones. The state's water guideline levels for PFOA and/or PFOS ranged from 13 to 1000 ng/L compared with 70 ng/L by the EPA for both compounds individually and combined (Table 7). For example, Minnesota established state guideline levels that were lower than the EPA guidelines of 35 ng/L PFOA and 27 ng/L PFOS. On the other hand, New Jersey has proposed 14 ng/L MCLs for PFOA and 13 ng/L for PFOS, the first lowest guideline standard in the US [5].

In USA, MCLs for any PFAS have not been identified by EPA, though they recently declared their intention to "initiate steps to evaluate the need for a maximum contaminant level (MCL) for PFOA and PFOS". Identifying the MCL would increase the ability of EPA's authority to study further on PFAS pollution [119]. In May 2016, USEPA lowered its drinking water health standard to 0.07 μg/L for the two most frequently found PFAS, PFOA and PFOS. A lifetime drinking water health advisory (HA) for PFOA was issued by the USEPA on the basis of a reference dose (RFD) of 0.07 micrograms per liter (μg/L) based on a developmental toxicity analysis in mice [7]. The toxicity values of PFAS are site specific with the highly likely temporal and spatial variation of these values. Moreover, the rapidly developed analytical methods represents another challenge with the rapid changing regulations.

Some toxicity values are varying from one standard to another and there are no uniform standard PFAS toxicity values across all countries.

Table 7. US PFAO/PFOS drinking water guideline levels (After [5]).

Guideline	Advisory Level		Reference Dose	
	PFAO (ng/L)	PFOS (ng/L)	PFAO (ng/kg-Day)	PFOS (ng/kg-Day)
U.S. EPAa, 2016, Health Advisory Level	70	70	20	20
Alaska DECb, 2016, Groundwater cleanup level	400	400	20	20
Maine DEPb, 2016, Remedial action guideline	130	560	6	80
Minnesota DOH, 2017, Noncancer health-based level	35	27	18	5.1
New Jersey DEP, 2017, Maximum contaminant level	14	13	2	1.8
North Carolina DENRb, 2012, Interim maximum allowable concentration	1000	-	N/A	NA
Texas CEQb, 2017, Protective concentration level	290	560	15	20
Vermonta DEC/DOH,6 2016, Primary groundwater enforcement standard	20	20	20	20

EU guidelines just recently has initiated a preliminary guideline on maximum allowable PFAS concentrations. In Germany, a health-based guidance of maximum PFAS level was proposed by the drinking water commission under the Ministry of Health. The proposed value is based on the safe lifelong exposure for all population groups of 300 ng/L for both PFOA and PFOS. In Germany, upon the detection of PFOA in drinking water at concentrations up to 0.64 g/L, the German Drinking Water Commission (TWK) established the first health based lifelong PFAO and PFOS exposure of 0.3 g/L in drinking water in June 2006 [40]. Until recently, Italy, has no PFAS guidelines in drinking water and the PFAS regulations were introduced upon the extreme detection of PFAS in water bodies in an area of the Veneto Region [38]. The highest amount of PFAS in drinking water was enforced by the Italian National Health Institute to protect human's health risk with PFOS $\leq$ 30 ng/L, PFOA $\leq$ 500 ng/L, and other PFAS $\leq$500 ng/L. In Spain, frequent PFAS monitoring programs were carried out and water samples were regularly tested for various PFAS substance. PFAS level varied across Spain with the conclusion of an unlikely health risk under the detected PFOS and PFOA levels where the maximum average levels of PFOS and PFOA were 1.81 and 2.40 ng/L, respectively [39].

Canada has developed federal guidelines for a few PFAS levels to avoid any potential human health affect, while values to safeguard ecological receptors are offered for PFOS (Table 8) [118]. An in-depth study was carried out on ecotoxicology and toxicology, environmental fate and behavior, and exposure. In order to initiate toxicological reference values (TRVs), adequate information on the PFAS impact was obtained, while ECCC assessed the suitability of obtainable non- or low-effect ecotoxicological data for the derivation of PFOS recommendations for multiple matrices for the safety of different trophic levels. The degree of PFAS in drinking water, soil, groundwater, and bird eggs are now available in the Canadian PFAS regulation.

Table 8. Health based guidance for usage in field investigation in Canada [110].

PFAS Name	Acronym	Drinking Water Screening Value (ng/L)
perfluorobutanoate	PFBA	30
perfluorobutane sulfonate	PFBS	15
perfluorohexanesulfonate	PFHxS	0.6
perfluoropentanoate	PFPeA	0.2
perfluorohexanoate	PFHxA	0.2
perfluoroheptanoate	PFHpA	0.2
perfluorononanoate	PFNA	0.02
fluorotelomer sulfonate	6:2 FTS	0.2
fluorotelomer sulfonate	8:2 FTS	0.2

In Australia, PFASs have been widely used in several industrial applications. PFAS health-based guidance values for PFOS, PFOA and PFHxS, were developed by The Department of Health, Food Standards Australia (Table 9). The inconsistent release of PFAS in the environment with the multiple PFAS sources have created additional barrier in PFAS management. The knowledge gap regarding the PFAS spread in the Australian environment made the process of setting definite guidelines a bit complex. Gallen et al. [24], for example, found that the level of PFAS in in the treated WWTP effluent was higher than the wastewater influence.

Table 9. Health based guidance for utilization in site investigation in Australia [122].

Health Based Guideline Value	PFOS and PFHxS (ng)	PFOA (ng)	PFAO (ng)
Tolerable daily intake	20	160	0.16
Guideline for drinking water quality	70	560	0.56
Guideline value for Recreational water quality	2000	10,000	10

The guideline values considered the health of the general community due to the PFAS exposure due water and food consumption [122]. Through an intergovernmental agreement (IGA), Australia has been forced to agree on a National Structure for Reacting to PFAS contamination to restrict the spread of PFAS contamination due to increasing concerns about the risk of PFAS. Under the IGA, Australians have followed a policy to respond to PFAS pollution, introduced national environmental management of PFAS, and implemented guidelines to advise government agencies involved in responding to PFAS contamination. The guidance values based on health indicate the threshold value of the amount of PFAS in food or drinking water that an individual can consume without being affected over a lifetime.

7. Factors Contributing to Variation in PFAS Guideline Levels

Since the guidelines are always driven by the toxicology and the risk values identified by the human body, the regulations for PFA substances were viewed to limit the potential health impact and risk. Data obtained from toxicity tests in China showed that the criteria maximum concentration (CMC) for protection of aquatic organisms were 3.78 and 45.54 mg·L^{-1} for PFOS and PFAO, respectively which is higher than values derived in North America [20]. This variation indicates the challenges in setting out uniform PFAS guideline values due to the prevailing uncertainty in risk assessment and the lack of solid scientific background. Moreover, PFAS regulations are also influenced by the technical capacity and other socio-economic factors involved. There is a developing concern towards PFASs regulations due to the growing frequent detection of PFAS in drinking water in the US. The USEPA has issued a long-term health advisory PFAs level in drinking water of 70 ng/L (for combination of PFOS and/or PFOA). This regulation is intended to lower the number of individual PFASs reported. However, these guidelines of PFAS in drinking water have been dropped to 70 ng/L by many states, as in the case of New Jersey where a maximum PFAS level for PFOS was set at 13 ng/L and a 14 ng/L target for PFOA proposed [7]. Other states including California have adopted the same while the State of New York made it even lower with a maximum allowable PFAS value of 10 ng/L for PFOS and PFOA. Some other states went even for lower maximum allowable PFAS as in the case of New Hampshire and Michigan with an MCL of 12 ng/L for PFOA and 16 ng/L health-based value for PFO, respectively [7]. Much more conservative and lower levels of PFAS were proposed by Denmark, with a temporary level of 3 ng/L for PFOS being considered. The challenge in setting out these regulatory values remains the technical capacity in providing proper detection levels as well as the applicability of meeting these targets with the development of industrial activities and the lack of knowledge in identifying the toxicology of PFAs at these levels.

8. Conclusions

The intensive and the widespread nature of industries that are heavily releasing PFAS substances has contributed to a PFAS build up in the environment which presents a serious threat to human life. Occasional mysterious PFAS release, fate, transport, and exposure by many industries are adding more complexity for the policymakers. While the pathway of PFAS is still not very clear, PFAS characterization and behavior need to be better explained, particularly in terms of occurrence, transformation, and degradation pathway.

Although PFAS has been linked to various health risks, such as cancer, liver damage, and hormone disruption, the extent of this risk remains uncertain. This is due to the poor understanding of the risk scale resulting from the exposure frequency as well as the severity and the exposure duration. Since the PFAS in soil can leach down to great depths (around 15 m), PFOS levels in soil have been advised by governments including Australia to have a maximum of 20 ng/g for land uses and industry, which vary according to the land use. This leaching potential is getting more evident in case of point source pollution areas including landfill and treatment plants. However, the retention of PFOS increases with the clay content and the organic matter as well as the decrease of soil pH.

There is an ongoing effort by many governments to phase out PFAS substances and find alternatives to PFAS substances. This effort has caused PFAS levels to decline particularly in surface water. Whilst the alternative chemicals should be less toxic and not persist in the environment, the phasing out process has not resulted in PFAS elimination or decrease where many new industries are prone to PFAS release. This may require extra effort in understanding the measures to phase out.

It is concluded that the longer the exposure to PFAS compound, the higher the risk is due to the ubiquitous uptake of PFAS. PFAS high doses uptake. There is still a lack of precise knowledge around the PFAs toxicology and the threshold values at which PFAs can pose severe health risk. Moreover, this link is still lacking the interpretation of the relationship between the extent of the PFAS exposure and the associated impact, where no specific data can explain the impact of longevity, frequency, and severity of this impact. While the guidelines agree on considering human health as a base for any regulatory values, there seems to be a significant variation across the global guidelines in setting unified PFAS standards since the PFAS level and rate are inconsistent within the same country and across the globe. These discrepancies stem from the differences in PFAS sources, toxicology decisions, and exposure rates where PFAS compound can transform from one compound to another as well as transport with sediment and water far from the source based on the surrounding environment and therefore creating temporal and spatial variation. The review highlighted the need for further research towards identifying the characteristics, fate, frequency, and the severity of PFAS represented by the exposure time and extent to better understand the nature of PFAS pathways and exposure.

Author Contributions: Idea, Z.A.; conceptualization, M.J.K.B., Z.A., M.Y.D.A.; writing—impact on drinking water, M.Y.D.A., writing—treatment technologies, M.J.K.B.; writing—regulations and measurement-based data collected from various countries, Z.A.; formatting and conclusion M.J.K.B., M.Y.D.A. All authors have read and agreed to the published version of the manuscript.

Funding: The authors declare this research has no funding.

Conflicts of Interest: The authors declare no conflict of interest.

References

1. Buck, R.C.; Franklin, J.; Berger, U.; Conder, J.M.; Cousins, I.T.; de Voogt, P.; Jensen, A.A.; Kannan, K.; Mabury, S.A.; van Leeuwen, S.P. Perfluoroalkyl and Polyfluoroalkyl Substances in the Environment: Terminology, Classification, and Origins. *Integr. Environ. Assess. Manag.* **2011**, *7*, 513–541. [CrossRef] [PubMed]
2. Fu, Y.; Wang, T.; Wang, P.; Fu, Q.; Lu, Y. Effects of age, gender and region on serum concentrations of perfluorinated compounds in general population of Henan, China. *Chemosphere* **2014**, *110*, 104–110. [CrossRef] [PubMed]
3. Clout, L.; Priddle, D.; Spafford, P.J. The Coalition against PFAS. 2018. Available online: https://cappfas.com/what-is-pfas/ (accessed on 19 December 2020).
4. ITRC. *PFAS Technichal and Regulatory Guidanace Document and Fact Sheet PFAS-1*; ITRC: Washington, DC, USA, 2020. [CrossRef]
5. Cordner, A.; de la Rosa, V.Y.; Schaider, L.A.; Rudel, R.A.; Richter, L.; Brown, P. Guideline levels for PFOA and PFOS in drinking water: The role of scientific uncertainty, risk assessment decisions, and social factors. *J. Expo. Sci. Environ. Epidemiol.* **2019**, *29*, 157–171. [CrossRef] [PubMed]
6. Giesy, J.P.; Kannan, K. Global Distribution of Perfluorooctane Sulfonate in Wildlife. *Environ. Sci. Technol.* **2001**, *35*, 1339–1342. [CrossRef] [PubMed]
7. USEPA. *Drinking Water Health Advisories for PFOA and PFOS.*; USEPA: Washington, DC, USA, 2016; pp. 1–4. Available online: https://www.epa.gov/sites/production/files/201605/documents/pfos_health_advisory_final-plain.pdf (accessed on 10 October 2020).
8. ATSDR. *Toxicological Profile for Perfluoroalkyls (Draft for Public Comment)*; Agency for Toxic Substances and Disease Registry: Atlanta, GA, USA, 2018. Available online: https://www.atsdr.cdc.gov/toxprofiles/tp.asp?id=1117&tid=237 (accessed on 28 October 2020).
9. Fàbrega, F.; Kumar, V.; Schuhmacher, M. PBPK modeling for PFOS and PFOA: Validation with human experimental data. *Toxicol. Lett.* **2014**, *230*, 244–251. [CrossRef]
10. Lake, N.; Cao, Y.; Cao, X.; Wang, H.; Wan, Y.; Wang, S. Assessment on the distribution and partitioning of perfluorinated compounds in the water and sediment. *Environ. Monit. Assess.* **2015**, *611*, 611. [CrossRef]
11. Zareitalabad, P.; Siemens, J.; Hamer, M.; Amelung, W. Perfluorooctanoic acid (PFOA) and perfluorooctanesulfonic acid (PFOS) in surface waters, sediments, soils and wastewater—A review on concentrations and distribution coefficients. *Chemosphere* **2013**, *91*, 725–732. [CrossRef]
12. Post, G.B.; Louis, J.B.; Cooper, K.R.; Boros-Russo, B.J.; Lippincott, R.L. Occurrence and Potential Significance of Perfluorooctanoic Acid (PFOA) Detected in New Jersey Public Drinking Water Systems. *Environ. Sci. Technol.* **2009**, *43*, 4547–4554. [CrossRef]
13. Jain, R.B. Time trends over 2003–2014 in the concentrations of selected perfluoroalkyl substances among US adults aged ≥ 20 years: Interpretational issues. *Sci. Total Environ.* **2018**, *645*, 946–957. [CrossRef]
14. Liu, B.; Zhang, H.; Li, J.; Dong, W. Perfluoroalkyl acids (PFAAs) in sediments from rivers of the Pearl River Delta, southern China. *Environ. Monit. Assess.* **2017**, *189*, 213. [CrossRef]
15. Dunn, A.M.; Hofmann, O.S.; Waters, B.; Witchel, E. Cloaking malware with the trusted platform module. In Proceedings of the 20th USENIX Security Symposium, San Francisco, CA, USA, 8–12 August 2011; pp. 395–410.
16. Vedagiri, U.K.; Anderson, R.H.; Loso, H.M.; Schwach, C.M. Ambient levels of PFOS and PFOA in multiple environmental media. *Remediat. J.* **2018**, *28*, 9–51. [CrossRef]
17. Goosey, E.; Harrad, S. Perfluoroalkyl compounds in dust from Asian, AustralianEuropean, and North American homes and UK cars, classrooms, and offices. *Environ. Int.* **2011**, *37*, 86–92. [CrossRef] [PubMed]
18. ITRC. *Environmental Fate and Transport for per-and Polyfluoroalkyl Substances*; ITRC: Washington, DC, USA, 2018; Available online: https://pfas-1.itrcweb.org/wp-content/uploads/2018/03/pfas_fact_sheet_fate_and_transport__3_16_18.pdf (accessed on 28 October 2020).
19. Kucharzyk, K.H.; Darlington, R.; Benotti, M.; Deeb, R.; Hawley, E. Novel treatment technologies for PFAS compounds: A critical review. *J. Environ. Manag.* **2017**, *204*, 757–764. [CrossRef] [PubMed]
20. Wang, T.; Wang, P.; Meng, J.; Liu, S.; Lu, Y.; Khim, J.S.; Giesy, J.P. A review of sources, multimedia distribution and health risks of perfluoroalkyl acids (PFAAs) in China. *Chemosphere* **2015**, *129*, 87–99. [CrossRef]

21. Page, D.; Vanderzalm, J.; Kumar, A.; Cheng, K.Y.; Kaksonen, A.H.; Simpson, S. Risks of perfluoroalkyl and polyfluoroalkyl substances (PFAS) for sustainable water recycling via aquifers. *Water* **2019**, *11*, 1737. [CrossRef]

22. Gallen, C.; Drage, D.; Eaglesham, G.; Grant, S.; Bowman, M.; Mueller, J.F. Australia-wide assessment of perfluoroalkyl substances (PFASs) in landfill leachates. *J. Hazard. Mater.* **2017**, *331*, 132–141. [CrossRef]

23. Cousins, I.T.; Vestergren, R.; Wang, Z.; Scheringer, M.; McLachlan, M.S. The precautionary principle and chemicals management: The example of perfluoroalkyl acids in groundwater. *Environ. Int.* **2016**, *94*, 331–340. [CrossRef]

24. Gallen, C.; Eaglesham, G.; Drage, D.; Nguyen, T.H.; Mueller, J.F. A mass estim Figure 2020 ate of perfluoroalkyl substance (PFAS) release from Australian wastewater treatment plants. *Chemosphere* **2018**, *208*, 975–983. [CrossRef]

25. Domingo, J.L.; Nadal, M. Human exposure to per-and polyfluoroalkyl substances (PFAS) through drinking water: A review of the recent scientific literature. *Environ. Res. J.* **2019**, *177*, 108648. [CrossRef]

26. Gonzalez, D.; Thompson, K.; Quiñones, O.; Dickenson, E.; Bott, C. Assessment of PFAS fate transport, and treatment inhibition associated with a simulated AFFF release within a WASTEWATER treatment plant. *Chemosphere* **2021**, *262*, 127900. [CrossRef]

27. Simon, J.A.; Cassidy, D.; Cherry, J.; Bryant, D.; Cox, D. PFAS Experts Symposium: Statements on regulatory policy, chemistry and analytics, toxicology, transport/fate, and remediation for per- and polyfluoroalkyl substances (PFAS) contamination issues. *Remediat. J.* **2019**, *29*, 31–48. [CrossRef]

28. Cerveny, D.; Grabic, R.; Fedorova, G.; Grabicova, K.; Turek, J.; Zlabek, V.; Randak, T. Fate of perfluoroalkyl substances within a small stream food web affected by sewage effluent. *Water Res.* **2018**, *134*, 226–233. [CrossRef] [PubMed]

29. Brusseau, M.L. Assessing the potential contributions of additional retention processes to PFAS retardation in the subsurface. *Sci. Total Environ.* **2018**, *613–614*, 176–185. [CrossRef] [PubMed]

30. Trudel, D.; Horowitz, L.; Wormuth, M.; Scheringer, M.; Cousins, I.T.; Hungerb, K. Estimating Consumer Exposure to PFOS and PFOA. *Risk Anal.* **2008**, *28*, 13–15. [CrossRef]

31. Sznajder-katarzy, K.; Surma, M.; Cie, I. A Review of Perfluoroalkyl Acids (PFAAs) in terms of Sources, Applications, Human Exposure, Dietary Intake, Toxicity, Legal Regulation, and Methods of Determination. *J. Chem.* **2019**, *2019*. [CrossRef]

32. Uebelacker, L.A. A Review of the Pathways of Human Exposure to Poly- and Perfluoroalkyl Substances (PFASs) and Present Understanding of Health Effects Elsie. *Physiol. Behav.* **2017**, *176*, 139–148. [CrossRef]

33. Pelch, K.E.; Reade, A.; Wol, T.A.M.; Kwiatkowski, C.F. Review article PFAS health e ff ects database: Protocol for a systematic evidence map. *Environ. Int. J.* **2019**, *130*, 104851. [CrossRef]

34. Zhao, L.; Zhu, L.; Zhao, S.; Ma, X. Sequestration and bioavailability of perfluoroalkyl acids (PFAAs) in soils: Implications for their underestimated risk. *Sci. Total Environ.* **2016**, *572*, 169–176. [CrossRef]

35. Yan, H.; Cousins, I.T.; Zhang, C.; Zhou, Q. Perfluoroalkyl acids in municipal landfill leachates from China: Occurrence, fate during leachate treatment and potential impact on groundwater. *Sci. Total Environ.* **2015**, *524–525*, 23–31. [CrossRef]

36. Eriksson, U.; Kärrman, A.; Rotander, A.; Mikkelsen, B.; Dam, M. Perfluoroalkyl substances (PFASs) in food and water from Faroe Islands, Environ. *Sci. Pollut. Res.* **2013**, *20*, 7940–7948. [CrossRef]

37. Lu, Z.; Lu, R.; Zheng, H.; Yan, J.; Song, L.; Wang, J.; Yang, H. Risk exposure assessment of per-and polyfluoroalkyl substances (PFASs) in drinking water and atmosphere in central eastern China. *Environ. Sci. Pollut. Res.* **2018**, *25*, 9311–9320. [CrossRef] [PubMed]

38. Mastrantonio, M.; Bai, E.; Uccelli, R.; Cordiano, V.; Screpanti, A.; Crosignani, P. Drinking water contamination from perfluoroalkyl substances (PFAS): An ecological mortality study in the Veneto Region, Italy. *Eur. J. Public Health* **2017**, *28*, 180–185. [CrossRef] [PubMed]

39. Domingo, J.L.; Ericson-jogsten, I.; Perello, G.; Nadal, M.; van Bavel, B.; Ka, A. Human Exposure to Perfluorinated Compounds in Catalonia, Spain: Contribution of Drinking Water and Fish and Shellfish. *Agric. Food Chem.* **2012**, *60*, 4408–4415. [CrossRef] [PubMed]

40. Wilhelm, M.; Bergmann, S.; Dieter, H.H. Occurrence of perfluorinated compounds (PFCs) in drinking water of North Rhine-Westphalia, Germany and new approach to assess drinking water contamination by shorter-chained C4–C7 PFCs. *Int. J. Hyg. Environ. Health* **2010**, *213*, 224–232. [CrossRef] [PubMed]

41. Schwanz, T.G.; Llorca, M.; Farré, M.; Barceló, D. Perfluoroalkyl substances assessment in drinking waters from Brazil, France and Spain. *Sci. Total Environ.* **2016**, *539*, 143–152. [CrossRef] [PubMed]

42. Quinete, N.; Wu, Q.; Zhang, T.; Yun, S.H.; Moreira, I.; Kannan, K. Specific profiles of perfluorinated compounds in surface and drinking waters and accumulation in mussels, fish, and dolphins from southeastern Brazil. *Chemosphere* **2009**, *77*, 863–869. [CrossRef]

43. Russell, M.H.; Berti, W.R.; Buck, R.C. Investigation of the Biodegradation Potential of a Fluoroacrylate Polymer Product in Aerobic Soils. *Environ. Sci. Technol.* **2008**, *42*, 800–807. [CrossRef]

44. Mak, Y.L.; Taniyasu, S.; Yeung, L.W.Y.; Lu, G.; Jin, L.; Yang, Y.; Lam, P.K.S.; Kannan, K.; Yamashita, N. Perfluorinated Compounds in Tap Water from China and Several Other Countries. *Environ. Sci. Technol.* **2009**, *43*, 4824–4829. [CrossRef]

45. Ghisi, R.; Manzetti, S. Accumulation of perfluorinated alkyl substances (PFAS) in agricultural plants: A review. *Environ. Res. J.* **2019**, *169*, 326–341. [CrossRef]

46. Bartolomé, M.; Gallego-picó, A.; Cutanda, F.; Huetos, O.; Esteban, M.; Pérez-Gómez, B.; Castaño, A. Perfluorinated alkyl substances in Spanish adults: Geographical distribution and determinants of exposure. *Sci. Total Environ.* **2017**, *603–604*, 352–360. [CrossRef]

47. Boiteux, V.; Dauchy, X.; Bach, C.; Colin, A.; Hemard, J.; Sagres, V.; Rosin, C.; Munoz, J. Science of the Total Environment Concentrations and patterns of perfluoroalkyl and polyfluoroalkyl substances in a river and three drinking water treatment plants near and far from a major production source. *Sci. Total Environ.* **2017**, *583*, 393–400. [CrossRef] [PubMed]

48. Loos, R.; Wollgast, J.; Huber, T. Polar herbicides, pharmaceutical products, perfluorooctanesulfonate (PFOS), perfluorooctanoate (PFOA), and nonylphenol and its carboxylates and ethoxylates in surface and tap waters around Lake Maggiore in Northern Italy. *Anal. Bioanal. Chem.* **2007**, *387*, 1469–1478. [CrossRef] [PubMed]

49. Lindstrom, A.B.; Strynar, M.J.; Libelo, E.L. Polyfluorinated compounds: Past, present, and future. *Environ. Sci. Technol.* **2011**, *45*, 7954–7961. [CrossRef] [PubMed]

50. Gellrich, V.; Brunn, H.; Stahl, T. Perfluoroalkyl and polyfluoroalkyl substances (PFASs) in mineral water and tap water. *J. Environ. Sci. Health Part A Toxic/Hazard. Subst. Environ. Eng.* **2013**, *48*, 129–135. [CrossRef] [PubMed]

51. Horst, J.; Mcdonough, J.; Ross, I.; Dickson, M.; Miles, J.; Hurst, J.; Storch, P. Advances in Remediation Solutions Water Treatment Technologies for PFAS: The Next Generation, Gr. *Water Monit. Remediat.* **2018**, *38*. [CrossRef]

52. Zhang, X.; Lohmann, R.; Dassuncao, C.; Hu, X.C.; Weber, A.K.; Vecitis, C.D.; Sunderland, E.M. Source Attribution of Poly- and Perfluoroalkyl Substances (PFASs) in Surface Waters from Rhode Island and the New York Metropolitan Area. *Environ. Sci. Technol. Lett.* **2016**, *3*, 316–321. [CrossRef]

53. Egeghy, P.P.; Judson, R.; Gangwal, S.; Mosher, S.; Smith, D.; Vail, J.; Cohen, E.A. The exposure data landscape for manufactured chemicals. *Sci. Total Environ.* **2012**, *414*, 159–166. [CrossRef]

54. Schultz, M.M.; Higgins, C.P.; Huset, C.A.; Luthy, R.G.; Barofsky, D.F.; Field, J.A. Fluorochemical mass flows in a municipal wastewater treatment facility. *Environ. Sci. Technol.* **2006**, *40*, 7350–7357. [CrossRef]

55. Yin, T.; Chen, H.; Reinhard, M.; Yi, X.; He, Y.; Gin, K.Y.H. Perfluoroalkyl and polyfluoroalkyl substances removal in a full-scale tropical constructed wetland system treating landfill leachate. *Water Res.* **2017**, *125*, 418–426. [CrossRef]

56. Bolan, N.; Sarkar, B.; Yan, Y.; Li, Q.; Wijesekara, H.; Kannan, K.; Tsang, D.C.W.; Schauerte, M.; Bosch, J.; Noll, H.; et al. Remediation of poly-and perfluoroalkyl substances (PFAS) contaminated soils–To mobilize or to immobilize or to degrade? *J. Hazard. Mater.* **2021**, *401*. [CrossRef]

57. Lee, H.; Tevlin, A.G.; Mabury, S.A.; Mabury, S.A. Fate of Polyfluoroalkyl Phosphate Diesters and Their Metabolites in Biosolids-Applied Soil: Biodegradation and Plant Uptake in Greenhouse and Field Experiments. *Environ. Sci. Technol.* **2014**, *48*. [CrossRef] [PubMed]

58. Karnjanapiboonwong, A.; Deb, S.K.; Subbiah, S.; Wang, D.; Anderson, T.A. Perfluoroalkylsulfonic and carboxylic acids in earthworms (Eisenia fetida): Accumulation and effects results from spiked soils at PFAS concentrations bracketing environmental relevance. *Chemosphere* **2018**, *199*, 168–173. [CrossRef] [PubMed]

59. Lechner, M.; Knapp, H. Carryover of Perfluorooctanoic Acid (PFOA) and Perfluorooctane Sulfonate (PFOS) from Soil to Plant and Distribution to the Different Plant Compartments Studied in Cultures of Carrots (*Daucus carota* ssp. *Sativus*), Potatoes (*Solanum tuberosum*), and cucumbers (*Cucumis sativus*). *J. Agric. Food Chem.* **2011**, 11011–11088. [CrossRef]

60. Stahl, T.; Heyn, J.; Thiele, H.; Hüther, J.; Failing, K.; Georgii, S.; Brunn, H. Carryover of Perfluorooctanoic Acid (PFOA) and Perfluorooctane Sulfonate (PFOS) from Soil to Plants. *Arch. Environ. Contam. Toxicol.* **2009**, *57*, 289–298. [CrossRef] [PubMed]

61. Cai, Y.; Chen, H.; Yuan, R.; Wang, F.; Chen, Z.; Zhou, B. Toxicity of perfluorinated compounds to soil microbial activity: Effect of carbon chain length, functional group and soil properties. *Sci. Total Environ.* **2019**, *690*, 1162–1169. [CrossRef] [PubMed]

62. Qiao, W.; Xie, Z.; Zhang, Y.; Liu, X.; Xie, S.; Huang, J.; Yu, L. Perfluoroalkyl substances (PFASs) influence the structure and function of soil bacterial community: Greenhouse experiment. *Sci. Total Environ.* **2018**, *642*, 1118–1126. [CrossRef]

63. Cai, Y.; Chen, H.; Yuan, R.; Wang, F.; Chen, Z.; Zhou, B. Metagenomic analysis of soil microbial community under PFOA and PFOS stress. *Environ. Res.* **2020**, *188*, 109838. [CrossRef]

64. Sun, Y.; Wang, T.; Peng, X.; Wang, P.; Lu, Y. Bacterial community compositions in sediment polluted by perfluoroalkyl acids (PFAAs) using Illumina high-throughput sequencing. *Environ. Sci. Pollut. Res.* **2016**, *23*, 10556–10565. [CrossRef]

65. Li, B.; Bao, Y.; Xu, Y.; Xie, S.; Huang, J. Vertical distribution of microbial communities in soils contaminated by chromium and perfluoroalkyl substances. *Sci. Total Environ.* **2017**, *599–600*, 156–164. [CrossRef]

66. Chen, H.; Wang, Q.; Cai, Y.; Yuan, R.; Wang, F.; Zhou, B.; Chen, Z. Effect of perfluorooctanoic acid on microbial activity in wheat soil under different fertilization conditions. *Environ. Pollut.* **2020**, *264*, 114784. [CrossRef]

67. Perkola, N.; Sainio, P. Survey of perfluorinated alkyl acids in Finnish effluents, storm water, landfill leachate and sludge. *Environ. Sci. Pollut. Res.* **2013**, *20*, 7979–7987. [CrossRef] [PubMed]

68. Busch, J.; Ahrens, L.; Sturm, R.; Ebinghaus, R. Polyfluoroalkyl compounds in landfill leachates. *Environ. Pollut.* **2010**, *158*, 1467–1471. [CrossRef] [PubMed]

69. Robey, N.M.; da Silva, B.F.; Annable, M.D.; Townsend, T.G.; Bowden, J.A. Concentrating Per- And Polyfluoroalkyl Substances (PFAS) in Municipal Solid Waste Landfill Leachate Using Foam Separation. *Environ. Sci. Technol.* **2020**, *54*, 12550–12559. [CrossRef] [PubMed]

70. Dauchy, X.; Boiteux, V.; Colin, A.; Hémard, J.; Bach, C.; Rosin, C.; Munoz, J.F. Deep seepage of per-and polyfluoroalkyl substances through the soil of a firefighter training site and subsequent groundwater contamination. *Chemosphere* **2019**, *214*, 729–737. [CrossRef]

71. Høisæter, Å.; Pfaff, A.; Breedveld, G.D. Leaching and transport of PFAS from aqueous film-forming foam (AFFF) in the unsaturated soil at a firefighting training facility under cold climatic conditions. *J. Contam. Hydrol.* **2019**, *222*, 112–122. [CrossRef]

72. Gao, Y.; Liang, Y.; Gao, K.; Wang, Y.; Wang, C.; Fu, J.; Wang, Y.; Jiang, G.; Jiang, Y. Levels, spatial distribution and isomer profiles of perfluoroalkyl acids in soil, groundwater and tap water around a manufactory in China. *Chemosphere* **2019**, *227*, 305–314. [CrossRef]

73. Wang, Y.; Fu, J.; Wang, T.; Liang, Y.; Pan, Y.; Cai, Y.; Jiang, G. Distribution of perfluorooctane sulfonate and other perfluorochemicals in the ambient environment around a manufacturing facility in china. *Environ. Sci. Technol.* **2010**, *44*, 8062–8067. [CrossRef]

74. Liu, B.; Zhang, H.; Yu, Y.; Xie, L.; Li, J.; Wang, X.; Dong, W. Perfluorinated Compounds (PFCs) in Soil of the Pearl River Delta, China: Spatial Distribution, Sources, and Ecological Risk Assessment. *Arch. Environ. Contam. Toxicol.* **2020**, *78*, 182–189. [CrossRef]

75. Chen, S.; Jiao, X.C.; Gai, N.; Li, X.J.; Wang, X.C.; Lu, G.H.; Piao, H.T.; Rao, Z.; Yang, Y.L. Perfluorinated compounds in soil, surface water, and groundwater from rural areas in eastern China. *Environ. Pollut.* **2016**, *211*, 124–131. [CrossRef]

76. Dalahmeh, S.; Tirgani, S.; Komakech, A.J.; Niwagaba, C.B.; Ahrens, L. Per-and polyfluoroalkyl substances (PFASs) in water, soil and plants in wetlands and agricultural areas in Kampala, Uganda. *Sci. Total Environ.* **2018**, *631–632*, 660–667. [CrossRef]

77. Armstrong, D.L.; Lozano, N.; Rice, C.P.; Ramirez, M.; Torrents, A. Temporal trends of perfluoroalkyl substances in limed biosolids from a large municipal water resource recovery facility. *J. Environ. Manag.* **2016**, *165*, 88–95. [CrossRef] [PubMed]

78. Herzke, D.; Olssonb, E.; Posner, S. Perfluoroalkyl and polyfluoroalkyl substances (PFASs) in consumer products in Norway—A pilot study. *Chemosphere* **2012**, *88*, 980–987. [CrossRef] [PubMed]

79. Clarke, B.O.; Anumol, T.; Barlaz, M.; Snyder, S.A. Investigating landfill leachate as a source of trace organic pollutants. *Chemosphere* **2015**, *127*, 269–275. [CrossRef] [PubMed]

80. Benskin, J.P.; Li, B.; Ikonomou, M.G.; Grace, J.R.; Li, L.Y. Per-and Polyfl uoroalkyl Substances in Landfill Leachate: Patterns, Time Trends, and Sources. *Environ. Sci. Technol.* **2012**, *46*, 11532–11540. [CrossRef]

81. Eggen, T.; Moeder, M.; Arukwe, A. Municipal landfill leachates: A signifi cant source for new and emerging pollutants. *Sci. Total Environ.* **2010**, *408*, 5147–5157. [CrossRef]

82. Fuertes, I.; Gómez-Lavín, S.; Elizalde, M.P.; Urtiaga, A. Perfluorinated alkyl substances (PFASs) in northern Spain municipal solid waste landfill leachates. *Chemosphere* **2017**, *168*, 399–407. [CrossRef]

83. Huset, C.A.; Barlaz, M.A.; Barofsky, D.F.; Field, J.A. Quantitative determination of fluorochemicals in municipal landfill leachates. *Chemosphere* **2011**, *82*, 1380–1386. [CrossRef]

84. Garg, S.; Kumar, P.; Mishra, V.; Guijt, R.; Singh, P.; Dumée, L.F.; Sharma, R.S. A review on the sources, occurrence and health risks of per-/poly-fluoroalkyl substances (PFAS) arising from the manufacture and disposal of electric and electronic products. *J. Water Process Eng.* **2020**, *38*, 101683. [CrossRef]

85. Loos, R.; Carvalho, R.; Anto, D.C.; Locoro, G.; Tavazzi, S.; Paracchini, B.; Ghiani, M.; Lettieri, T.; Blaha, L.; Jarosova, B.; et al. EU-wide monitoring survey on emerging polar organic contaminants in wastewater treatment plant effluents. *Water Res.* **2013**, *7*, 6475–6487. [CrossRef]

86. Zhang, Q.; Zhang, W.L.; Liang, Y.N. Adsorption of perfluoroalkyl and polyfluoroalkyl substances (PFASs) from aqueous solution—A review. *Sci. Total Sci. Total Environ.* **2020**, *748*, 142354. [CrossRef]

87. Xiao, F. Emerging poly-and perfluoroalkyl substances in the aquatic environment: A review of current literature. *Water Res.* **2017**, *124*, 482–495. [CrossRef] [PubMed]

88. Ochoa-Herrera, V.; Sierra-Alvarez, R. Removal of perfluorinated surfactants by sorption onto granular activated carbon, zeolite and sludge. *Chemosphere* **2008**, *72*, 1588–1593. [CrossRef]

89. Tang, H.; Xiang, Q.; Lei, M.; Yan, J.; Zhu, L.; Zou, J. Efficient degradation of perfluorooctanoic acid by UV–Fenton process. *Chem. Eng. J.* **2012**, *184*, 156–162. [CrossRef]

90. Gorenflo, A.; Veliizquez-Padon, D.; Frimmel, F.H. Nanofiltration of a German groundwater of high hardness and NOM content: Performance and costs. *Desalination* **2002**, *151*, 253–265. [CrossRef]

91. Hori, H.; Yamamoto, A.R.I. Efficient Decomposition of Environmentally Persistent Perfluorocarboxylic Acids by Use of Persulfate as a Photochemical Oxidant. *Environ. Sci. Technol.* **2005**, *39*, 2383–2388. [CrossRef] [PubMed]

92. Vecitis, C.D.; Wang, Y.; Cheng, J.I.E.; Park, H.; Mader, B.T. Sonochemical Degradation of Perfluorooctanesulfonate in Aqueous Film-Forming Foams. *Environ. Sci. Technol.* **2010**, *44*, 432–438. [CrossRef]

93. Franke, V.; McCleaf, P.; Lindegren, K.; Ahrens, L. Efficient removal of per-And polyfluoroalkyl substances (PFASs) in drinking water treatment: Nanofiltration combined with active carbon or anion exchange. *Environ. Sci. Water Res. Technol.* **2019**, *5*, 1836–1843. [CrossRef]

94. Zhang, D.; Luo, Q.; Gao, B.; Chiang, S.D.; Woodward, D.; Huang, Q. Sorption of perfluorooctanoic acid, perfluorooctane sulfonate and perfluoroheptanoic acid on granular activated carbon. *Chemosphere* **2016**, *144*, 2336–2342. [CrossRef]

95. Li, J.; Li, Q.; Li, L.; Xu, L. Removal of perfluorooctanoic acid from water with economical mesoporous melamine-formaldehyde resin microsphere. *Chem. Eng. J.* **2017**, *320*, 501–509. [CrossRef]

96. Du, Z.; Deng, S.; Chen, Y.; Wang, B.; Huang, J.; Wang, Y.; Yu, G. Removal of perfluorinated carboxylates from washing wastewater of perfluorooctanesulfonyl fluoride using activated carbons and resins. *J. Hazard. Mater.* **2015**, *286*, 136–143. [CrossRef]

97. Gao, Y.; Deng, S.; Du, Z.; Liu, K.; Yu, G. Adsorptive removal of emerging polyfluoroalky substances F-53B and PFOS by anion-exchange resin: A comparative study. *J. Hazard. Mater.* **2017**, *323*, 550–557. [CrossRef] [PubMed]

98. Franke, V.; Schäfers, D.; Lindberg, J.J.; Ahrens, L. Removal of per-and polyfluoroalkyl substances (PFASs) from tap water using heterogeneously catalyzed ozonation. *Environ. Sci. Water Res. Technol.* **2019**, *5*, 1887–1896. [CrossRef]

99. Andres, V.; Espana, A.; Mallavarapu, M.; Naidu, R. Treatment technologies for aqueous perfluorooctanesulfonate (PFOS) and perfluorooctanoate (PFOA): A critical review with an emphasis on field testing. *Environ. Technol. Innov.* **2015**, *4*, 168–181. [CrossRef]

100. Collings, A.F.; Gwan, P.B. Large scale environmental applications of high power ultrasound. *Ultrason. Sonochem.* **2010**, *17*, 1049–1053. [CrossRef] [PubMed]

101. Fernandez, N.A.; Rodriguez-Freire, L.; Keswani, M.; Sierra-Alvarez, R. Effect of chemical structure on the sonochemical degradation of perfluoroalkyl and polyfluoroalkyl substances (PFASs)[†]. *Environ. Sci. Water Res. Technol.* **2016**, 975–983. [CrossRef]

102. Zhao, L.; Bian, J.; Zhang, Y.; Zhu, L.; Liu, Z. Comparison of the sorption behaviors and mechanisms of perfluorosulfonates and perfluorocarboxylic acids on three kinds of clay minerals. *Chemosphere* **2014**, *114*, 51–58. [CrossRef]

103. Zhang, R.; Yan, W.; Jing, C. Mechanistic study of PFOS adsorption on kaolinite and montmorillonite, Colloids Surfaces A Physicochem. *Eng. Asp.* **2014**, *462*, 252–258. [CrossRef]

104. Wu, T.; Wu, Z.; Ma, D.; Xiang, W.; Zhang, J.; Liu, H.; Deng, Y.; Tan, S.; Cai, X. Fabrication of Few-Layered Porous Graphite for Removing Fluorosurfactant from Aqueous Solution. *Langmuir* **2018**, *34*, 15181–15188. [CrossRef]

105. Chen, X.; Xia, X.; Wang, X.; Qiao, J.; Chen, H. A comparative study on sorption of perfluorooctane sulfonate (PFOS) by chars, ash and carbon nanotubes. *Chemosphere* **2011**, *83*, 1313–1319. [CrossRef]

106. Zhang, Q.; Deng, S.; Yu, G.; Huang, J. Removal of perfluorooctane sulfonate from aqueous solution by crosslinked chitosan beads: Sorption kinetics and uptake mechanism. *Bioresour. Technol.* **2011**, *102*, 2265–2271. [CrossRef]

107. Fagbayigbo, B.O.; Opeolu, B.O.; Fatoki, O.S. Adsorption of perfluorooctanoic acid (PFOA) and perfluorooctane sulfonate (PFOS) from water using leaf biomass (Vitis vinifera) in a fixed-bed column study. *J. Environ. Health Sci. Eng.* **2020**, *18*, 221–233. [CrossRef] [PubMed]

108. Stebel, E.K.; Pike, K.A.; Nguyen, H.; Hartmann, H.A.; Klonowski, M.J.; Lawrence, M.G.; Collins, R.M.; Hefner, C.E.; Edmiston, P.L. Absorption of short-chain to long-chain perfluoroalkyl substances using swellable organically modified silica. *Environ. Sci. Water Res. Technol.* **2019**, *5*, 1854–1866. [CrossRef]

109. Qian, J.; Shen, M.; Wang, P.; Wang, C.; Hu, J.; Hou, J.; Ao, Y.; Zheng, H.; Li, K.; Liu, J. Co-adsorption of perfluorooctane sulfonate and phosphate on boehmite: Influence of temperature, phosphate initial concentration and pH. *Ecotoxicol. Environ. Saf.* **2017**, *137*, 71–77. [CrossRef] [PubMed]

110. Jian, J.M.; Zhang, C.; Wang, F.; Lu, X.W.; Wang, F.; Zeng, E.Y. Effect of solution chemistry and aggregation on adsorption of perfluorooctanesulphonate (PFOS) to nano-sized alumina. *Environ. Pollut.* **2019**, *251*, 425–433. [CrossRef] [PubMed]

111. Sunderland, E.M.; Hu, X.C.; Dassuncao, C.; Tokranov, A.K.; Wagner, C.C.; Allen, J.G. A review of the pathways of human exposure to poly-and perfluoroalkyl substances (PFASs) and present understanding of health effects. *J. Expo. Sci. Environ. Epidemiol.* **2019**, *29*, 131–147. [CrossRef] [PubMed]

112. Gebbink, W.A.; Berger, U.; Cousins, I.T. Estimating human exposure to PFOS isomers and PFCA homologues: The relative importance of direct and indirect (precursor) exposure Estimating human exposure to PFOS isomers and PFCA homologues: The relative importance of direct and indirect (precursor). *Environ. Int.* **2015**, *74*, 160–169. [CrossRef] [PubMed]

113. Skaar, J.S.; Ræder, E.M.; Lyche, J.L.; Ahrens, L.; Kallenborn, R. Elucidation of contamination sources for poly-and perfluoroalkyl substances (PFASs) on Svalbard (Norwegian Arctic). *Environ. Sci. Pollut. Res.* **2019**, *26*, 7356–7363. [CrossRef]

114. Andersen, M.E.; Butenhoff, J.L.; Chang, S.; Farrar, D.G.; Kennedy, G.L.; Lau, C.; Olsen, G.W.; Seed, J.; Wallace, K.B. Perfluoroalkyl Acids and Related Chemistries—Toxicokinetics and Modes of Action. *Toxicol. Sci.* **2008**, *102*, 3–14. [CrossRef]

115. Hoffman, K.; Webster, T.F.; Bartell, S.M.; Weisskopf, M.G.; Fletcher, T. Private Drinking Water Wells as a Source of Exposure to Perfluorooctanoic Acid Private Drinking Water Wells as a Source of Exposure to Perfluorooctanoic Acid (PFOA) in Communities Surrounding a Fluoropolymer Production Facility. *Environ. Health Perspect.* **2010**, *119*, 92–97. [CrossRef]

116. Longpré, D.; Lorusso, L.; Levicki, C.; Carrier, R.; Cureton, P. PFOS, PFOA, LC-PFCAS, and certain other PFAS: A focus on Canadian guidelines and guidance for contaminated sites management. *Environ. Technol. Innov.* **2020**, *18*, 100752. [CrossRef]

117. Seyoum, A.; Pradhan, A.; Jass, J.; Olsson, P. Perfluorinated alkyl substances impede growth, reproduction, lipid metabolism and lifespan in Daphnia magna. *Sci. Total Environ.* **2020**, *737*, 139682. [CrossRef] [PubMed]

118. Skogheim, T.S.; Villanger, G.D.; Weyde, K.V.F.; Engel, S.M.; Surén, P.; Øie, M.G.; Skogan, A.H.; Biele, G.; Zeiner, P.; Øvergaard, K.R.; et al. Prenatal exposure to perfluoroalkyl substances and associations with symptoms of attention-deficit/hyperactivity disorder and cognitive functions in preschool children. *Int. J. Hyg. Environ. Health* **2020**, *223*, 80–92. [CrossRef] [PubMed]

119. Niu, J.; Liang, H.; Tian, Y.; Yuan, W.; Xiao, H.; Hu, H.; Sun, X.; Song, X.; Wen, S.; Yang, L.; et al. Prenatal plasma concentrations of Perfluoroalkyl and polyfluoroalkyl substances and neuropsychological development in children at four years of age. *Environ. Health* **2019**, *18*, 53. [CrossRef] [PubMed]
120. McCarthy, C.; Kappleman, W.; DiGuiseppi, W. Ecological Considerations of Per-and Polyfluoroalkyl Substances (PFAS). *Curr. Pollut. Reports.* **2017**, *3*, 289–301. [CrossRef]
121. Cordner, A.; College, W.; Ave, B.; Walla, W.; States, U.; Richter, L.; Brown, P. Can Chemical Class Approaches Replace Chemical-by-Chemical Strategies? Lessons from Recent U.S. FDA Regulatory Action on per-and Polyfluoroalkyl Substances. *Environ. Sci. Technol.* **2016**, *50*, 12584–12591. [CrossRef] [PubMed]
122. Australian-Government. *Health Based Guidance Values for PFAS*; Australian-Government, Department of Health: Canberra, Australia, 2019.

Publisher's Note: MDPI stays neutral with regard to jurisdictional claims in published maps and institutional affiliations.

Article

Removal of Polycyclic Aromatic Hydrocarbons (PAHs) from Produced Water by Ferrate (VI) Oxidation

Tahir Haneef [1], Muhammad Raza Ul Mustafa [1,2,*], Khamaruzaman Wan Yusof [1], Mohamed Hasnain Isa [3], Mohammed J.K. Bashir [4], Mushtaq Ahmad [5] and Muhammad Zafar [5]

[1]   Department of Civil and Environmental Engineering, Universiti Teknologi PETRONAS, Seri Iskandar 32610, Perak, Malaysia; tahirhanifuaf@gmail.com (T.H.); khamaruzaman.yusof@utp.edu.my (K.W.Y.)

[2]   Centre for Urban Resource Sustainability, Institute of Self-Sustainable Building, Universiti Teknologi PETRONAS, Seri Iskandar 32610, Perak, Malaysia

[3]   Civil Engineering Programme, Faculty of Engineering, Universiti Teknologi Brunei, Tungku Highway, Gadong BE1410, Brunei; hasnain_isa@yahoo.co.uk

[4]   Department of Environmental Engineering, Faculty of Engineering and Green Technology (FEGT), Universiti Tunku Abdul Rahman, Kampar 31900, Perak, Malaysia; jkbashir@utar.edu.my

[5]   Department of Plant Sciences, Quaid-i-Azam University, Islamabad 45320, Pakistan; mushtaqflora@hotmail.com (M.A.); zafar@qau.edu.pk (M.Z.)

*   Correspondence: raza.mustafa@utp.edu.my

Received: 5 August 2020; Accepted: 21 September 2020; Published: 9 November 2020

Abstract: Polycyclic aromatic hydrocarbons (PAHs) are mutagenic and carcinogenic contaminants made up of fused benzene rings. Their presence has been reported in several wastewater streams, including produced water (PW), which is the wastewater obtained during oil and gas extraction from onshore or offshore installations. In this study, ferrate (VI) oxidation was used for the first time for the treatment of 15 PAHs, with the total concentration of 1249.11 µg/L in the produced water sample. The operating parameters viz., ferrate (VI) dosage, pH, and contact time were optimized for maximum removal of PAHs and chemical oxygen demand (COD). Central composite design (CCD) based on response surface methodology (RSM) was used for optimization and modeling to evaluate the optimal values of operating parameters. PAH and COD removal percentages were selected as the dependent variables. The study showed that 89.73% of PAHs and 73.41% of COD were removed from PW at the optimal conditions of independent variables, i.e., ferrate (VI) concentration (19.35 mg/L), pH (7.1), and contact time (68.34 min). The high values of the coefficient of determination (R^2) for PAH (96.50%) and COD (98.05%) removals show the accuracy and the suitability of the models. The results showed that ferrate (VI) oxidation was an efficient treatment method for the successful removal of PAHs and COD from PW. The study also revealed that RSM is an effective tool for the optimization of operating variables, which could significantly help to reduce the time and cost of experimentation.

Keywords: Fe (VI) oxidation; chemical oxygen demand; polycyclic aromatic hydrocarbons; central composite design; RSM

---

1. Introduction

Polycyclic aromatic hydrocarbons (PAHs) are hazardous organic micropollutants that are colorless or pale-yellow and widely present in the ecosystem. Chemically, these micropollutants consist of two or more fused benzene rings [1]. All PAHs, both low and high molecular weight, are stable and resistant to biodegradation [2]. These compounds are generated by both anthropogenic (industrial discharge, waste incineration, and biomass burning) and natural sources (natural oil seeps, forest fires, and volcanic eruptions). PAHs can cause cancer (i.e., lung, bladder, and skin cancer) in human beings and also severe health problems in aquatic life by inhalation and ingestion even in very low concentrations (ng/L–µg/L) [1,3]. The United States Environmental Protection Agency (USEPA) has categorized 16 PAHs as priority micropollutants because of their mutagenic and carcinogenic effects [4]. Produced water (PW) is also one of the largest anthropogenic sources that contains a considerable amount of PAHs [2,5]. PW is a byproduct of oil and gas industries, which is generated during the oil and gas extraction process. Many studies have confirmed that acute and chronic toxicity of PW is mainly because of PAHs, phenols, and high amounts of COD [6]. Globally, the production of PW increases day by day, and its production has been reported to have reached up to 250 million barrels per day [7]. Moreover, almost 40% of PW is directly discharged into water bodies without any treatment [6]. The direct discharge of untreated PW into the environment contaminates surface and groundwater. Several treatment methods, such as volatilization, combined microfiltration and biological processes, chlorination, biochar adsorption, ozonation, electrodialysis, reverse osmosis, electrocoagulation, ion exchange, membrane-based technology, and conventional phase separation, have been employed for the treatment of PW [8,9]. Although more than 90% of organic pollutants removal from PW was attained by electrocoagulation, it is not cost-effective, produces a large amount of sludge, and consumes more energy [10]. Membrane-based technologies are efficient for PW treatment; however, these technologies have several problems such as membrane fouling, high energy consumption, and the lack of potential to degrade refractory organic pollutants [11]. Combined microfiltration and biological process reported almost 65% of COD removal from oilfield wastewater; however, it is also a time-consuming process [12]. More than 50% removal of total dissolved solids from PW was obtained by electrodialysis, but it utilized a large amount of energy to accomplish the treatment [13]. Constructed wetland is an efficient method for organic matter removal from PW, but its maintenance cost is very high [14]. Moreover, most of the techniques are very expensive. In many cases, they just transfer organic pollutants from one phase to another and cannot remove dissolved organic contaminants from PW. Some of the available technologies produce toxic byproducts that limit their practical use [15]. Therefore, it is imperative to explore economic and ecotechnological solutions for PAH reduction from PW.

Different materials/chemicals such as zeolites, metal oxides, and nanoparticles have been applied for the remediation of contaminants from water and wastewater [16]. In recent years, ferrate (VI) (Fe (VI)) with high valent iron (VI) has gained attention due to its high oxidation/reduction potential [17–19]. The Fe (VI) oxidation method is a promising technique due to its environment-friendly nature, low cost, and high efficiency for organic pollutant removal [20]. Surprisingly, Fe (VI) acts as a coagulant, disinfectant, and oxidizer at the same time. Fe (VI) is considered as one of the most efficient oxidants for the treatment of wastewater due to its strong oxidizing power [3]. The redox potential of the Fe (VI) ion in both acidic and neutral environments is higher than many other oxidizing agents, as shown in Table 1, making it a favorable oxidant for the treatment of wastewater [21]. In an acidic environment, Fe (VI)'s electrode potential is 2.20 V, and, in an alkaline medium, it is 0.72 V. However, the Fe (VI) cations' structure can be modified by adjusting the pH value to control the oxidation activity, so as to achieve high selectivity [22]. Fe (VI) is a robust multifunctional oxidizer with a tetrahedral structure (FeO_4^{2-}). Fe (VI) is converted into Fe^{3+} and $Fe(OH)_3$ during disinfection and oxidation processes and acts as coagulant and oxidizer, as shown in Equations (1) and (2). Fe (VI) produces molecular oxygen during spontaneous oxidation, as shown in Equation (3) [23]. The reactivity of Fe (VI) with refractory organic and inorganic pollutants shows its usefulness for the removal of pollutants from industrial effluents [19,21].

$$FeO_4^{2-} + 8H^+ + 3e^- \rightarrow Fe^{3+} + 4H_2O \tag{1}$$

$$FeO_4^{2-} + 4H_2O + 3e^- \rightarrow Fe(OH)_3 + 5OH^- \tag{2}$$

$$2FeO_4^{2-} + 5H_2O \rightarrow 2Fe^{3+} + 3/2O_2 + 10OH^- \tag{3}$$

Table 1. Redox potential of different oxidants.

Oxidant	E° (Volt)	Oxidant	E° (Volt)
Permanganate	1.67	Dissolved oxygen	1.22
Ozone	2.07	Perchlorate	1.38
Chlorine dioxide	0.95	Hypochlorite	1.48
Chlorine	1.35	Ferrate (VI)	2.20

Response surface methodology (RSM) using Design-Expert software is a powerful statistical and mathematical tool that is commonly used for the systematic design and analysis of experiments. It provides optimization and validation of a system based on its statistical modeling. RSM is far better than the conventional one-factor-at-a-time optimization technique because it helps to reduce the vast amount of laboratory experimental work. Traditional methods are complicated, time-consuming, and expensive for multivariable experiments [24]. Moreover, the influence of multiple variables on responses during the optimization process can be studied in RSM [25]. It also addresses the interaction between different independent variables and can be practiced in multivariable analyses for the optimization of functional variables [24].

Various researchers have studied the application of Fe (VI) for remediation of different organic pollutants in several types of wastewater, such as municipal wastewater secondary effluent [18], oil sands process-affected water [26], fracturing wastewater [27], dyeing effluent [28], textile wastewater [29], coking wastewater [30], and tannery wastewater [31]. In addition, a few studies have explored the potential of Fe (VI) for the removal of just one or two PAHs from synthetic wastewater rather than from real wastewater, especially PW [22,30,32]. Guan et al. [22] investigated the separate removal of only three PAHs, i.e., phenanthrene, pyrene and naphthalene, from synthetic water using Fe (VI) oxidation. Li et al. [30] studied the potential of Fe (VI) for the removal of just one PAH (phenanthrene) in coking wastewater. Similarly, Tan et al. [32] evaluated the potential of Fe (VI) just for one PAH, i.e., phenanthrene removal in synthetic wastewater. Each PAH has different characteristics based on its number of benzene rings; each PAH may react differently against Fe (VI) oxidation. More studies are required to investigate the potential of Fe (VI) for combination of PAHs, especially priority PAHs in aqueous media, so that consistent performance is achieved. It appears from literature that, so far, no study has explored the application of Fe (VI) oxidation for 15 PAHs and COD removal from PW. Secondly, integrated optimization of 15 PAHs and COD removal using RSM is also yet to be explored. Therefore, understanding of the performance of Fe (VI) for combined PAH removal is important to fill the gap in the previous studies. The objectives of this study are (i) to evaluate the potential of Fe (VI) for PAHs and COD removal from PW and (ii) to optimize the independent parameters viz., Fe (VI) concentration, pH, and contact time using RSM.

2. Experimental Work

2.1. Materials/Chemicals

All chemicals used in this work were analytical grade and were utilized without purification. H_2SO_4 (95–98%) and NaOH (30% *w/v*) were purchased from R&M Chemicals Malaysia. Commercially available potassium Fe (VI) was purchased from NANO IRON (Židlochovice, Czech Republic) and utilized as received. Syringe filters (25 mm dia, cat. no. 6874–2504) by Whatman U.S.A. were used. PW samples were collected from an oil and gas exploration site in the South East Asia region and stored in cold storage at 4 °C in compliance with the standard protocol of the American Public Health Association (APHA) [33].

2.2. Fe(VI) Oxidation and GC-MS Analysis

For the oxidation process, a 500 mL glass beaker was used as a reactor. Aluminum foil was used to cover the reactor for shielding the water sample from light. The pH of PW was 8.2, and it was adjusted according to experimental requirements during the oxidation process using 1 M solutions of NaOH and H_2SO_4. The temperature and magnetic stirrer speed during the oxidation process were kept constant at 25 °C and 250 rpm, respectively. The reactor was positioned on a hot/magnetic plate, and a magnetic stirrer inside the reactor was used to blend reagent homogeneously in PW. The oxidation process was initiated by adding a measured quantity of Fe (VI) into the water sample. The oxidation process was performed for different contact times, varying from 10–120 min. The reaction was stopped after a pre-decided contact time by increasing the pH of the solution up to pH 12.0 with the help of a 1 M NaOH solution. The treated water sample was filtered using WhatmanTM filter paper. The filtered PW samples were utilized for COD and PAH analysis using HACH vial and gas chromatography–mass spectrometry (GC–MS), respectively. All the experiments were performed in triplicate, and average values were taken.

Polycyclic aromatic hydrocarbons in untreated and treated PW samples were quantified using GC–MS analysis. Liquid–liquid extraction (LLE) was performed before the GC–MS analysis to concentrate the PAHs present in PW samples. The USEPA 3510C LLE technique, with some modifications, was employed for PAH extraction. Methylene chloride (DCM) was used as a solvent extractor for PAH extraction in LLE. After the LLE, the samples were re-concentrated by a rotary evaporator. In the rotary evaporator, the pressure of the condenser was set at 789 bar, and bath temperature was held at 40 °C. The rotation of the receiving flask was set at 30 rpm. Then, a water sample was taken in the receiving flask and evaporated until the volume of the sample reached 1.0 to 0.5 mL. After that, the water sample was transferred into a GC–MS vial (1.5 mL) for GC–MS analysis. In GC–MS, a column of 30 m (Elite 5MS) with 0.25 mm inner diameter (ø) and 0.25 μm film thickness was used, while helium (He) gas was utilized as a transporter gas. The column temperature was raised from 60 to 175 °C (at 6 °C/min) and then raised up to 240 °C (at 3 °C/min) and then finally held at 300 °C for 7 min. Injector and transition line temperatures were 280 and 300 °C, respectively [34]. For PAH quantification, a standard solution with 16 PAHs (2000 mg/L concentration) was utilized. Equations (4) and (5) were applied to calculate percentage removal of COD and PAHs, respectively.

$$X = \frac{D_i - D_f}{D_i} \times 100 \tag{4}$$

where X shows the percentage of COD removal; D_i and D_f represent the concentration of COD before and after treatment, respectively.

$$Z = \frac{P_i - P_f}{P_i} \times 100 \tag{5}$$

where Z represents the percentage of PAHs removal; P_i and P_f show the concentration of PAHs before and after treatment, respectively.

2.3. Development of Experimental Design Using RSM

Response surface methodology (RSM) is an efficient statistical and mathematical tool that is usually used for developing experimental designs. It optimizes the operating parameters and predicts the responses based on polynomial quadratic models. Furthermore, it also validates the interaction between independent and dependent variables. The entire process consists of four stages, i.e., (i) the experiment layout, (ii) the response selection by conducting experiments, (iii) the design of an RSM numerical model, and (iv) developed model testing for validation and optimization [35]. In RSM, the most effective and widely applied method for the development of experimental design is the central composite design (CCD) method [25]. This method helps to reduce the number of experiments and evaluate the interaction between the variables. The selection of data points in a two-level factorial design (i.e., CCD) includes minimum (−1), maximum (+1), and middle (0) points for all parameters [24]. A quadratic model (Equation (6)) developed by CCD is applied for response estimation [25].

$$Y = \beta_0 + \sum \beta_i x_i + \sum \beta_{ii} x_i^2 + \sum \beta_{ij} x_i x_j \tag{6}$$

where Y indicates the responses (i.e., COD/PAHs removal); β_0, β_i, β_{ii}, and β_{ij} stand for the constant-coefficients, i.e., linear coefficients, quadratic coefficient, and interaction coefficient, respectively; x_i and x_j are independent parameters. The defined independent variables, with their codes and levels (low, medium, and high), are presented in Table 2. The values for the three independent parameters were carefully chosen by conducting preliminary tests after a comprehensive literature review [7,36,37].

Table 2. Ranges of independent parameters and coded values for the experimental design.

Factors	Independent Parameters	Low	Medium	High
		−1	**0**	**0**
A	Fe (VI) concentration (mg/L)	10	20	30
B	pH	5.0	7.5	10.0
C	Contact time (min)	10	50	90

The experimental design was developed using Design-Expert software (version 11). In CCD design, k, 2^k, 2k, and n stand for the number of independent factors, the number of factorial experiments, the number of axial point experiments, and the number of center point experiments, respectively. The total number of tests can be determined using the formula ($2^k + 2k + n$) [38]. Thus, a total of 20 experimental runs were designed by CCD based on three independent variables (k = 3), including 8 factorial point experiments, 6 axial point experiments, and 6 central point experiments (Table 3).

Table 3. Experimental trials designed by response surface methodology (RSM) for the reduction of chemical oxygen demand (COD) and polycyclic aromatic hydrocarbons (PAHs).

Experiment No.	Parameters		
	Fe (VI) Concentration (mg/L)	Contact Time (min)	pH
1	10	10	10.0
2	20	50	7.5
3	10	50	7.5
4	10	90	10.0
5	20	50	5.0
6	20	50	7.5
7	30	50	7.5
8	30	90	5.0
9	10	90	5.0
10	30	10	5.0
11	20	10	7.5
12	20	50	7.5
13	20	50	7.5
14	20	50	7.5
15	30	10	10.0
16	20	90	7.5
17	10	10	5.0
18	20	50	7.5
19	20	50	10.0
20	30	90	10.0

3. Results and Discussion

3.1. Concentration of PAHs and COD in PW

In PW, 15 PAHs were quantified via GC–MS analysis, with a total concentration of PAHs (ΣPAHs) of 1249.11 µg/L, as shown in Figure 1. It was noticed that the concentration of naphthalene (196.46 µg/L) was higher than the other PAH concentrations in PW, while benzo (g, h, i) perylene concentration (27.15 µg/L) was the lowest. Fluorene and carbazole were both 90.00 µg/L. In addition, the concentration of COD in PW was found to be 2213 mg/L. The high concentrations of COD and PAHs in PW indicate that PW contains a high amount of organic pollutants that needed to be removed before discharging PW into the environment.

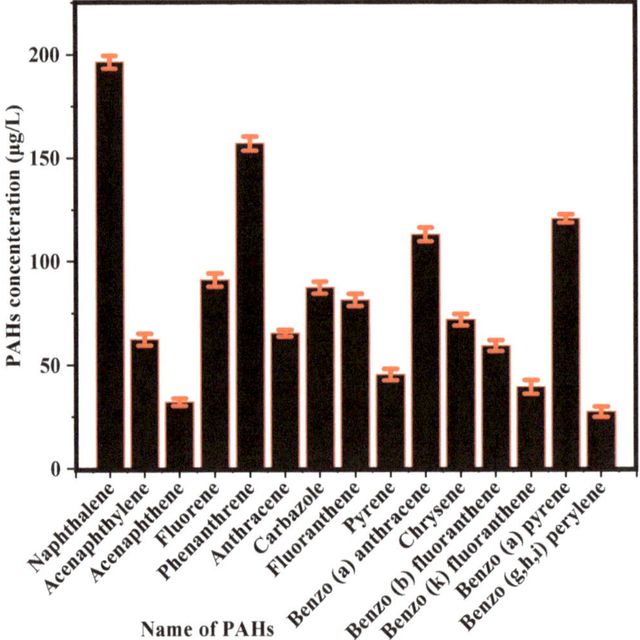

Figure 1. Quantified 15 PAHs in produced water (before treatment) via GC–MS analysis.

3.2. Screening Batch Experiments

The screening experiments were conducted to study the impact of the three independent parameters: the concentration of Fe (VI) (*A*), pH (*B*), and contact time (*C*) on COD removal. Different ranges of these variables were examined for the oxidation process, as shown in Figure 2.

As shown in Figure 2a, various pH values were tested in the range of 3.0 to 10.0 at constant Fe (VI) concentration (20 mg/L) and contact time (60 min) for the removal of COD. Maximum COD reduction (68.0%) was obtained at pH 8.0. A rising trend of COD removal was observed from 35.3% to 68.0%, with an increase in pH value from 3.0–8.0. Moreover, a sudden decline in COD removal was noted as the pH value was increased beyond 8.0. It may have happened due to the generation of reactive byproducts at the higher pH, which adversely affected the COD removal and also affected the stability and oxidation potential of ferrate ions [39]. Subsequently, the COD removal reached 50.4% at pH 10.0. Moreover, after observing the effect of pH on COD removal, the pH range of 5.0 to 10.0 was chosen for further comprehensive analysis of PAHs and COD removal via RSM optimization.

Figure 2b shows the effect of different Fe (VI) concentrations (1 to 35 mg/L) at constant pH (8.0) and contact time (60 min) on COD removal. The highest COD reduction (70.2%) was attained at 25 mg/L of

Fe (VI) concentration. An increasing trend of COD removal was observed from 29.0% to 70.2% with an increase in the concentration of Fe (VI) from 1 to 25 mg/L. However, a sudden decline in COD removal efficiency was noticed as the concentration of Fe (VI) was increased beyond 25 mg/L. It appeared that as the concentration of Fe (VI) was increased, more Fe^{3+} ions were generated, which may have raised Fe (VI) decomposition instead of organic pollutant decomposition and adversely affected the COD removal [40]. Subsequently, the COD removal at 35 mg/L of Fe (VI) concentration was reduced to 49.9%. After examining the impact of Fe (VI) concentration on COD removal, the Fe (VI) concentration range of 10 to 30 mg/L was selected for the optimization process.

As presented in Figure 2c, different contact times from 10 to 120 min were used for COD removal at constant pH (8.0) and Fe (VI) concentration (25 mg/L). The peak removal of COD (70.9%) was achieved at 50 min of contact time. A rising trend for COD removal was noticed from 38.0% to 70.9% with an increase of contact time from 10 to 50 min. COD removal remained constant, and no further change was observed after the contact time of 50 min. Thus, the contact time range of 10 to 90 min was adopted for optimization analysis using RSM.

The ranges of independent variables obtained via preliminary experiments were used for RSM optimization. The values of these three independent variables were used as minimum and maximum in the Design-Expert software (CCD/RSM) for both PAHs and COD removal. PAHs are organic compounds that are also part of COD, so the same values (ranges) of independent variables attained from the screening test can also be used for the optimization study of PAHs and COD via CCD.

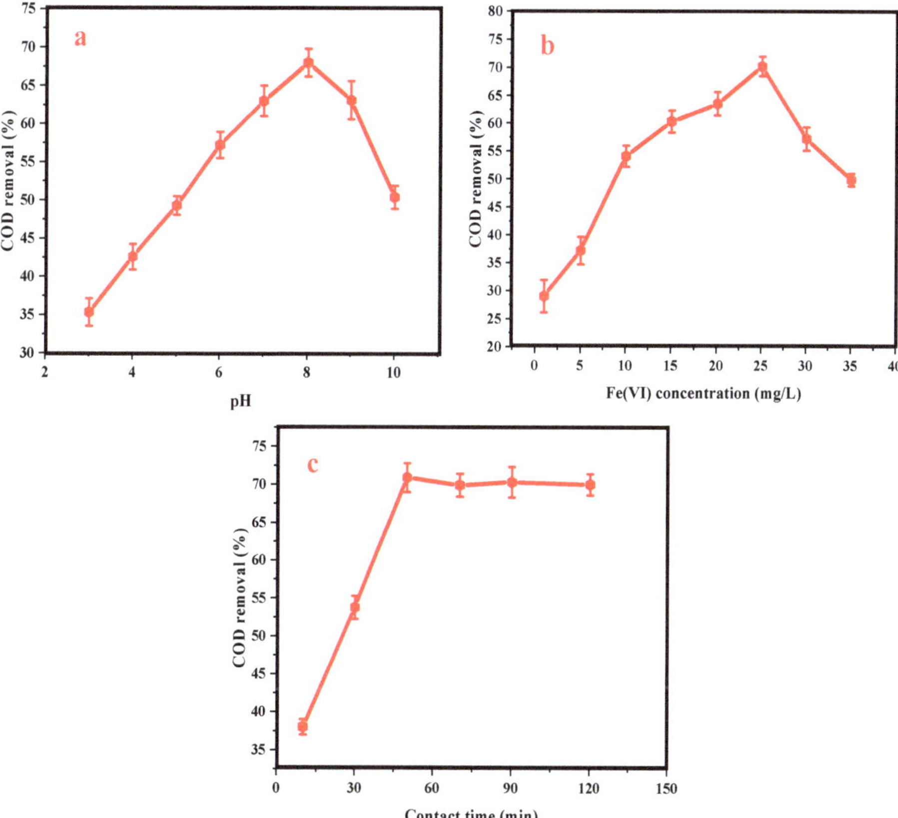

Figure 2. Screening test results for COD removal: (**a**) removal of COD (%) against pH, (**b**) removal of COD (%) against Fe (VI), (**c**) removal of COD (%) against contact time.

3.3. Analysis of Variance (ANOVA)

Analysis of variance (ANOVA) results obtained from CCD based on a second-order model (Equation (6)) were used to evaluate the adequacy and significance of the models, as shown in Tables 4 and 5. The significance of the models was determined based on p-values and F-values of coefficient terms. From Table 4, it can be noticed that the F-value and p-value of the model for COD removal were 55.95 and 0.001, respectively, which indicated that the model was significant and could be utilized for COD removal optimization. Model terms, Fe (VI) concentration *(A)*, pH *(B)*, contact time *(C)*, the two-level interaction of Fe (VI) concentration and pH *(A × B)*, pH and contact time *(B × C)*, and the quadratic effect of Fe (VI) concentration (A^2), pH (B^2), and contact time (C^2) were significant. The lack of fit of the model was nonsignificant with a 0.211 p-value. In addition, the adequacy of a model can be verified by R-squared values (R^2). The high value of R^2 (98.05%) implied that the model could not describe only 1.95% of the overall sample variation. The value of adjusted R^2 (96.30%) was also high, which illustrated that the model represented the system well. The high value of predicted R^2 (89.23%) indicated that the current model was able to predict appropriate responses for new observations.

Table 4. Results from the ANOVA table for COD removal.

Source	* SS	* DF	* MS	F-Value	p-Value	Model's Status
Model	4446.85	9	494.09	55.95	0.0001	Significant
A	60.02	1	60.02	6.80	0.0262	
B	67.60	1	67.60	7.65	0.0199	
C	276.60	1	276.68	31.33	0.0002	
AB	45.13	1	45.13	5.11	0.0473	
AC	32.00	1	32.00	3.62	0.0861	
BC	72.00	1	72.00	8.15	0.0171	
A^2	646.00	1	646.04	73.16	0.0001	
B^2	349.70	1	349.74	39.60	0.0001	
C^2	85.54	1	85.54	9.69	0.0110	
Residual	88.31	10	8.83			
Lack of Fit	60.20	5	12.04	2.14	0.2116	Non-significant
Model summary:	R^2	$R^2_{adj.}$	$R^2_{pred.}$			
	98.05%	96.30%	89.23%			

Note: * SS = sum of square, * DF = degree of freedom, * MS = mean square.

Table 5. Results from the ANOVA table for PAH removal.

Source	* SS	* DF	* MS	F-Value	p-Value	Model's Status
Model	5104.18	9	567.13	30.93	0.0001	Significant
A	110.89	1	110.89	6.05	0.0337	
B	158.40	1	158.40	8.64	0.0148	
C	305.81	1	305.81	16.68	0.0022	
AB	113.25	1	113.25	6.18	0.0323	
AC	150.51	1	150.51	8.21	0.0168	
BC	0.01	1	0.01	0.00	0.9807	
A^2	295.88	1	295.88	16.13	0.0025	
B^2	723.74	1	723.74	39.47	0.0001	
C^2	141.48	1	141.48	7.71	0.0195	
Residual	183.39	10	18.34			
Lack of Fit	73.30	5	14.66	0.6669	0.6669	Non-significant
Model summary:	R^2	$R^2_{adj.}$	$R^2_{pred.}$			
	96.50%	93.40%	79.72%			

Note: * SS = sum of square, * DF = degree of freedom, * MS = mean square.

The F-value and p-value of the PAH model were 30.93 and less than 0.001, respectively, as shown in Table 5. These values showed that the model was significant and can be utilized for PAH removal optimization. Model terms, Fe (VI) concentration *(A)*, pH *(B)*, contact time *(C)*, the two-level interaction of Fe (VI) concentration and pH $(A \times B)$, Fe(VI) concentration and contact time $(A \times C)$, and the quadratic effect of Fe (VI) concentration (A^2), pH (B^2), and contact time (C^2) were significant. The lack of fit of the model was nonsignificant, with a 0.666 p-value. Moreover, the adequacy of the model can be evaluated by R squared (R^2) values. The high value of R^2 (96.53%) suggested that the model could not describe only 3.47% of the overall sample variation. The value of adjusted R^2 (93.41%) was also high, which demonstrated the suitability of the model to describe the system. The high value of predicted R^2 (79.72%) indicated that the model can predict appropriate responses for new observations. The quadratic models for COD and PAH removal are shown in Equations (7) and (8).

$$Y1 = 68.26 + 2.45A - 2.60B + 5.26C + 2.37AB - 2.00AC - 3.00BC - 15.33A^2 - 11.28B^2 - 5.58C^2 \tag{7}$$

$$Y2 = 89.54 - 3.33A - 3.98B + 5.53C - 3.76AB - 4.34AC + 0.03BC - 10.37A^2 - 16.22B^2 - 7.17C^2 \tag{8}$$

where Y1 and Y2 are the removal of COD and PAHs, respectively; A, B, and C show the concentration of Fe (VI), pH, and contact time, respectively. The predicted (attained from Equations (7) and (8)) and experimental responses are presented in Table 6. It revealed that the proposed polynomial quadratic models were appropriate for the prediction of responses and showed a remarkably good agreement between the results.

Table 6. Experimental and predicted results of PAHs and COD in percentage (%).

Experimental Run	Actual PAHs Removal	Predicted PAHs Removal	Actual COD Removal	Predicted COD Removal
1	45.50	48.98	27.00	24.39
2	82.30	89.54	71.30	68.26
3	85.30	82.50	47.00	50.48
4	70.00	68.79	33.00	32.91
5	75.20	77.30	60.30	59.58
6	94.30	89.54	67.30	68.26
7	72.30	75.84	56.50	55.38
8	65.00	61.34	43.00	45.01
9	67.80	69.15	49.00	48.86
10	58.00	59.03	33.00	32.49
11	79.00	76.84	53.20	57.42
12	90.10	89.54	65.30	68.26
13	91.70	89.54	68.90	68.26
14	86.30	89.54	70.50	68.26
15	45.00	43.47	38.50	38.04
16	85.00	87.90	69.80	67.94
17	50.30	49.49	29.00	28.34
18	94.00	89.54	71.00	68.26
19	70.70	69.34	51.30	54.38
20	45.30	45.93	38.50	38.56

The COD and PAH removal predicted by the polynomial quadratic models agreed well with the experimental values. Furthermore, the distribution of all data points was close to the 45° straight line, as shown in Figure 3. These graphs showed a satisfactory agreement between experimental and predicted values. The model adequacy was also determined by the normal plot of residuals from the least square fit. The assumption of normality was checked by creating a plot between normal percentage of probability and studentized residuals, which was found to be satisfactory for COD and PAH removal efficiencies, as shown in Figure 4. All the data points in the graphs of residuals were approximately along the straight line, which is an indication of normal distribution.

3.4. Response Surface and Contour Plots

Three-dimensional (3D) surface and contour plots of the quadratic models obtained from Equations (7) and (8) were used for the graphical representation of independent variables' impact on COD and PAH removal. These plots also showed the interaction among independent variables, as illustrated in Figures 5–7. In these graphs, two independent variables were continually varied for the responses, and the remaining variable was fixed. Correlations between the independent parameters were significant, so the peaks of the response surface plots were prominent, as shown in Figures 5–7. All response plots had noticeable peaks, which implied that all the variables in the design space were given optimum conditions for the highest response.

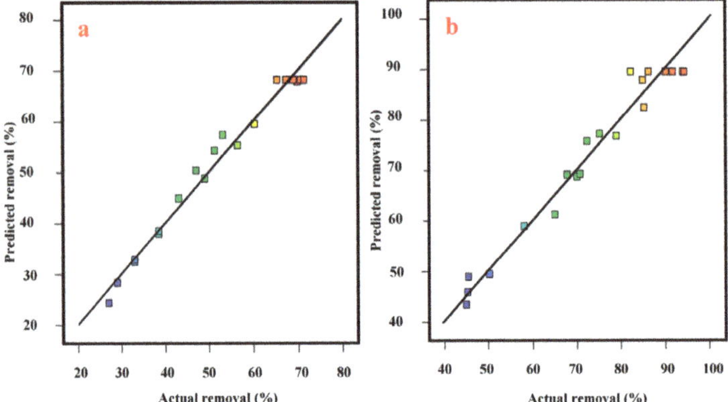

Figure 3. Comparison between predicted and experimental removal efficiencies. (**a**) Predicted COD removal vs. experimental COD removal (%) (**b**) Predicted PAH removal vs. experimental PAH removal (%).

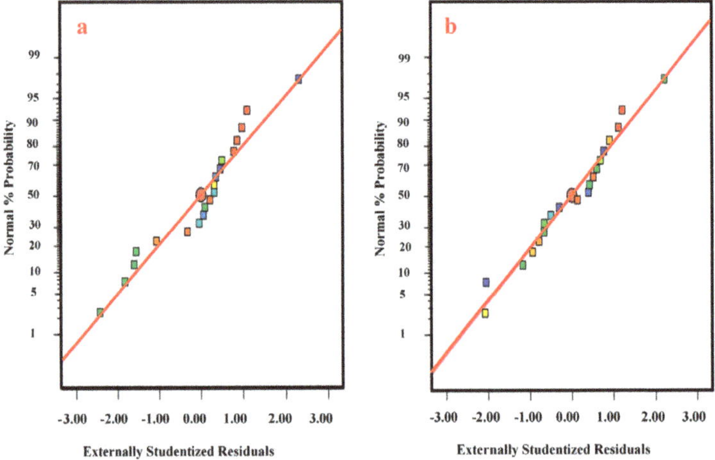

Figure 4. Residual normal plots for (**a**) COD removal efficiency and (**b**) PAH removal efficiency.

3.4.1. pH Effect on COD and PAH Removal

The interaction between pH and Fe (VI) concentration at 50 min of contact time (fixed) is presented in Figure 5; the peaks in both plots were prominent. The COD and PAH removal efficiencies increased by increasing the pH of the aqueous solution and concentration of Fe (VI) until specific values of both factors in Fe (VI) oxidation. A strong interaction was observed between both variables (Fe (VI) and pH), where the p-values of both parameters' interaction were found to be significant for COD (p-value 0.0473) and PAH (p-value 0.0323) removal. Moreover, more information on the interaction between pH and Fe (VI) concentration can be attained from the plots, as shown in Figure 5. The COD and PAH removal in Fe (VI) oxidation were highly dependent on pH. It was observed at pH 5.0, the COD and PAH removal efficiencies were 45.0% and 67.0% (respectively) and it increased further when pH was increased from pH 5.0 to pH 7.5. The maximum removal of COD (71.3%) and PAHs (94.3%) were obtained at pH 7.5, which showed that near-neutral pH was more suitable for Fe (VI) and micropollutants reaction in PW. Additionally, based on the working mechanism, Fe (VI) is converted to $Fe^{3+}/Fe(OH)_3$ in an aqueous solution, and these products slightly increase the pH of the solution, which affects the decomposition process [41]. The removal efficiencies were decreased when the pH of the water sample was increased from pH 7.5 to pH 10.0, as shown in Figure 5. It was expected that Fe (VI) stability would increase with the pH increment, whereas the redox potential would decrease. In addition, Fe (VI) has high oxidation potential and is highly unstable (rapidly reduced) in an acidic environment due to the high concentration of H^+ existence in aqueous solution [42]. Hence, Fe (VI) quickly reduced in an acidic environment and could not completely react with organic pollutants, which resulted in incomplete oxidation of these hazardous pollutants, while in a neutral environment, the Fe (VI) stability and redox potential were comparatively higher, and oxidation process was relatively more favorable for the degradation of micropollutants. As a result, higher COD and PAH removal efficiencies were attained under a neutral environment. In comparison with the previous studies, Karaatli et al. [43] and Song et al. [44] obtained maximum removal of pollutants from lake water at 6.5 pH. Similarly, Ciabatti et al. [28] achieved maximum remediation (70.0%) of organic pollutants from dyeing effluents at 8.5 pH. Jiang et al. [17] reported almost pH 10.0 as the optimum value for the highest degradation of contaminates from domestic sewage effluent by Fe (VI) oxidation. The optimum value of pH of the current study was different from the previous studies because, in previous studies, they employed Fe (VI) for the treatment of different types of source water/wastewater that contained different kind of pollutants, which influenced the Fe (VI) oxidation accordingly. In the present study, the maximum removal of pollutants was attained at pH 7.5, which indicates that PW influenced the Fe (VI) oxidation. Optimum values of operating parameters may vary depending on the characteristic of the wastewater.

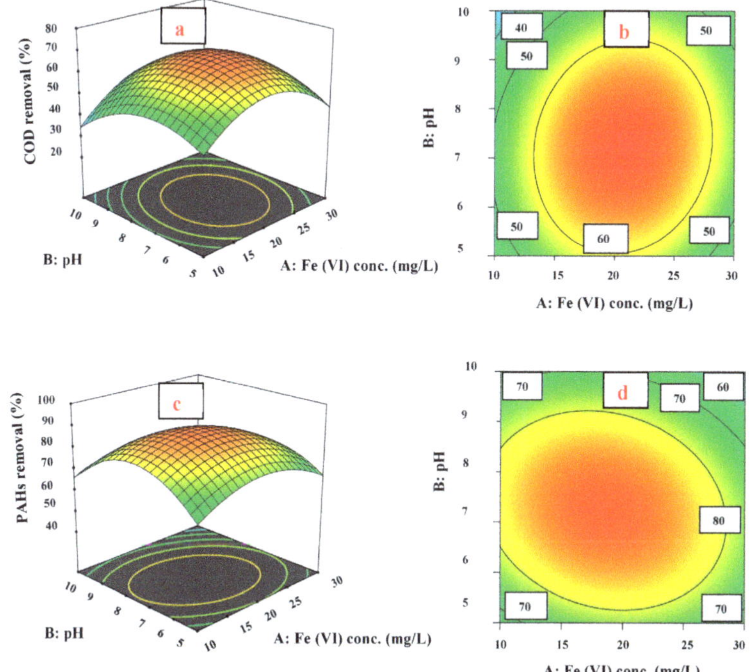

Figure 5. Effect of pH on COD and PAH removal. (**a**) 3D plot and (**b**) couture plot for COD removal; (**c**) 3D plot and (**d**) couture plot for PAH removal.

3.4.2. Fe (VI) Concentration Effect on COD and PAH Removal

The correlation between contact time and Fe (VI) concentration at pH 7.5 (fixed) is presented in Figure 6. The plots showed that the interaction between both parameters (contact time and Fe (VI) concentration) was nonsignificant (*p*-value 0.0861) for COD reduction while it was significant (*p*-value 0.0168) for PAH removal. The response surface graphs indicated that COD reduction was significantly affected by the individual impact, and there was some interaction between both parameters. When pH was fixed at 7.5, the reaction time showed a trend where it was proportional to the COD removal, while a reverse effect was observed when increasing the Fe (VI) dosage beyond optimum value. It may be because of more Fe^{3+} ions, which adversely affected the COD reduction and caused insignificant interaction between both parameters.

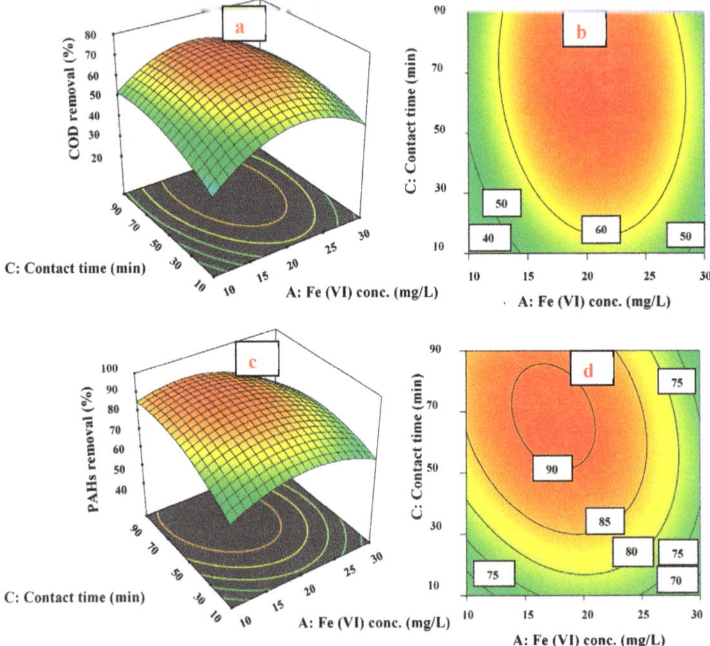

Figure 6. Effect of Fe (VI) concentration on COD and PAH removal. (**a**) 3D plot and (**b**) couture plot for COD removal; (**c**) 3D plot and (**d**) couture plot for PAH removal.

In addition, the concentration of Fe (VI) played a significant role in COD and PAH removal, as shown in Figure 6. At 10 mg/L of Fe (VI) concentration, the removal percentage of COD and PAHs were 38.0% and 65.0%, respectively. The rise in COD and PAH removal was observed when the concentration of Fe (VI) was increased from 10 to 20 mg/L. The maximum removal of COD and PAHs at 58.0% and 78.0%, respectively (at 10 min of contact time), was observed at 20 mg/L of Fe (VI) concentration. As the Fe (VI) concentration was enhanced, the degradation was predominant in approximately neutral environment, which increased the ability of $HFeO_4^-$ (Fe (VI) in the protonated form) to react with organic pollutants [30,45]. Thus, $HFeO_4^-$ resulted in surging the COD and PAH removal efficiencies. Moreover, a reverse phenomenon was observed as the concentration of Fe (VI) was increased above 20 mg/L; the COD and PAH removal efficiencies declined. The COD and PAH removal percentage at 30 mg/L of Fe (VI) concentration decreased to 47.0% and 67.7% (at 10 min of contact time), respectively. Micropollutant removal did not improve by increasing the Fe (VI) concentration beyond 20 mg/L. It appeared that as the concentration of Fe (VI) was increased, more Fe^{3+} ions were generated, which may have increased Fe (VI) decomposition instead of organic pollutant decomposition [40].

Beyond the optimum value of Fe (VI), the oxidized products generated during the reaction of Fe (VI) and organic pollutants consumed an extra amount of Fe (VI) and decreased the removal efficiencies of COD and PAHs, as shown in Figure 6. Wang et al. [46] reported 20 mg/L Fe (VI) concentration as the optimal value for maximum removal of organic pollutants from papermaking wastewater in line with the current study. Some studies, however, reported different optimal values of Fe (VI) depending on the characteristics of the wastewater [28,32], as PW characteristics affected the Fe (VI) oxidation as well.

3.4.3. Contact Time Effect on COD and PAH Removal

In Figure 7, the effect of pH and contact time on COD and PAH removal have been shown at fixed Fe (VI) concentration (20 mg/L). The interaction between both parameters significantly affected the COD removal in PW. While the interaction between both parameters was insignificant for PAH removal, the response surface graphs indicated that PAH reduction was significantly affected by the individual impact, and there was some interaction between both parameters. It may have been due to the sensitive nature of Fe^{3+} to pH value in aqueous media. The plots portrayed that increasing the amount of both parameters up to a specific value increased the removal efficiencies of PAHs and COD. The contact time impact on organic compound degradation was studied by varying the time range from 10 to 90 min. As shown in Figure 7, it was noticed that by increasing the contact time, COD and PAH removal efficiencies also increased up to a specific value of contact time. Initially, after 10 min of contact time, the reduction percentage of COD and PAHs was 53.5% and 73.7%, respectively, at pH 6.0. The highest reduction of COD and PAHs of 66.4% and 87.8% was obtained at 50 min of contact time. After 50 min, the degradation of organic pollutants was constant. It could be expected that the Fe (VI) amount was completely consumed within 50 min of contact time, so it was not available after that time for further reaction. In this study, the COD and PAH removal were stable after 50 min of reaction time, which indicated that Fe (VI) was completely degraded within 50 min.

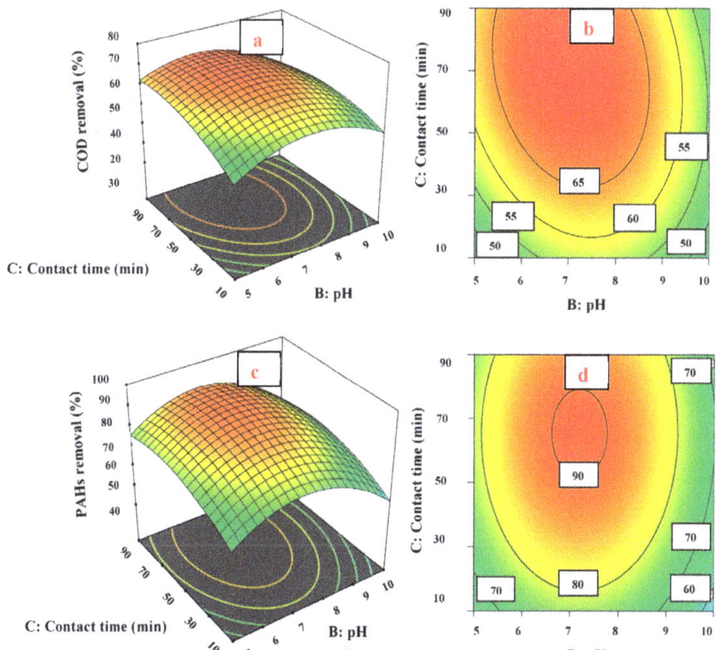

Figure 7. Effect of contact time on COD and PAH removal. (**a**) 3D plot and (**b**) couture plot for COD removal; (**c**) 3D plot and (**d**) couture plot for PAH removal.

3.5. Optimization and Validation

Central composite design (CCD)-based numerical optimization was utilized to evaluate the optimal conditions for maximum removal of COD and PAHs in PW. The optimal parameters were obtained for COD and PAH removal depending on the desirability functions. In numerical optimization, for all independent parameters, Fe (VI) concentration, pH, and the contact time "in range" option were chosen. In contrast, "maximum range" was selected for COD and PAH removal. Optimized conditions were acquired under maximum desirability for the defined conditions. The COD and PAH removal efficiencies were predicted under these specified conditions by CCD-based numerical optimization. An additional test under optimal conditions was carried out to check the predicted models' precision and efficiency. The experimental removal efficiencies were in good agreement with predicted removal efficiencies, as shown in Table 7. The small difference between observed and predicted results indicated a strong agreement between responses obtained from experiments and the ones proposed by the quadratic models. In addition, the PAH removal by Fe (VI) oxidation was higher than COD removal. As COD of PW comprises of several organic and inorganic pollutants [45] it could be expected that some pollutants (which were part of COD) were resistant to Fe (VI) oxidation. On the other hand, all PAHs in PW were entirely removed by the Fe (VI) oxidation process except naphthalene and phenanthrene. Fe (VI) oxidation only partially removed naphthalene and phenanthrene. Both PAHs were quantified in very low concentrations in PW after treatment of PW at optimal conditions, as shown in Figure 8. From the results, it can be observed that significant removal of COD and PAHs from PW was attained. However, Fe (VI) oxidation could not attain 100% removal of pollutants from PW. A few pollutants such as naphthalene, phenanthrene, 2,5,5-trimethyl-1-hexen-3-yne, and 2-propenamide were identified after the treatment of PW at optimal conditions. Additionally, based on the identified pollutants after PW treatment, the Fe (VI) oxidation mechanism for PAH removal could be proposed. It is expected that the reduction of PAHs in the treated PW occurred by the transformation of aromatic rings. It might occur through 1,3-dipolar cycloaddition and result in two aldehyde groups, which can be further oxidized by Fe (VI) to acid groups [26]. A future study may be conducted on the Fe (VI) oxidation mechanism for PAH removal from PW.

Table 7. Optimum values of independent variables found by Design-Expert software and verification.

Dependent Variables	Fe (VI) Conc. (mg/L)	pH	Contact Time (min)	Predicted Removal (%)	Observed Removal (%)
COD	19.35	7.1	68.34	69.69	73.41
PAHs	19.35	7.1	68.34	91.00	89.73

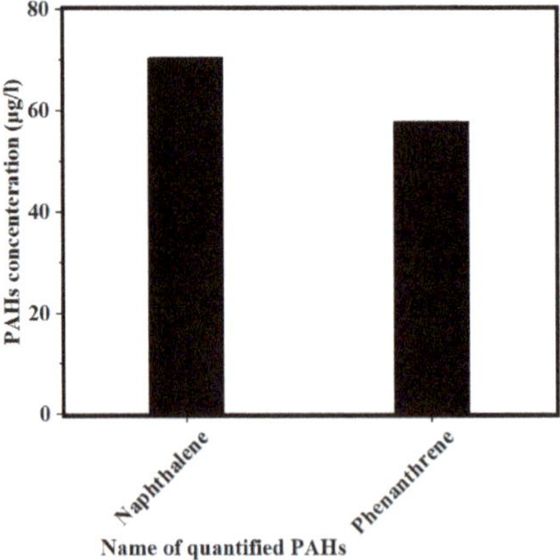

Figure 8. Name/concentration of PAHs quantified in PW after treatment of PW at optimal conditions.

In comparison with previously conducted scientific studies, Guang et al. [22] reported the maximum removal of PAHs (82.6%) was achieved at 10 mg/L of Fe (VI) concentration and pH 7.1 from synthetic wastewater using Fe (VI) oxidation. Duan et al. [30] reported 84.0% removal of organic contaminants from coking wastewater by Fe (VI) oxidation. Amirreza et al. [46] reported a maximum of 48.0% of micropollutant removal from municipal wastewater was achieved via Fe (VI) oxidation. The results obtained by Ciabatti et al. [28] showed a maximum of 80.0% removal of organic compounds at 70 mg/L of Fe (VI) concentration. The removal efficiencies in this present study were higher than all studies discussed above, which indicates the strong applicability of Fe (VI) to oxidize micropollutants in PW.

4. Conclusions

In this study, the remediation of COD and PAHs in PW by Fe (VI) oxidation was investigated systematically. The effectiveness of Fe (VI) oxidation depends on the operating parameters viz. Fe (VI) concentration, pH, and contact time. Initially, the ranges of operating parameters were evaluated by a screening test. Based on three operating parameters, the experimental design, consisting of 20 runs, was developed by CCD/RSM. Optimum values of independent variables for the current oxidation study were observed, such as 19.35 mg/L, 7.1, and 68.34 min for Fe (VI) concentration, pH, and contact time, respectively. The maximum COD (73.41%) and PAH (89.73%) removal were achieved under tested operating conditions. A satisfactory agreement was confirmed between experimental data and predicted data (obtained from the quadratic regression models). The high values of the coefficient of determination ($R^2 > 0.90$) by the analysis of variance verified the adequacy of the selected model. The findings of this study indicate an excellent efficiency of Fe (VI) for the remediation of micropollutants in PW. The study recommends that it would be advantageous to conduct a continuous flow study for the assessment of field applications of Fe (VI) for the treatment of produced water.

Author Contributions: Conceptualization, M.R.U.M. and M.H.I.; data curation, T.H.; formal analysis, T.H. and M.J.K.B.; funding acquisition, M.R.U.M.; investigation, T.H.; methodology, T.H.; project administration, M.R.U.M.; resources, M.R.U.M.; supervision, M.R.U.M. and K.W.Y.; validation, T.H.; writing—original draft, T.H.; writing—review and editing, M.R.U.M., M.H.I., M.J.K.B., M.A., and M.Z. All authors have read and agreed to the published version of the manuscript.

Funding: This research was funded by the YUTP research project (Cost center 015LC0-044).

Acknowledgments: The authors would like to acknowledge Universiti Teknologi PETRONAS for providing facilities and financial support under the YUTP grant (cost center; 015LC0-044) for this study and the first author also acknowledges the Graduate Assistantship (GA) scheme by the Centre for Graduate Studies of Universiti Teknologi PETRONAS (UTP).

Conflicts of Interest: The authors declare no conflict of interest.

References

1. Yaqub, A.; Isa, M.H.; Kutty, S.R.M.; Ajab, H.E. lectrochemical degradation of PAHs in produced water using Ti/Sb$_2$O$_5$-SnO$_2$-IrO$_2$ anode. *Electrochem. Commun.* **2014**, *82*, 979–984. [CrossRef]

2. Vela, N.; Martínez-Menchón, M.; Navarro, G.; Pérez-Lucas, G.; Navarro, S. Removal of polycyclic aromatic hydrocarbons (PAHs) from groundwater by heterogeneous photocatalysis under natural sunlight. *J. Photochem. Photobiol.* **2012**, *232*, 32–40. [CrossRef]

3. Liu, B.; Chen, B.; Zhang, B.; Song, X.; Zeng, G.; Lee, K. Photocatalytic ozonation of offshore produced water by TiO2 nanotube arrays coupled with UV-LED irradiation. *J. Hazard. Mater.* **2020**, *402*, 123456. [CrossRef] [PubMed]

4. Jiménez, S.; Micó, M.M.; Arnaldos, M.; Medina, F.; Contreras, S. State of the art of produced water treatment. *Chemosphere* **2018**, *192*, 186–208.

5. Nawaz, R.; Kait, C.F.; Chia, H.Y.; Isa, M.H.; Huei, L.W. Glycerol-mediated facile synthesis of colored titania nanoparticles for visible light photodegradation of phenolic compounds. *Nanomaterials* **2019**, *9*, 1586. [CrossRef]

6. Haneef, T.; Ul Mustafa, M.R.; Rasool, K.; Ho, Y.C.; Mohamed Kutty, S.R. Removal of polycyclic aromatic hydrocarbons in a heterogeneous Fenton like oxidation system using nanoscale zero-valent iron as a catalyst. *Water* **2020**, *12*, 2430. [CrossRef]

7. Haneef, T.; Mustafa, M.R.U.; Farhan Yasin, H.M.; Farooq, S.; Hasnain Isa, M. Study of Ferrate(VI) oxidation for COD removal from wastewater. *IOP Conf. Ser. Earth Environ. Sci.* **2020**, *442*, 1–8. [CrossRef]

8. Smol, M.; Włodarczyk, M.M. The effectiveness in the removal of PAHs from aqueous solutions in physical and chemical processes: A review. *Polycycl. Aromat. Comp.* **2017**, *37*, 292–313. [CrossRef]

9. Sahrin, N.T.; Nawaz, R.; Fai Kait, C.; Lee, S.L.; Wirzal, M.D.H. Visible light photodegradation of formaldehyde over TiO_2 nanotubes synthesized via electrochemical anodization of titanium foil. *Nanomaterials* **2020**, *10*, 128. [CrossRef]

10. Manilal, A.M.; Soloman, P.A.; Basha, C.A. Removal of Oil and Grease from Produced Water Using Electrocoagulation. *J. Hazard. Toxic Radioact. Waste* **2020**, *24*, 04019023. [CrossRef]

11. Chang, H.; Li, T.; Liu, B.; Vidic, R.D.; Elimelech, M.; Crittenden, J.C. Potential and implemented membrane-based technologies for the treatment and reuse of flowback and produced water from shale gas and oil plays: A review. *Desalination* **2019**, *455*, 34–57. [CrossRef]

12. Campos, J.C.; Borges, R.M.H.; Oliveira Filho, A.M.D.; Nobrega, R.; Sant'Anna, G.L., Jr. Oilfield wastewater treatment by combined microfiltration and biological processes. *Water Res.* **2002**, *36*, 95–104. [CrossRef]

13. Hao, H.; Huang, X.; Gao, C.; Gao, X. Application of an integrated system of coagulation and electrodialysis for treatment of wastewater produced by fracturing. *Desalin. Water Treat.* **2015**, *55*, 2034–2043. [CrossRef]

14. Clay, L.; Pichtel, J. Treatment of simulated oil and gas produced water via pilot-scale rhizofiltration and constructed wetlands. *Int. J. Environ. Sci. Technol.* **2019**, *13*, 185–198. [CrossRef]

15. Rasool, K.; Pandey, R.P. Water treatment and environmental remediation applications of two-dimensional metal carbides (MXenes). *Mater. Today* **2019**, *30*, 80–102. [CrossRef]

16. Jiang, J.Q.; Stanford, C.; Petri, M. Practical application of ferrate (VI) for water and wastewater treatment–site study's approach. *Water-Energy Nexus* **2018**, *1*, 42–46. [CrossRef]

17. Gombos, E.; Barkács, K.; Felföldi, T.; Vértes, C.; Makó, M.; Palkó, G.; Záray, G. Removal of organic matters in wastewater treatment by ferrate (VI)-technology. *Microchem. J.* **2013**, *107*, 115–120. [CrossRef]

18. Alsheyab, M.; Jiang, J.Q.; Stanford, C. On-line production of ferrate with an electrochemical method and its potential application for wastewater treatment—A review. *J. Environ. Manag.* **2009**, *90*, 1350–1356. [CrossRef]

19. Sun, S.; Pang, S.; Jiang, J.; Ma, J.; Huang, Z.; Zhang, J.; Liu, Y.; Xu, C.; Liu, Q.; Yuan, Y. The combination of ferrate (VI) and sulfite as a novel advanced oxidation process for enhanced degradation of organic contaminants. *Chem. Eng. J.* **2018**, *333*, 11–19. [CrossRef]

20. Jiang, J.Q. Research progress in the use of ferrate(VI) for the environmental remediation. *J. Hazard. Mater.* **2007**, *146*, 617–623. [CrossRef]

21. Guan, W.; Xie, Z.; Zhang, J. Preparation and aromatic hydrocarbon removal performance of potassium Ferrate. *Int. J. Spectrosc.* **2014**, *22*, 1–8. [CrossRef]

22. Sharma, V.K.; Rivera, W.; Smith, J.O.; O'Brien, B. Ferrate (VI) oxidation of aqueous cyanide. *J. Environ. Sci. Technol.* **1998**, *32*, 2608–2613. [CrossRef]

23. Ahmad, A.L.; Ismail, S.; Bhatia, S. Optimization of coagulation-flocculation process for palm oil mill effluent using response surface methodology. *Environ. Sci. Technol.* **2005**, *39*, 2828–2834. [CrossRef] [PubMed]

24. Saeed, M.O.; Azizli, K.; Isa, M.H.; Bashir, M.J. Application of CCD in RSM to obtain optimize treatment of POME using Fenton oxidation process. *J. Water Process. Eng.* **2015**, *8*, 7–16. [CrossRef]

25. Wang, C.; Klamerth, N.; Huang, R.; Elnakar, H.; Gamal El-Din, M. Oxidation of oil sands process-affected water by potassium Ferrate(VI). *Environ. Sci. Technol.* **2016**, *50*, 4238–4247. [CrossRef]

26. Han, H.; Li, J.; Ge, Q.; Wang, Y.; Chen, Y.; Wang, B. Green Ferrate(VI) for multiple treatments of fracturing wastewater: Demulsification, visbreaking, and chemical oxygen demand removal. *Int. J. Mol. Sci.* **2019**, *20*, 1857. [CrossRef]

27. Ciabatti, I.; Tognotti, F.; Lombardi, L. Treatment and reuse of dyeing effluents by potassium ferrate. *Desalination* **2010**, *250*, 222–228. [CrossRef]

28. Malik, S.N.; Ghosh, P.C.; Vaidya, A.N.; Waindeskar, V.; Das, S.; Mudliar, S.N. Comparison of coagulation, ozone and ferrate treatment processes for color, COD and toxicity removal from complex textile wastewater. *Water Sci. Technol.* **2017**, *76*, 1001–1010. [CrossRef]

29. Li, Y.N.; Duan, Z.H.; Wang, Y.F.; Yuan, Z.J.; Wang, G.Y. Preliminary treatment of phenanthrene in coking wastewater by a combined potassium ferrate and Fenton process. *Int. J. Sci. Environ. Technol.* **2018**, *16*, 4483–4492. [CrossRef]
30. Kozik, V.; Barbusinski, K.; Thomas, M.; Sroda, A.; Jampilek, J.; Sochanik, A.; Smolinski, A.; Bak, A. Taguchi method and response surface methodology in the treatment of highly contaminated tannery wastewater using commercial potassium Ferrate. *Materials* **2019**, *12*, 3784. [CrossRef]
31. Tan, X.M.; Ji, F.Y.; Wang, X.D.; Zhang, G.Z. Preparation of potassium Ferrate and Ferrate(VI) oxidation of phenanthrene. *Adv. Mater. Res.* **2012**, *523*, 784–789. [CrossRef]
32. Eaton, A.D. *Standard Methods for the Examination of Water and Wastewater*; American Public Health Association: Washington, DC, USA, 1915.
33. Perez, S.; Guillamon, M.; Barceló, D. Quantitative analysis of polycyclic aromatic hydrocarbons in sewage sludge from wastewater treatment plants. *J. Chromatogr. A* **2001**, *938*, 57–65. [CrossRef]
34. Zahid, M.; Shafiq, N.; Isa, M.H.; Gil, L. Statistical modeling and mix design optimization of fly ash based engineered geopolymer composite using response surface methodology. *J. Clean. Prod.* **2018**, *194*, 483–498. [CrossRef]
35. Rai, P.K.; Lee, J.; Kailasa, S.K.; Kwon, E.E.; Tsang, Y.F.; Ok, Y.S.; Kim, K.H. A critical review of ferrate (VI)-based remediation of soil and groundwater. *Environ. Res.* **2018**, *160*, 420–448. [CrossRef]
36. Jiang, J.Q.; Lloyd, B. Progress in the development and use of ferrate(VI) salt as an oxidant and coagulant for water and wastewater treatment. *Water Res.* **2001**, *36*, 1397–1408. [CrossRef]
37. Matin, A.R.; Yousefzadeh, S.; Ahmadi, E.; Mahvi, A.; Alimohammadi, M.; Aslani, H.; Nabizadeh, R. A comparative study of the disinfection efficacy of H_2O_2/Ferrate and UV/H_2O_2/Ferrate processes on inactivation of Bacillus subtilis spores by response surface methodology for modeling and optimization. *Food Chem. Toxicol.* **2018**, *116*, 129–137. [CrossRef]
38. Karim, A.V.; Krishnan, S.; Pisharody, L.; Malhotra, M. Application of Ferrate for Advanced Water and Wastewater Treatment. In *Advanced Oxidation Processes-Applications, Trends, and Prospects*; IntechOpen: London, UK, 2020.
39. Sun, X.; Zhang, Q.; Liang, H.; Ying, L.; Xiangxu, M.; Sharma, V.K. Ferrate(VI) as a greener oxidant: Electrochemical generation and treatment of phenol. *J. Hazard. Mater.* **2016**, *319*, 130–136. [CrossRef]
40. Graham, N.; Jiang, C.C.; Li, X.Z.; Jiang, J.Q.; Ma, J. The influence of pH on the degradation of phenol and chlorophenols by potassium ferrate. *Chemosphere* **2004**, *56*, 949–956. [CrossRef]
41. Karaatli, T. Disinfection of Surface Waters by Ferrate. Master's Thesis, Middle East Technical University, Ankara, Turkey, 1998.
42. Song, Y.; Men, B.; Wang, D.; Ma, J. On line batch production of ferrate with an chemical method and its potential application for greywater recycling with Al(III) salt. *Int. J. Environ. Sci. Technol.* **2017**, *52*, 1–7. [CrossRef]
43. Jiang, J.Q. Advances in the development and application of ferrate (VI) for water and wastewater treatment. *J. Chem. Technol.* **2014**, *89*, 165–177. [CrossRef]
44. Miao, Z.C.; Wang, F.; Deng, D.; Wang, L.; Yang, J.Z. Removal effect of potassium Ferrate to COD in different wastewater. *Adv. Mater. Res.* **2012**, *553*, 2288–2290. [CrossRef]
45. Duraisamy, R.T.; Beni, A.H.; Henni, A. State of the art treatment of produced water. *Water Treat.* **2013**, *233*, 199–222.
46. Talaiekhozani, A.; Eskandari, Z.; Bagheri, M.; Talaie, M.R. Removal of H_2S and COD using UV, Ferrate and UV/Ferrate from municipal wastewater. *J. Hum. Environ. Health Prom.* **2016**, *2*, 1–8. [CrossRef]

Publisher's Note: MDPI stays neutral with regard to jurisdictional claims in published maps and institutional affiliations.

Article

Presence and Reduction of Anthropogenic Substances with UV Light and Oxidizing Disinfectants in Wastewater—A Case Study at Kuopio, Finland

Jenni Ikonen [1,*], Ilpo Nuutinen [2], Marjo Niittynen [1], Anna-Maria Hokajärvi [1], Tarja Pitkänen [1,3], Eero Antikainen [2] and Ilkka T. Miettinen [1]

[1] Water Microbiology Laboratory, Department of Health Security, Finnish Institute for Health and Welfare, P.O. Box 95, FI-70701 Kuopio, Finland; marjo.niittynen@thl.fi (M.N.); anna-maria.hokajarvi@thl.fi (A.-M.H.); tarja.pitkanen@thl.fi (T.P.); ilkka.miettinen@thl.fi (I.T.M.)

[2] School of Engineering and Technology, Savonia University of Applied Sciences, P.O. Box 6 (Microkatu 1), FI-70201 Kuopio, Finland; ilpo.nuutinen@savonia.fi (I.N.); eero.antikainen@savonia.fi (E.A.)

[3] Faculty of Veterinary Medicine, Department of Food Hygiene and Environmental Health, University of Helsinki, P.O. Box 66, FI-00014 Helsinki, Finland

[*] Correspondence: jenni.ikonen@thl.fi; Tel.: +358-29-524-6375

Citation: Ikonen, J.; Nuutinen, I.; Niittynen, M.; Hokajärvi, A.-M.; Pitkänen, T.; Antikainen, E.; Miettinen, I.T. Presence and Reduction of Anthropogenic Substances with UV Light and Oxidizing Disinfectants in Wastewater—A Case Study at Kuopio, Finland. *Water* **2021**, *13*, 360. https://doi.org/10.3390/w13030360

Academic Editor: Amin Mojiri

Received: 18 December 2020

Accepted: 27 January 2021

Published: 30 January 2021

Abstract: Anthropogenic substances are a major concern due to their potential harmful effects towards aquatic ecosystems. Because wastewater treatment plants (WWTPs) are not designed to remove these substances from wastewater, a part of the anthropogenic substances enter nature via WWTP discharges. During the spring 2019, the occurrence of anthropogenic substances in the municipal wastewater effluent in Kuopio, Finland, was analysed. Furthermore, the capacity of selected disinfection methods to reduce these substances from wastewater was tested. The disinfection methods were ozonation (760 mL min^{-1}) with an OxTube hermetic dissolution method (1), the combined usage of peracetic acid (PAA) (<5 mg L^{-1}) and ultraviolet (UV) disinfection (12 mJ/cm^2) (2), and the combined usage of hydrogen peroxide (H_2O_2) (<10 mg L^{-1}) and UV disinfection (12 mJ/cm^2) (3). The substances found at the concentrations over 1 µg L^{-1} in effluent (N = 3) were cetirizine (5.2 ± 1.3 µg L^{-1}), benzotriazole (BZT) (2.1 ± 0.98 µg L^{-1}), hydrochlorothiazide (1.7 ± 0.2 µg L^{-1}), furosemide (1.6 ± 0.2 µg L^{-1}), lamotrigine (1.5 ± 0.06 µg L^{-1}), diclofenac (DCF) (1.4 ± 0.2 µg L^{-1}), venlafaxine (1.0 ± 0.13 µg L^{-1}) and losartan (0.9 ± 0.2 µg L^{-1}). The reduction (%) with different methods (1, 2, 3) were: cetirizine (99.9, 5.0, NR = no removal), benzotriazole (67.9, NR, NR), hydrochlorothiazide (91.1, 5.9, NR), furosemide (99.7, 5.9, NR), lamotrigine (46.4, NR, 6.7), diclofenac (99.7, 7.1, 16.7), venlafaxine (91.3, NR, 1.1), losartan (99.6, 13.8, NR). Further research concerning the tested disinfection methods is needed in order to fully elucidate their potential for removing anthropogenic substances from purified wastewater.

Keywords: anthropogenic substances; disinfection; wastewater

1. Introduction

Emerging anthropogenic pollutants are a permanent global challenge to freshwater quality and safety [1–3]. A major group of emerging pollutants in the aquatic environmental consists of pharmaceuticals [4]. After being used for human or animal medication [5], pharmaceuticals are mainly excreted in urine and faeces as such or as metabolites [6]. Subsequently, they are distributed in the environment via wastewater treatment plants (WWTPs) [7] where they pass through various treatment processes and, therefore, are easily transferred to the receiving waters. In Finland, the legislation does not require the removal of anthropogenic substances from wastewater before discharge into the environment and most of the treated wastewaters are discharged to the receiving waters without disinfection. The Water Framework Directive (WFD) [8] aimed to improve the status of all European

Union (EU) inland waters, coastal waters, and groundwater by 2015. The deadline has been extended until 2027 at the latest under the WFD derogation rules.

In the future, the use of pharmaceuticals is likely to increase due to the ageing population. Unless efforts are made to reduce emissions, more pharmaceutical residues will end up in the environment. To compare the effectiveness of different wastewater treatment methods, more research data on the existence and harmfulness of these substances in the environment is needed. EU Member States are required to monitor the concentrations of 45 substances or groups of substances in the aquatic environment [9]. These substances are listed in the directive 2013/39/EU (amending Directives 2000/60/EC and 2008/105/EC as regards priority substances in the field of water policy). Moreover, some of the substances are listed as priority hazardous substances. Furthermore environmental quality standards (EQS) are included in the directive for these 45 substances or groups of substances that EU Member States are required to monitor. The concentrations of the substances in water or biota must not exceed the EQS set for them. With the aim of achieving good surface water chemical status, the revised EQS for existing priority substances should be met by the end of 2021 and the EQS for newly identified priority substances by the end of 2027.

At the WWTPs pharmaceuticals may transform, retain in sewage sludge, or end up in receiving water. In a recent risk assessment study concerning Finnish surface waters, the calculated environmental risk was assessed by a so called risk component; or risk quotient (RQ). A risk quotient > 1 was found for 29 of the evaluated substances, suggesting that these substances potentially pose a risk in Finnish surface waters. Four substances: diclofenac (DCF) (0.022 μg L^{-1}), azithromycin (0.0015 μg L^{-1}), ciprofloxacin (0.034 μg L^{-1}), and 17α- ethinylestradiol (0.00018 μg L^{-1}) concentrations measured in Finnish surface waters exceeded concentrations assessed as harmful [10].

As the pharmaceuticals and other anthropogenic substances are not properly removed in current WWTP processes, alternative, tentatively more efficient removal options such as novel disinfection methods to remove these substances from wastewater have been studied. For example in Mexico, Mejía-Morales et al. studied a post treatment with advanced oxidation process (AOP) based on an ultraviolet (UV)/H_2O_2/O_3 system in hospital wastewaters [11]. In addition, various other methods have been tested to remove inorganic and organic impurities from water such as porous ceramic disk filter (PVDF) ultrafiltration membrane [12] and porous ceramic disk filter coated with Fe/TiO2 nano-composites [13]. Here we studied the efficiency of three disinfection methods, i.e., ozonation (760 mL min^{-1}) with OxTube mixing; a combination of peracetic acid (PAA) (<5 mg L^{-1}) and UV disinfection (12 mJ/cm^2), and a combination of hydrogen peroxide (<10 mg L^{-1}) and UV disinfection (12 mJ/cm^2) in order to reduce the amount of anthropogenic substances in treated wastewater. This study was one part of a project in which we studied the removal of certain microbes and chemicals in different water matrices with different disinfection methods.

2. Materials and Methods

Treated wastewater samples (N = 3) were collected from the municipal WWTP of the city of Kuopio (Lehtoniemi WWTP) in the spring of 2019. The wastewater that was used in the tests was from the channel where purified wastewater is discharged into the surface water. The population of the service area of the Lehtoniemi WWTP is 90,697; and the total population of the city of Kuopio is 118,000. The disinfection methods tested herein were an ozone purification process with the OxTube hermetic dissolution method, a combination of the usage of PAA (<5 mg L^{-1}) and UV disinfection (12 mJ/cm^2) processes, and a combination of hydrogen peroxide (<10 mg L^{-1}) and UV disinfection (12 mJ/cm^2) processes. A wide set of chemical substances were analysed (N = 121). Wastewater samples were taken before and after each disinfection treatment. All wastewater samples were frozen and stored at $-20\,^\circ$C and subsequently sent to a commercial laboratory (Eurofins Environment Testing Finland Oy, Lahti) for analysis. Analysed substances are listed in the Table S1. The substances were analysed with the U.S. Environmental Protection Agency

(EPA) method 1694. Method 1694 is used for determination of pharmaceuticals and personal care products (PPCPs) in multi-media environmental samples by high-performance liquid chromatography combined with tandem mass spectrometry (HPLC/MS/MS) using isotope dilution and internal standard quantitation techniques [14].

Experimental Design

All disinfection experiments were carried out in the Savonia Water laboratory (Savonia University of Applied Sciences, Kuopio, Finland). The experimental design is shown in the Supplementary Material (Figure S1). First, a 1000-litre food-grade plastic container was filled with 500 L of wastewater. The treated wastewater was mixed with an electric motor-operated water mixer to ensure the homogeneity of the sample water. After the wastewater was disinfected, it was collected into a plastic container with a capacity of 45 L (Curtec Ltd., Denmark). The pipe material used was a plastic water pipe with an inner diameter of 15 mm (Uponor Aqua Pipe, PEX 15/18 mm polyethylene, PE) and the connectors were acid-resistant stainless steel water pipe fittings (various manufacturers).

For the pumping of the tested wastewater, a 24 V resistance-adjustable gear pump designed for drinking water systems in boats with a maximum flow of 26 L min^{-1} (Marco UP/Em, Castenedolo, Italy,) was used. Acid-resistant steel valves (EGO, stainless) were used in the test system due to the oxidizing peroxide chemicals used in the experiments.

For the supply of peroxide chemicals (PAA and H_2O_2) (Lamor water technology, Finland), a chemical pump (Grundfos DDA 12-10 AR-PP/E/C-F-31U2U2FG, Bjerringbro, Denmark) with proven chemical supply and adjustability for a wide feed rate between 12 mL h^{-1} and 12 L h^{-1} was obtained. A rotameter (Kobold, Germany) with a flow scale of 100 to 1000 L h^{-1} was obtained to measure the water flow. A tube UV lamp (Wedeco Aquada 1, Xylem, Herford) was used in the disinfection experiments. A Faraday Ozone L 40 G (Farady Ozone, Coimbatore, India) device, which is capable of producing ozone with a capacity of 40.000 mg h^{-1}, was used as the ozone generator. The flow of ozone gas was controlled with the mass flow controller (Brooks GF040). An OxTube water treatment tube (OxTubeDN20, Sansox Oy, Lahti, Finland) was used for mixing and dissolving ozone hermetically in the test water. The OxTube hermetic dissolution method (Figure 1) treats the water in flowing condition in its hermetic tube. The air gases are sucked by the vacuum effect in the nozzle zone and led directly into the middle of the main flow. Other gases like pure oxygen, ozone, CO_2 as well as chemicals can be fed and dispensed through the same channel. The water and gases are mixed evenly and the meeting probability of the molecules is high. Chemical reactions follow immediately in the hermetic condition. There are four main functions following each other seamlessly in one tube or in separate modules by function. The water is clarified and dissolved with desirable ingredients, e.g., air gases in the tube within a second or less.

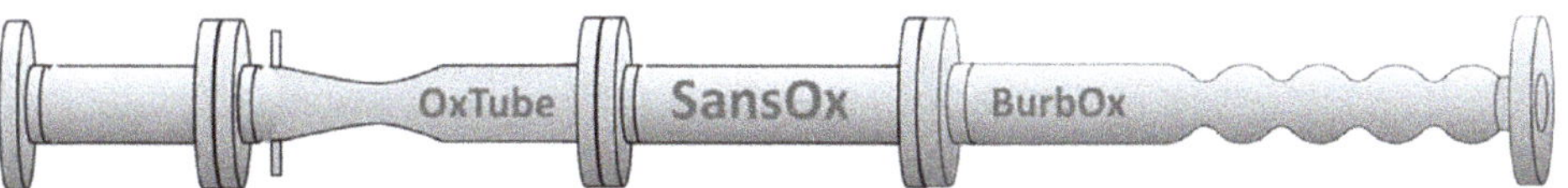

Figure 1. OxTube hermetic dissolution method (Sansox Oy, Lahti, Finland).

Chemical concentrations were measured using a Chemetrics Inc. V-2000 spectrometer (Chemetrics Inc., Midland, VA, USA) and suitable measuring ampoules including a Chemetrics K7913 for peracetic acid, K-5543 for hydrogen peroxide and K-7423 for ozone.

3. Results and Discussion

The anthropogenic substances with detected concentrations over 1 µg L^{-1} in the wastewater are shown in Table 1, as well as the removal efficiencies for the chemicals with the tested disinfection methods. The measurement uncertainty is between 45–51% in the analyses presented here. Each of the substances is discussed later in this manuscript.

Anthropogenic substances detected concentrations below concentrations of 1 µg L^{-1} in the wastewater are presented in the Supplemental Materials (Table S1).

Table 1. Anthropogenic substances with detected concentrations over 1 µg L^{-1} in wastewater effluent and their removal efficiencies with the depicted methods.

Anthropogenic Substance (Intended Use)	Ozone and Ox Tube Device (N = 1)		Peracetic Acid (PAA) and Ultraviolet (UV) Disinfection (N = 1)		H$_2$O$_2$ and UV Disinfection (N = 1)		EQS
	Initial Concentration (µg L^{-1})	Reduction (%)	Initial Concentration (µg L^{-1})	Reduction (%)	Initial Concentration (µg L^{-1})	Reduction (%)	
Cetirizine (antihistamine)	5.8	99.9	6.0	5.0	3.7	–	NR
Benzotriazole (chemical, anticorrosive)	2.8	67.9	2.5	–	0.98	–	NR
Hydrochlorothiazide (diuretic)	1.8	91.1	1.7	5.9	1.5	–	NR
Furosemide (diuretic)	1.8	99.7	1.7	5.9	1.4	–	NR
Lamotrigine (antiepileptic/ antidepressant)	1.4	46.4	1.5	–	1.5	6.7	NR
Diclofenac (DCF) (anti-inflammatory medicine)	1.5	99.7	1.4	7.1	1.2	16.7	NR
Venlafaxine (antidepressant)	1.1	91.3	1.1	–	0.87	1.1	NR
Losartan (used to treat high blood pressure)	1.2	99.6	0.8	13.8	0.84	–	NR

– = no removal detected; NR = not regulated in Directive 2013/39/EU.

3.1. Cetirizine

Cetirizine is an ingredient that is used in the treatment of symptoms of seasonal allergic rhinitis [15], perennial allergic rhinitis, and chronic idiopathic urticarial in adults [16]. Unfortunately, cetirizine has been shown to induce adverse biochemical effects in the mussel *Mytilus galloprovincialis* and is thus problematic from the ecotoxicological point of view [17]. In our study, the concentration of cetirizine detected in the Kuopio wastewater effluent (5.2 ± 1.3 µg L^{-1}) clearly exceeds the concentrations reported earlier in Finland. The appearance and level of cetirizine in municipal wastewaters has previously been studied in the city of Turku in 2007 [7]. In that study cetirizine was found to be the dominating antihistamine in nearly all samples. The sampling included 12 influent and 12 effluent samples, and it was conducted between March and September. The highest detected concentration of cetirizine in that study was 0.22 µg L^{-1} (influent) and elimination rate of cetirizine in the sewage treatment process was 16% [7]. Concentrations ranging from 0.1 µg L^{-1} to 0.7 µg L^{-1} have been detected in the influent of WWTPs in Berlin. Cetirizine levels were significantly increased between the hay season [18]. The concentration of this chemical is only slightly degraded during the wastewater treatment process [7]. For instance cetirizine removal from wastewater has been tested with granular activated carbon (GAC). Only 30.4% of cetirizine was removed even when the contact time was 15 min with a GAC column [19]. The purification process with ozone and the OxTube hermetic dissolution method removed 99.9% of the cetirizine. The method consisting of PAA with UV disinfection was clearly less efficient, as it removed only 5%. The third method (with a combination of H_2O_2 and UV) did not remove any of the cetirizine. The result obtained indicates that wastewater effluent disinfection with ozone is a very efficient method to remove cetirizine.

3.2. Benzotriazoles (BZTs)

Benzotriazoles (BZTs) are heterocyclic aromatic compounds that are widely used in industrial applications due to their excellent properties as corrosion inhibitors [20], antifreeze agents, and UV radiation stabilizers [21]. Another cause for the occurrence of BZTs in municipal wastewaters is their use in dishwasher products; tablets and powders [22]. BZTs are highly water soluble and highly polar compounds. In addition, they are moderately resistant against biological and photochemical degradation processes in the aquatic environment [23]. Moreover, BZTs have been identified in river water, groundwater, drinking water, wastewater as well as in soil, and in human samples. This is due to the low volatility of these compounds, their strong resistance to oxidation, and limited degradation under environmental conditions [24]. In our tests, the ozone purification process with the OxTube hermetic dissolution method removed 67.9% of the found BZT. PAA or H_2O_2 together with UV disinfection did not achieve removal efficiency. Loos et al. (2013) found that median concentrations of BZTs in an EU-wide wastewater treatment plant study were 2.7 μg L^{-1} for 1H-benzotriazole, and 2.1 μg/L for methylbenzotriazole (mixture of 4- and 5-isomers, also called tolyltriazoles), with maximum values up to 221 μg L^{-1} and 24.3 μg L^{-1}, respectively. In our study concentrations of 2.1 $\pm$ 0.98 μg L^{-1} of BZT in wastewater were detected (BZTs were not specified), which were median concentrations compared to concentrations measured in the EU [25]. Our study addresses the fact that when using ozone disinfecting for wastewater effluent, significant removal of BZT can be achieved.

3.3. Hydrocholorothiazide

Hydrochlorothiazide is a diuretic and an antihypertensive drug that is widely used by itself or in combination with other drugs for the treatment of edema and hypertension, as well as for other disorders such as diabetes insipidus, hypoparathyroidism, or hypercalciuria [26,27]. In Finland, hydrochlorothiazide concentrations of 1.8–6.7 μg L^{-1} have been reported in effluent wastewaters [28], which were on average higher than concentrations found in this study (1.7 $\pm$ 0.2 μg L^{-1}). This compound has been frequently detected in the influents and effluents of WWTPs in Europe. In the Netherlands [29] concentrations of 1.27 $\pm$ 0.26 μg L^{-1} (effluent) of hydrochlorothiazide were detected in municipal wastewater samples. In Spain, detected concentrations ranged between 2.5 μg L^{-1} and 14 μg L^{-1} in raw wastewater [30].

The removal of hydrochlorothiazide from wastewater has been studied using biological membranes in the laboratory and reduction percentages between 56% and 85% have been achieved. Slightly better removal has been achieved by conventional wastewater treatment [31]. In the present study, a removal rate of 91.1% was achieved with ozone purification using an OxTube hermetic dissolution method. With PAA and UV disinfection treatment, the removal was only 5.9%. The combination of H_2O_2 and UV did not remove hydrochlorothiazide at all. Therefore, ozone disinfection was a superior method in terms of hydrochlorothiazide removal.

3.4. Furosemide

Furosemide, a diuretic that has been widely used since the 1960s, is poorly metabolized by humans [32]. In 2016, it was the most used diuretic in Finland [33]. The average concentration of furosemide detected in the Kuopio wastewater effluent (1.6 $\pm$ 0.2 μg L^{-1}) is slightly higher than concentrations detected before in Finland. A furosemide concentration of 1.4 μg L^{-1} in wastewater effluent has previously been reported in another Finnish WWTP, at the Turku Kakolanmäki WWTP (Turku, Finland) [28]. It has previously been detected at concentrations of 0.615 and 0.2 μg L^{-1} in the Viskan river at Jössabron Borås, in Sweden [34]. In Norwegian surface water sample concentrations of up to 0.05 μg L^{-1} and up to 1.9 μg L^{-1}

in treated wastewater [35] were detected. Deblonde et al. [1] presented in their review furosemide concentrations of 0.413 µg L^{-1} and 0.166 µg L^{-1} in wastewater influent and effluent, respectively. Among human pharmaceuticals divided into six categories (IA, IB, IIA, IIB, III, IV), furosemide belongs to the highest risk group (IA) as it has been shown to pose a risk to the aquatic environment already at concentrations of potential exposure (PEC) > 0.1 µg L^{-1} [36]. In the study by Jelic et al. (2011) removal rates for furosemide were found between WWTPs to be 30%, 60% and 80% [37]. In our experiment, 99.7% of furosemide was removed with ozone purification with the OxTube hermetic dissolution method and 5.9% using PAA and UV disinfection. With the treatment of H$_2$O$_2$ and UV disinfection, no reduction was detected. Thus furosemide was successfully removed from the wastewater effluent with using ozone as the disinfection method.

3.5. Lamotrigine

Lamotrigine is an anticonvulsant medication used to treat epilepsy and bipolar disorder. Lamotrigine has recently been recognized as a persistent pharmaceutical in the water environment and in wastewater effluent. Bollman et al. [38], found N2-glucuronide conjugates of lamotrigine cleaved to form lamotrigine and that the concentration of lamotrigine increased from 1.1 to 1.6 µg L^{-1} in WWTPs. In this study, the lamotrigine concentration in wastewater was within the range of 1.5 ± 0.06 µg L^{-1}. In a previous study [39], it was found to be present in 94% of the studied wastewater samples, with a mean concentration of 0.488 µg L^{-1}. The same study found lamotrigine in two drinking water samples. As lamotrigine has also been detected in groundwater, it has been suggested that lamotrigine could be used as an indicator for the presence of treated wastewater in raw water used for drinking water production [38]. Lamotrigine is very persistent chemically and physically and can resist UV photolysis and ozone, but it reacts rapidly with hydroxyl radicals. Therefore, advanced oxidation processes might be effective for removing this compound during water treatment [40]. In this study, 46.4% of lamotrigine was removed by ozone purification with the OxTube hermetic dissolution method. With PAA and UV disinfection there was no reduction and with H$_2$O$_2$ and UV disinfection the reduction was 6.7%. The removal capacity of the ozone disinfection was less efficient for lamotrigine than for other anthropogenic substances studied. However, ozone disinfection was more efficient for removal of lamotrigine than any other tested disinfection method.

3.6. Diclofenac

Diclofenac (2-2-2,6-dichlorophenylaminophenylacetic acid; DCF) is a common non-steroidal anti-inflammatory drug that is used as oral tablets or as a topical gel. It is especially known for its harmful effects on vultures [41,42]. DCF is commonly found in municipal wastewater in Finland [43]. In 2002, the average concentration of DCF in wastewater influents was 0.35 ± 0.1 µg L^{-1} [44]. In 2013, DCF was selected for inclusion on the watch list of the WFD in order to collect data on it for the determination of risk reduction measures. According to the proposed EQS document, the maximum allowable concentrations of DCF is 0.1 µg L^{-1} in fresh waters and 0.01 µg L^{-1} in marine waters [45].

In recent years, the highest detected concentration of DCF in wastewater effluents in Finland has been 0.62 µg L^{-1} [46] and in surface waters 0.022 and 0.05 µg L^{-1} [10]. Furthermore, Lindholm-Lehto et al. [43] found some high concentrations of DCF. In this study, 1.4 ± 0.2 µg L^{-1} DCF was detected in the Kuopio wastewater effluent. Even though DCF is removed by natural processes such as photodegradation, the residues still remain in the environment as potential toxic metabolites and as the original compound. In the environment DCF is detected in lower concentrations, such as ng L^{-1} to mg L^{-1}, than in wastewater. It has been stated that DCF has adverse effects on several environmental species already at concentrations of ≤1 µg L^{-1} [47]. It has been suggested that DCF used as an NSAID (non-steroidal anti-inflammatory drug) and as a pain gel cannot be removed effectively in WWTPs. The removal efficiencies of diclofenac in WWTPs varied from 0% up to 80%, but were in mainly in the range of 21–40% in the study by Zhang et al.

(2008) [48]. In our current study, DCF removal from the tested wastewater was 99.7% using ozone purification with the OxTube hermetic dissolution method. PAA and UV disinfection removed 7.1% and with H_2O_2 and UV disinfection the reduction of DCF was 16.7%. Diclofenac was efficiently removed from the wastewater effluent by using the ozone disinfection. Also, the combined H_2O_2 and UV disinfection was able to remove diclofenac more efficiently compared to the other studied substances.

3.7. Venlafaxine

Venlafaxine is one of the most abundant antidepressants in municipal wastewaters where concentrations of the substance have been generally shown to range between 0.003 and 0.743 µg L^{-1} wastewater effluent receiving waters [49]. In this study concentrations of 1.0 ± 0.13 µg L^{-1} were found. Venlafaxine has also been detected at very low (<0.005 µg L^{-1}) concentrations in untreated drinking water [50]. More than 60% of venlafaxine has successfully been removed with anaerobic biological reactors [51]. Ozone purification with the OxTube hermetic dissolution method removed 91.3% of the detected venlafaxine. H_2O_2 and UV disinfection removed 1.1%, and with PAA and UV disinfection there was no reduction. In this study, venlafaxine was removed efficiently from the wastewater effluent by using ozone disinfection.

3.8. Losartan

Losartan, an antihypertensive, was one of the 10 most used medicines in Finland in 2018 [52]. Losartan can undergo structural modification resulting in formation of valsartan acid, which is a persistent pollutant ending up into activated sludge [53]. Losartan can also be found in various water matrices such as surface water and rivers [54]. It has been shown to be present in municipal wastewaters, e.g., in Colombia losartan has been detected in wastewater effluent at concentrations of 1.97 and 1.00 µg L^{-1} [55]. When studying pharmaceutical residues, Kot-Wask et al. (2016) found signs of losartan in wastewater from the Pomerania area in Poland [56]. In this study, a mean concentration of 0.9 ± 0.2 µg L^{-1} was detected. The removal of losartan has been studied with a WWTP that was designed for biological nitrogen removal and chemical precipitation of phosphorus. The removal efficiency of losartan in the system varied between 50–80% [57]. In our current study 99.6% of losartan was removed using ozone purification with the OxTube hermetic dissolution method. With PAA and UV disinfection, and H_2O_2 and UV disinfection the reduction was 13.8% and zero, respectively. Thus, the disinfection method using ozone as a disinfectant worked well in removal of losartan. Partial losartan reduction was also achieved with combined PAA and UV disinfection.

3.9. The Most Efficient Removal of Anthropogenic Substances Achieved by Using Ozone Purification with OxTube Hermetic Dissolution Method

Dissolved ozone has been used for years to disinfect and purify water [58]. Ozone is produced by separating oxygen from the air with an oxygen generator or industrial bottled instrument oxygen gas O_2. Pure oxygen is passed through a strong electric field with continuous corona discharge. When ozone decomposes in water, the hydrogen peroxy (HO_2) and hydroxyl (OH) are formed and they have great oxidizing capacity [59,60]. The half-life of ozone in aqueous solution depends, among other things, on pH and temperature of the water. In our study, the usage of ozone with the OxTube hermetic dissolution method was relatively efficient in removing of the detected anthropogenic substances. The reason for the achieved reduction capacities could be due to free radicals that are formed.

The use of ozone-based cleaning and disinfecting agents has increased in recent years in industry and water treatment sectors. The advantage of ozone compared to chlorine or other disinfectants is that ozone is very reactive, degrades rapidly and leaves no toxic or unwanted end products. It is an exceptionally good disinfectant with faster disinfection kinetics and more potency to eliminate most microorganisms than other chemical disinfectants in use. Ozonation followed by chlorination is proved to be better in terms of producing less disinfection byproducts than the sole use of chlorination [61].

Wastewater is a complex mixture of water and various substances, its viscosity is usually higher than water, the movement between substances is slow and thus its handling differs greatly from e.g., domestic water. This may be one reason why using OxTube hermetic dissolution method produced such good results in our case although we did not test the efficiency of ozone disinfection without this device.

4. Conclusions

Many anthropogenic substances are harmful to the environment. Out of the 121 analysed substances 44 were detected (Table S1) in the treated wastewater samples collected from the Kuopio (Lehtoniemi) WWTP. Eight substances (cetirizine, BZT, hydrocholorothiazide, furosemide, lamotrigine, DFC, venlafaxine, and losartan) were detected at concentrations over 1 μg L^{-1}. Among these eight substances, DCF is the only one that appears on the European Union's WFD monitoring list. In 2013, it was included on the first watch list to gather monitoring data for the purpose of facilitating the determination of appropriate measures to address the risk posed by the substance.

The results from this study showed that ozone disinfection using an OxTube hermetic dissolution method can efficiently reduce the concentration of pharmaceuticals in wastewater effluent. In future work, the OxTube hermetic dissolution method should be compared to other ozone mixing devices to prove the performance and capacity of this novel dissolution technique.

Supplementary Materials: The following are available online at https://www.mdpi.com/2073-4441/13/3/360/s1, Figure S1: Equipment used in the experiments; Table S1: List of the concentrations of analysed anthropogenic substances

Author Contributions: J.I.; writing—original draft preparation, I.N.; project administration, writing—review and editing, M.N. and A.-M.H. writing—review and editing, T.P. and I.T.M. supervision, writing—review and editing E.A. and I.T.M.; funding acquisition. All authors have read and agreed to the published version of the manuscript.

Funding: This study was funded by the European Regional Development (A72637).

Institutional Review Board Statement: Not applicable.

Informed Consent Statement: Not applicable.

Data Availability Statement: Not applicable.

Acknowledgments: This study was part of the project "Fostering market penetration and implementation of combined use of low energy Led-UV -technologies and PAA-chemicals in water disinfection". The authors thank the staff of the Savonia University of Applied Sciences, in Kuopio, Finland, and the staff of the Finnish Institute for Health and Welfare.

Conflicts of Interest: The authors declare no conflict of interest.

References

1. Deblonde, T.; Cossu-Leguille, C.; Hartemann, P. Emerging pollutants in wastewater: A review of the literature. *Int. J. Hyg. Environ. Health* **2011**, *214*, 442–448. [CrossRef] [PubMed]
2. Martín-Pozo, L.; De Alarcón-Gómez, B.; Rodríguez-Gómez, R.; García-Córcoles, M.T.; Çipa, M.; Zafra-Gómez, A. Analytical methods for the determination of emerging contaminants in sewage sludge samples. A review. *Talanta* **2019**, *192*, 508–533. [CrossRef] [PubMed]
3. Thomaidis, N.S.; Asimakopoulos, A.G.; Bletsou, A. Emerging contaminants: A tutorial mini-review. *Glob. NEST J.* **2012**, *14*, 72–79.
4. Gros, M.; Petrović, M.; Ginebreda, A.; Barceló, D. Removal of pharmaceuticals during wastewater treatment and environmental risk assessment using hazard indexes. *Environ. Int.* **2010**, *36*, 15–26. [CrossRef]
5. Song, P.; Huang, G.; An, C.; Xin, X.; Zhang, P.; Chen, X.; Ren, S.; Xu, Z.; Yang, X. Exploring the decentralized treatment of sulfamethoxazole-contained poultry wastewater through vertical-flow multi-soil-layering systems in rural communities. *Water Res.* **2021**, *188*, 116480. [CrossRef]
6. Lienert, J.; Güdel, K.; Escher, B.I. Screening Method for Ecotoxicological Hazard Assessment of 42 Pharmaceuticals Considering Human Metabolism and Excretory Routes. *Environ. Sci. Technol.* **2007**, *41*, 4471–4478. [CrossRef]

7. Kosonen, J.; Kronberg, L. The occurrence of antihistamines in sewage waters and in recipient rivers. *Environ. Sci. Pollut. Res.* **2009**, *16*, 555–564. [CrossRef]
8. European Union. *Water Framework Directive (WFD) 2000/60/EC: Directive 2000/60/EC of the European Parliament and of the Council of 23 October 2000 Establishing a Framework for Community Action in the Field of Water Policy*; European Union: Brussels, Belgium, 2012.
9. European Union. *Directive 2013/39/EU of the European Parliament and of the Council of 12 August 2013 Amending Directives 2000/60/EC and 2008/105/EC as Regards Priority Substances in the Field of Water Policy*; European Union: Brussels, Belgium, 2013.
10. Vieno, N.; Äystö, L.; Mehtonen, J.; Sikanen, T.; Karlsson, S.; Fjäder, P.; Nystén, T. Lääkejäämien vesistöriskien arviointi Suomessa. *Vesitalous* **2020**, *1*, 25–28. (In Finnish)
11. Mejía-Morales, C.; Hernández-Aldana, F.; Cortes-Hernandez, D.M.; Rivera-Tapia, J.A.; Castañeda-Antonio, D.; Bonilla, N. Assessment of Biological and Persistent Organic Compounds in Hospital Wastewater After Advanced Oxidation Process UV/H2O2/O3. *Waterairsoil Pollut.* **2020**, *231*, 1–10. [CrossRef]
12. Chen, X.; Huang, G.; Li, Y.; An, C.; Feng, R.; Wu, Y.; Shen, J. Functional PVDF ultrafiltration membrane for Tetrabromobisphenol-A (TBBPA) removal with high water recovery. *Water Res.* **2020**, *181*, 115952. [CrossRef]
13. Zhao, Y.; Huang, G.; An, C.; Huang, J.; Xin, X.; Chen, X.; Hong, Y.; Song, P. Removal of Escherichia Coli from water using functionalized porous ceramic disk filter coated with Fe/TiO2 nano-composites. *J. Water Process. Eng.* **2020**, *33*, 101013. [CrossRef]
14. EPA. *Method 1694: Pharmaceuticals and Personal Care Products in Water, Soil, Sediment, and Biosolids by HPLC/MS/MS*; Environmental Protection Agency: Washington, DC, USA, 2007.
15. Zhang, L.; Cheng, L.; Hong, J. The Clinical Use of Cetirizine in the Treatment of Allergic Rhinitis. *Pharmacology* **2013**, *92*, 14–25. [CrossRef] [PubMed]
16. Curran, M.P.; Scott, L.J.; Perry, C.M. Cetirizine. *Drugs* **2004**, *64*, 523–561. [CrossRef]
17. Teixeira, M.; Almeida, Â.; Calisto, V.; Esteves, V.I.; Schneider, R.J.; Wrona, F.J.; Soares, A.M.; Figueira, E.; Freitas, R. Toxic effects of the antihistamine cetirizine in mussel Mytilus galloprovincialis. *Water Res.* **2017**, *114*, 316–326. [CrossRef]
18. Bahlmann, A.; Carvalho, J.J.; Weller, M.G.; Panne, U.; Schneider, R.J. Immunoassays as high-throughput tools: Monitoring spatial and temporal variations of carbamazepine, caffeine and cetirizine in surface and wastewaters. *Chemosphere* **2012**, *89*, 1278–1286. [CrossRef]
19. Ma, D.; Chen, L.; Liu, R. Removal of novel antiandrogens identified in biological effluents of domestic wastewater by activated carbon. *Sci. Total. Environ.* **2017**, *595*, 702–710. [CrossRef]
20. Cantwell, M.G.; Sullivan, J.C.; Burgess, R.M. Benzotriazoles: History, environmental distribution, and potential ecological effects. In *Anonymous Comprehensive Analytical Chemistry*; Elsevier: Amsterdam, The Netherlands, 2015; pp. 513–545.
21. Muschietti, A.; Serrano, N.; Ariño, C.; Diaz-Cruz, M.S.; Cruz, J.M.D. Screen-Printed Electrodes for the Voltammetric Sensing of Benzotriazoles in Water. *Sensors* **2020**, *20*, 1839. [CrossRef]
22. Janna, H.; Scrimshaw, M.D.; Williams, R.J.; Churchley, J.; Sumpter, J.P. From Dishwasher to Tap? Xenobiotic Substances Benzotriazole and Tolyltriazole in the Environment. *Environ. Sci. Technol.* **2011**, *45*, 3858–3864. [CrossRef]
23. Hart, D.; Davis, L.; Erickson, L.; Callender, T. Sorption and partitioning parameters of benzotriazole compounds. *Microchem. J.* **2004**, *77*, 9–17. [CrossRef]
24. Alotaibi, M.D.; McKinley, A.J.; Patterson, B.M.; Reeder, A.Y. Benzotriazoles in the Aquatic Environment: A Review of Their Occurrence, Toxicity, Degradation and Analysis. *Waterairsoil Pollut.* **2015**, *226*, 226. [CrossRef]
25. Loos, R.; Carvalho, R.; António, D.C.; Comero, S.; Locoro, G.; Tavazzi, S.; Paracchini, B.; Ghiani, M.; Lettieri, T.; Blaha, L.; et al. EU-wide monitoring survey on emerging polar organic contaminants in wastewater treatment plant effluents. *Water Res.* **2013**, *47*, 6475–6487. [CrossRef] [PubMed]
26. Fernández-Perales, M.; Sánchez-Polo, M.; Rozalen, M.; López-Ramón, M.V.; Mota, A.; Rivera-Utrilla, J. Degradation of the diuretic hydrochlorothiazide by UV/Solar radiation assisted oxidation processes. *J. Environ. Manag.* **2020**, *257*, 109973. [CrossRef] [PubMed]
27. Ranjan, S.; Devarapalli, R.; Kundu, S.; Vangala, V.R.; Ghosh, A.; Reddy, C.M. Three new hydrochlorothiazide cocrystals: Structural analyses and solubility studies. *J. Mol. Struct.* **2017**, *1133*, 405–410. [CrossRef]
28. Äystö, L.; Mehtonen, J.; Kalevi, K. *Kartoitus Lääkeaineista Yhdyskuntajätevedessä ja Pintavedessä*; Finnish Environment Institute: Helsinki, Finland, 2014. (In Finnish)
29. Oosterhuis, M.; Sacher, F.; Ter Laak, T.L. Prediction of concentration levels of metformin and other high consumption pharmaceuticals in wastewater and regional surface water based on sales data. *Sci. Total. Environ.* **2013**, *442*, 380–388. [CrossRef] [PubMed]
30. Bueno, M.M.; Gomez, M.; Herrera, S.; Hernando, M.; Agüera, A.; Fernández Alba, A.R.R. Occurrence and persistence of organic emerging contaminants and priority pollutants in five sewage treatment plants of Spain: Two years pilot survey monitoring. *Environ. Pollut.* **2012**, *164*, 267–273. [CrossRef]
31. Radjenovic, J.; Petrovic, M.; Barceló, D. Analysis of pharmaceuticals in wastewater and removal using a membrane bioreactor. *Anal. Bioanal. Chem.* **2006**, *387*, 1365–1377. [CrossRef]
32. Laurencé, C.; Rivard, M.; Martens, T.; Morin, C.; Buisson, D.; Bourcier, S.; Sablier, M.C.; Oturan, M.A. Anticipating the fate and impact of organic environmental contaminants: A new approach applied to the pharmaceutical furosemide. *Chemosphere* **2014**, *113*, 193–199. [CrossRef]

33. Fimea, K. Finnish Statistics on Medicines. 2016. Available online: http://urn.fi/URN:NBN:fi-fe2017111750773 (accessed on 2 December 2020).

34. Khalaf, H.; Salste, L.; Karlsson, P.; Ivarsson, P.; Jass, J.; Olsson, P.-E. In vitro analysis of inflammatory responses following environmental exposure to pharmaceuticals and inland waters. *Sci. Total. Environ.* **2009**, *407*, 1452–1460. [CrossRef]

35. TemaNord. *PPCP Monitoring in the Nordic Countries—Status Report*; Nordic Council of Ministers: Copenhagen, Denmark, 2012; p. 519.

36. Besse, J.-P.; Garric, J. Human pharmaceuticals in surface waters: Implementation of a prioritization methodology and application to the French situation. *Toxicol. Lett.* **2008**, *176*, 104–123. [CrossRef]

37. Jelic, A.; Gros, M.; Ginebreda, A.; Cespedes-Sánchez, R.; Ventura, F.; Petrovic, M.; Barcelo, D. Occurrence, partition and removal of pharmaceuticals in sewage water and sludge during wastewater treatment. *Water Res.* **2011**, *45*, 1165–1176. [CrossRef]

38. Bollmann, A.F.; Seitz, W.; Prasse, C.; Lucke, T.; Schulz, W.; Ternes, T. Occurrence and fate of amisulpride, sulpiride, and lamotrigine in municipal wastewater treatment plants with biological treatment and ozonation. *J. Hazard. Mater.* **2016**, *320*, 204–215. [CrossRef] [PubMed]

39. Ferrer, I.; Thurman, E.M. Identification of a New Antidepressant and its Glucuronide Metabolite in Water Samples Using Liquid Chromatography/Quadrupole Time-of-Flight Mass Spectrometry. *Anal. Chem.* **2010**, *82*, 8161–8168. [CrossRef] [PubMed]

40. Keen, O.S.; Ferrer, I.; Thurman, E.M.; Linden, K.G. Degradation pathways of lamotrigine under advanced treatment by direct UV photolysis, hydroxyl radicals, and ozone. *Chemosphere* **2014**, *117*, 316–323. [CrossRef]

41. Richards, N.; Gilbert, M.; Taggart, M.; Naidoo, V. A Cautionary Tale: Diclofenac and Its Profound Impact on Vultures. *Encycl. Anthr.* **2018**, 247–255. [CrossRef]

42. Sathishkumar, P.; Meena, R.A.A.; Palanisami, T.; AshokKumar, V.; Palvannan, T.; Gu, F.L. Occurrence, interactive effects and ecological risk of diclofenac in environmental compartments and biota - a review. *Sci. Total. Environ.* **2020**, *698*, 134057. [CrossRef]

43. Lindholm-Lehto, P.C.; Ahkola, H.S.J.; Knuutinen, J.S.; Herve, S.H. Widespread occurrence and seasonal variation of pharmaceuticals in surface waters and municipal wastewater treatment plants in central Finland. *Environ. Sci. Pollut. Res.* **2016**, *23*, 7985–7997. [CrossRef] [PubMed]

44. Lindqvist, N.; Tuhkanen, T.; Kronberg, L. Occurrence of acidic pharmaceuticals in raw and treated sewages and in receiving waters. *Water Res.* **2005**, *39*, 2219–2228. [CrossRef]

45. Lonappan, L.; Brar, S.K.; Das, R.K.; Verma, M.; Surampalli, R.Y. Diclofenac and its transformation products: Environmental occurrence and toxicity - A review. *Environ. Int.* **2016**, *96*, 127–138. [CrossRef] [PubMed]

46. Vieno, N. Occurrence of pharmaceuticals in Finnish sewage treatment plants, surface waters, and their elimination in drinking water treatment processes. 2007. Available online: http://urn.fi/URN:NBN:fi:tty-200810021012 (accessed on 11 December 2020).

47. Vieno, N.; Sillanpää, M. Fate of diclofenac in municipal wastewater treatment plant — A review. *Environ. Int.* **2014**, *69*, 28–39. [CrossRef]

48. Zhang, Y.; Geissen, S.-U.; Gal, C. Carbamazepine and diclofenac: Removal in wastewater treatment plants and occurrence in water bodies. *Chemosphere* **2008**, *73*, 1151–1161. [CrossRef]

49. Thompson, W.A.; Vijayan, M.M. Environmental levels of venlafaxine impact larval behavioural performance in fathead minnows. *Chemosphere* **2020**, *259*, 127437. [CrossRef] [PubMed]

50. Metcalfe, C.D.; Chu, S.; Judt, C.; Li, H.; Oakes, K.D.; Servos, M.R.; Andrews, D.M. Antidepressants and their metabolites in municipal wastewater, and downstream exposure in an urban watershed. *Environ. Toxicol. Chem.* **2010**, *29*, 79–89. [CrossRef] [PubMed]

51. Falås, P.; Wick, A.; Castronovo, S.; Habermacher, J.; Ternes, T.A.; Joss, A. Tracing the limits of organic micropollutant removal in biological wastewater treatment. *Water Res.* **2016**, *95*, 240–249. [CrossRef] [PubMed]

52. Fimea/Kela, 2018. Finnish statistics of medicines. Available online: http://urn.fi/URN:NBN:fi-fe2019123149481 (accessed on 2 December 2020).

53. Kaur, B.; Dulova, N. UV-assisted chemical oxidation of antihypertensive losartan in water. *J. Environ. Manag.* **2020**, *261*, 110170. [CrossRef] [PubMed]

54. Carpinteiro, I.; Castro, G.; Rodríguez, I.; Cela, R. Free chlorine reactions of angiotensin II receptor antagonists: Kinetics study, transformation products elucidation and in-silico ecotoxicity assessment. *Sci. Total. Environ.* **2019**, *647*, 1000–1010. [CrossRef]

55. Botero-Coy, A.; Martínez-Pachón, D.; Boix, C.; Rincón, R.; Castillo, N.; Arias-Marín, L.; Manrique-Losada, L.; Torres-Palma, R.; Moncayo-Lasso, A.; Hernández, F.H. An investigation into the occurrence and removal of pharmaceuticals in Colombian wastewater. *Sci. Total. Environ.* **2018**, *642*, 842–853. [CrossRef] [PubMed]

56. Kot-Wasik, Á.; Jakimska, A.; Śliwka-Kaszyńska, M. Occurrence and seasonal variations of 25 pharmaceutical residues in wastewater and drinking water treatment plants. *Environ. Monit. Assess.* **2016**, *188*, 661. [CrossRef]

57. Gurke, R.; Rößler, M.; Marx, C.; Diamond, S.; Schubert, S.; Oertel, R.; Fauler, J. Occurrence and removal of frequently prescribed pharmaceuticals and corresponding metabolites in wastewater of a sewage treatment plant. *Sci. Total. Environ.* **2015**, *532*, 762–770. [CrossRef]

58. Bustos, Y.; Vaca, M.; López, R.; Bandala, E.; Torres, L.; Rojas-Valencia, N. Disinfection of Primary Municipal Wastewater Effluents Using Continuous UV and Ozone Treatment. *J. Water Resour. Prot.* **2014**, *6*, 16–21. [CrossRef]

59. Tchobanoglous, G. Wastewater Engineering: Treatment and Resource Recovery. *McGraw-Hill* **2014**, *2*, 1367–1377.

60. US EPA. *Wastewater Technology Fact Sheet, Ozone Disinfection*; US EPA: Washington, DC, USA, 2009.
61. Hu, J.Y.; Wang, Z.S.; Ng, W.J.; Ong, S.L. Disinfection By-Products in Water Produced by Ozonation and Chlorination. *Environ. Monit. Assess.* **1999**, *59*, 81–93. [CrossRef]

Review

Application of Natural Coagulants for Pharmaceutical Removal from Water and Wastewater: A Review

Motasem Y. D. Alazaiza [1,*], Ahmed Albahnasawi [2], Gomaa A. M. Ali [3], Mohammed J. K. Bashir [4], Dia Eddin Nassani [5], Tahra Al Maskari [1], Salem S. Abu Amr [6] and Mohammed Shadi S. Abujazar [6,7]

[1] Department of Civil and Environmental Engineering, College of Engineering, A'Sharqiyah University, Ibra 400, Oman; tahra.almaskari@asu.edu.om
[2] Department of Environmental Engineering-Water Center (SUMER), Gebze Technical University, Kocaeli 41400, Turkey; ahmedalbahnasawi@gmail.com
[3] Chemistry Department, Faculty of Science, Al-Azhar University, Assiut 71524, Egypt; gomaasanad@azhar.edu.eg
[4] Department of Environmental Engineering, Faculty of Engineering and Green Technology (FEGT), Universiti Tunku Abdul Rahman, Kampar 31900, Malaysia; jkbashir@utar.edu.my
[5] Department of Civil Engineering, Hasan Kalyoncu University, Gaziantep 27500, Turkey; diaeddin.nassani@hku.edu.tr
[6] Faculty of Engineering, Karabuk University, Demir Campus, Karabuk 78050, Turkey; sabuamr@hotmail.com (S.S.A.A.); mohammedshadi@hotmail.com (M.S.S.A.)
[7] Al-Aqsa Community Intermediate College, Al-Aqsa University, Gaza P.B.4051, Palestine
* Correspondence: my.azaiza@gmail.com

Abstract: Pharmaceutical contamination threatens both humans and the environment, and several technologies have been adapted for the removal of pharmaceuticals. The coagulation-flocculation process demonstrates a feasible solution for pharmaceutical removal. However, the chemical coagulation process has its drawbacks, such as excessive and toxic sludge production and high production cost. To overcome these shortcomings, the feasibility of natural-based coagulants, due to their biodegradability, safety, and availability, has been investigated by several researchers. This review presented the recent advances of using natural coagulants for pharmaceutical compound removal from aqueous solutions. The main mechanisms of natural coagulants for pharmaceutical removal from water and wastewater are charge neutralization and polymer bridges. Natural coagulants extracted from plants are more commonly investigated than those extracted from animals due to their affordability. Natural coagulants are competitive in terms of their performance and environmental sustainability. Developing a reliable extraction method is required, and therefore further investigation is essential to obtain a complete insight regarding the performance and the effect of environmental factors during pharmaceutical removal by natural coagulants. Finally, the indirect application of natural coagulants is an essential step for implementing green water and wastewater treatment technologies.

Keywords: natural coagulation; chemical coagulation; pharmaceuticals; *Moringa oleifera*; green treatment technology

Citation: Alazaiza, M.Y.D.; Albahnasawi, A.; Ali, G.A.M.; Bashir, M.J.K.; Nassani, D.E.; Al Maskari, T.; Amr, S.S.A.; Abujazar, M.S.S. Application of Natural Coagulants for Pharmaceutical Removal from Water and Wastewater: A Review. *Water* **2022**, *14*, 140. https://doi.org/10.3390/w14020140

Academic Editor: Laura Bulgariu

Received: 5 December 2021
Accepted: 5 January 2022
Published: 6 January 2022

Publisher's Note: MDPI stays neutral with regard to jurisdictional claims in published maps and institutional affiliations.

1. Introduction

The discharge of pharmaceutical waste into the environment poses a threat to both humans and environmental systems. The disposal of these contaminants without proper treatment has resulted in pharmaceuticals being widespread in ecosystems [1]. The presence and accumulation of these emerging compounds harm the ecosystem. Human drugs such as ibuprofen and acetaminophen are continuously accumulating in the environment, resulting in pollutants in water bodies and causing harmful effects [2]. In addition, the mineralization rate of pharmaceuticals such as diclofenac and ibuprofen through photocatalysis is low, resulting in the accumulation of these compounds in the environment [2].

63

The effluent of wastewater treatment plants is the typical source of pharmaceutical compounds, since the conventional wastewater treatment methods are not designed to remove these micropollutants [3]. Therefore, these harmful chemicals accumulate and contaminate soil, rivers, oceans, and groundwater [4].

Recently, several studies reported the efficiency of the coagulation-flocculation treatment method for pharmaceuticals' removal, especially in rich organic wastewater [1]. Coagulation-flocculation consists of two steps: (1) the tendency of colloidal particles to form large flocs by destabilization, and (2) settling these large flocs by precipitation. The removal of pharmaceuticals directly by means of the coagulation process is not reported in the literature. The mechanism of pharmaceuticals' removal by coagulation process is indirect by using colloidal particles as a vehicle for pharmaceuticals [3,5,6].

For many years, chemical-based coagulants such as aluminum sulfate (alum) and poly-aluminum [7,8] have had different environmental effects by producing highly toxic sludge. In addition, the consumption of water contaminated by the residual chemical coagulants may cause neurodegenerative diseases [9]. Thus, the transition towards natural-based coagulants for water and wastewater treatments has gained increasing attention in recent years [10].

Natural coagulants can be produced from natural sources such as plants and animals. Many studies reported several natural sources for extracting natural-based coagulants [11,12]. Natural resources that possess a higher molecular weight may contain a more extended polymer that increases these natural coagulants' efficiency [13–15]. These sources have been extensively studied to treat different types of wastewater, such as textile wastewater, dairy wastewater, and domestic wastewater [16,17]. In addition, coagulants can also be obtained from animal waste such as banes and shells [18]. The main challenge of using natural coagulants in general, especially animal-based coagulants, is their continuous availability for large-scale treatment [18].

Natural coagulants perform better at a wide pH range [19–21]. In addition, using natural coagulants does not change the pH of water compared to chemical coagulants. In addition, natural coagulants positively affect the ecosystem and the environment [10,22,23]

The application of natural coagulants has been reported in many studies for domestic and industrial wastewater [24]. However, fewer studies investigated the performance of natural coagulants for emergency pollutant removal. In addition, fewer reviews discussed the use of natural coagulant for pharmaceuticals removal. In line with the aforementioned gaps. These reviews present and discuss the recent natural coagulation method for pharmaceutical removal from water and wastewater. A comprehensive comparison between natural coagulants and chemical coagulants is also presented. Finally, this review highlights the required future research to overcome the shortcomings of using natural coagulants.

2. Fundamental of Coagulation Processes

The coagulation process is used wildly in water and wastewater treatment, as it is effective for removing suspended solids, turbidity, organic matter, oil, chemical oxygen demand (COD), and color [25]. The coagulation process is mainly conducted by adding a coagulant that allows small agglomerate particles (unsettleable fine particles) to form larger flocs that can settle. Coagulation and flocculation are interlinked. Coagulation is the clustering process under high-speed mixing, whereas flocculation is the settling process under gentle mixing. Generally, colloidal particles are negatively charged particles. Thus, coagulation is a chemical process that involves neutralizing these particles in water and wastewater, whereas flocculation is a physical process involving the formation of flakes from neutralized particles during the coagulation process. Thus, large flocs form during coagulation, and they aggregate and settle during flocculation [26].

Generally, the coagulation process depends on operating conditions such as settling time, mixing rate, coagulant type, and dosage. These factors determine the quality of the produced water. In addition, the coagulant dosage must be suitable for a decent suspension of particles, and the mixing speed should be high. The other coagulant properties, such as life span and quality, determine the coagulant's stability during storage. Following the coagulation-flocculation, the large flocs sink through gravitational settling; this process depends on the settling rate of the particles [27].

The colloidal particles' sizes range from 0.001 to 1.0 µm due to them being negatively charged and the small size being suspended in water. Four mechanisms are used to destabilize these fine particles using a coagulant; charge neutralization, polymer bridging, sweep flocculation, and double-layer compression (Figure 1).

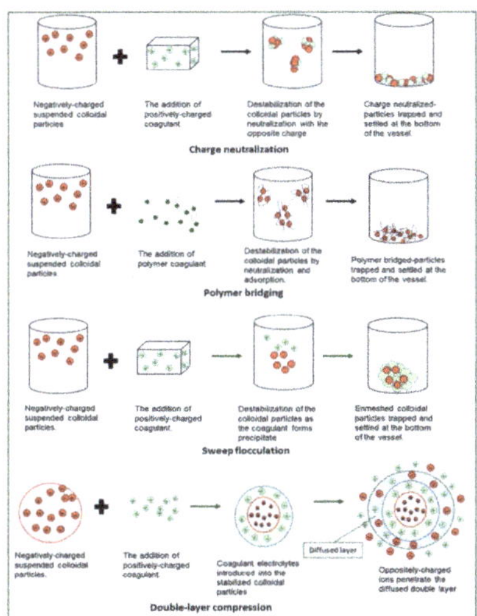

Figure 1. Coagulation mechanisms diagram showing charge naturalization, polymer bridging, sweep flocculation, and double-layer compression, copied with permission from Ref. [10], Copyright, 2021, Elsevier.

In the charge neutralization mechanism, the oppositely charged ions are used to attract colloidal particles, and coagulants added to the wastewater will further neutralize the electrical load until it reaches zero zeta potential; as a result, the colloidal particle charge neutralizes, and the electrostatic repulsion decreases or is almost eliminated [28]. Generally, when a chemical coagulant is added to water, a hydrolysis process occurs, producing cationic species which react colloidally. The polymer bridging mechanism takes place when a polymer or polyelectrolyte coagulant with a long chain destabilizes the colloidal particles by making a bridge that forms a connection between them. The polymer coagulant adsorbs multiple particles to the polymer molecule surface [29]. Thus, strong clusters of macro flocs are produced and tied together by bridges. The flocs formed by polymer bridging are flaky with irregular void spaces. The sweep flocculation coagulant traps the colloidal particles and forces them to sink to the bottom. A net-like structure is formed by the hydrolysis process that makes up precipitation of amorphous metal hydroxide. The double-layer compression includes a coagulant that helps the colloidal particles to reduce the repulsion force and assemble. This mechanism works by means of the presence of a high concentration of electrolyte ions around the colloidal; thus, an opposite charge enters the diffused double layer which surrounds the colloids; as a result, the density is increased [30].

The strongest flocs are those formed through polymer bridging, followed by those formed through charge neutralization and sweep flocculation. The flocs formed through charge neutralization are compacted but not strong, because they depend on the physical rather than chemical bonds. Analysis such as initial floc aggregation, the flocculation index, and the relative settling factor indicated that flocs produced by sweep flocculation have good settling behavior but have a slower formation rate. Flocs produced by double-layer compression are bigger due to the high aggregation rate, but their settling behavior is affected by the unnecessary friction force formed between flocs. Moreover, the coagulant's ionic charge significantly affects the strength of flocs. Divalent ions produce flocs stronger than monovalent ions and require less time to settle. Generally, the dominant coagulation mechanism for natural coagents is charge neutralization [10].

3. Factor Affecting Coagulation Process

Determination of optimum operating conditions is crucial, as an added coagulant is utilized thoroughly to remove the contaminants. Deferent optimal conditions can be achieved for different coagulants. A deep understanding of the reaction between pollutant and coagulant is needed to achieve high performance in addition to decreasing the cost and sludge volume. Many parameters affect the efficiency of the coagulation process for water and wastewater treatments, and these parameters are varied to control the optimal conditions for the highest efficiency. Coagulant dosage, pH, turbidity, mixing speed and time, and temperature are the main operating factors affecting coagulation speed [31]. These factors significantly impact the coagulation process, affecting the effectiveness and efficiency of coagulants in water and wastewater purification processes.

3.1. Coagulant Dosage

The optimal coagulant dosage is an important parameter that entirely controls coagulation reactions. The influence of coagulant dosage can be discussed for three different levels. The optimal coagulant dosage effectively aggregates the colloidal particles in water and wastewater. An underdosage inhabits the proper assembly of colloidal particles, whereas an overdosage pollutes the wastewater and causes an increase in organic load, turbidity, and higher slurry volume, which leads to an increase in the treatment cost [10].

3.2. pH

pH is an acidity/alkalinity measurement that varies between 1 and 14. The pH of water and wastewater is an essential environmental factor, as it affects chemical reactions during the treatment process [32,33]. The amphoteric coagulant molecules' charge highly influences the pH during the treatment process. In addition, alkalinity, which is defined as the capacity to neutralize acidity, controls the efficiency of the coagulation process. Most chemical-based coagulants, especially ferric-based coagulants, absorb a high percentage of alkalinity. Thus, adding a coagulant to wastewater with low alkalinity produces poor flocs. Additional alkaline agents such as caustic soda, slime, or soda ash should be added to the wastewater to overcome this problem. A pH value differing from the optimum pH produces a mixture of negative and positive charges of amino acids, which decreases the coagulant's cationic efficiency [29]. Moreover, pH determines the optimum coagulant dosage as it affects the protein molecule ionic charge. Therefore, the optimum pH's determination and adjustment must be performed before implementing the coagulation process.

3.3. Initial Turbidity

Initial turbidity is an essential factor that affects the coagulation process. The presence of a colloidal particle in water causes turbidity that affects the clarity of the water. Soil, abundant microorganisms, organic matter, decaying matter, colored compounds (pigment and dye), algae, and plankton induce turbidity in water, making it look murky, cloudy, and undesirable. Colloidal particles and turbidity are a challenge in water and wastewater treatment, as the increased rate of turbidity means more pollutant molecules are available, which means a higher number of collisions between the coagulant and pollutants may be produced [34]. More collisions result in sturdier and larger flocs, which settle faster.

On the other hand, a low initial turbidity decreases the collision chance between coagulants and pollutants. As a result, small flocs are formed, which settle slowly. Moreover, a low initial turbidity forms a flake-like structure that needs more time to sink.

3.4. Mixing Speed and Time

The mixing speed and time is an essential operation condition that affects the efficiency of the coagulation process. Rapid mixing is used during the coagulant's addition to evenly enhance the distribution of coagulant through the wastewater and destabilize the suspended particle, whereas gentle mixing is required to increase the collision between particles to form macro flocs [29]. These two speeds control the entire coagulation process as the efficiency of the coagulation process depends on the speed and time of mixing. Inadequate speed and time may inhibit the homogeneous agglomeration of the particles and increase the floc shear and tear.

4. Natural Coagulants

Recently, natural or green coagulants and their application for water and wastewater treatment have received attention, as they do not conserve alkalinity and maintain pH. In addition, natural coagulants do not add metals to the effluent, as chemical coagulants do; a lower sludge volume is produced, and thus, the cost of disposal is lower [10]. Natural coagulants are classified into plant-based coagulants and non-plant-based coagulants. Plant-based coagulants can be prepared from leaves, seeds, fruit wastes, the bark of trees, and other sources. Plant-based coagulants have been more widely investigated than non-plant-based coagulants due to their greater affordability [22]. A wide range of natural coagulants, such as moringa seeds, banana peel, jatropha curcas, cassava peel starch, watermelon, pawpaw, beans, nirmali seeds, and okra have been studied previously [35].

Natural coagulants in powder forms are usually added directly to wastewater. The preparation methods of natural coagulants depend on their source [36–38]. Figure 2 shows the preparation stages for natural powder coagulants from seeds. Oil extraction is an essential step for high oil-content seeds such as *Moringa oleifera*, which contain 30–40% oil, as when a coagulant made from high oil-content-based seeds is used without oil extraction, the organic matter in the treated wastewater will increase. Table 1 illustrates the main application forms of natural coagulants.

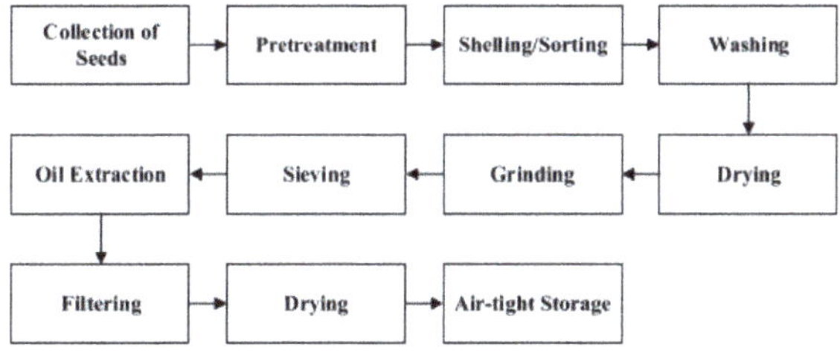

Figure 2. Flow chart of natural coagulant preparation from seeds, copied with permission from Ref. [10], Copyright, 2021, Elsevier.

4.1. Plant-Based Coagulant

Natural coagulants are used for water treatment; however, they are not used for industrial wastewater due to their higher costs than chemical coagulants. Generally, natural coagulants effectively treat water or wastewater with low turbidity ranging from 50 to 500 NTU (Nephelometric Turbidity Units). The primary sources of plant-based coagulants are *Moringa oleifera*, Nirmali seeds, cactus, and tannin. The extracted natural polymers from these seeds are biodegradable and eco-friendly [36]. Anionic polyelectrolytes are extracted from Nirmali seeds; this extract has hydroxyl (-OH) and carboxylic (-COOH) groups, increasing coagulation efficiency. The combination of galactan and polysaccharides extracted from Strychnospotatorum seeds may increase the turbidity removal efficiency up to 80%. The availability of the hydroxyl group (-OH) in the galactan and galactomannan enhances the adsorption process between the surface of colloidal and these polymers; thus, the polymers' bridging action may increase. The polyelectrolytes neutralize the negative colloidal particles and adsorb onto the surface particles. Natural coagulants possess several functional and charged groups such as -COOH, -NH$_2$, and -OH. Generally, the action of natural polymeric coagulants combines polymer bridging and charge neutralization.

Table 1. The main application forms of natural coagulants.

Natural Coagulant	Application Form	Reference
Moringa oleifera	Seed paste	[38]
	Press cake (solid)	[39]
	Powder	[40]
Chitosan	Powder	[41]
	Stock solution (0.1 M HCl)	[42]
	Solution (1% acetic acid)	[43]
	Stock solution (0.1 M HCl and distilled water)	[44]
Rice starch	Starch solution	[45]
Jatropha curcas	Press cake (solid)	[46]
Watermelon seeds	Oil-free powder	[47]
Banana pith	Powder	[48]
Ocimum basilicum	Mucilage	[49]

4.2. Animal Base

The source of animal-based coagulants is usually obtained from the exoskeleton of shellfish extracts, animal bone shell extracts, and chitosan. Chitosan is a polymer (cellulose-like biopolymer) with a high molecular weight produced from the deacetylation of chitin, extracted from the shells of crabs, lobsters, shrimps, diatoms, fungi, insects, freshwater and marine sponges, and mollusks. The applicability of using chitosan as a natural coagulant has been studied intensively for wastewater treatment in the agricultural industry, textile industry, food processing industry, paper mills, soap and detergent industry, and other industries [34]. The main advantages of using chitosan as a coagulant are that when added to acidic wastewater, it reacts and produces positive charges that destabilize colloidal particles' negative charge [37].

5. Pharmaceutical in Water and Wastewater

Pharmaceuticals are a set of developed chemicals used for human and veterinary medication. Recently, they have been classified as ecological contaminants that threaten both humans and environments [50]. Pharmaceuticals include antibiotics, analgesics, both legal and illicit, beta-blockers, steroids, etc., and they have been detected in wastewater treatment plants' effluents, sediments, sludge, natural waters, groundwater, and drinking water. The presence of pharmaceuticals in the soil may trigger the development of antibiotic-resistant genes [51].

Currently, pharmaceuticals and their biotransformative compounds are bioaccumulating and harmfully affecting the ecosystem. However, these chemicals have been discharged to the environment for a long time, their environmental effects have only been considered recently. Many pharmaceuticals (around 160) have been detected in water bodies in low concentrations. These chemicals are classified as pseudo-persistent pollutants which environmentally persistent and are continuously discharged into the environment at low concentrations. These pharmaceuticals' eco-toxicological impacts on aquatic and terrestrial life are unknown [52].

5.1. Pharmaceuticals' Consumption and Fate

The consumption of pharmaceutical compounds has increased dramatically due to many reasons, such as a decrease in production cost and chronic disease treatment demand. As a result, the presence of these compounds has increased. Currently, the environmental management of pharmaceuticals is challenging, as these substances are found in wastewater treatment plants' effluents in low concentrations (usually in ng/L). Sophisticated analytical apparatuses and complex methods are needed to quantify pharmaceuticals at this low concentration [53].

Pharmaceuticals are generally moved and transported by the demonstrated routes in Figure 3 [54]. After consumption, metabolism in the human body, and extraction, pharmaceuticals usually reach aquatic environments by being discharged in treated domestic wastewater effluents [55]. During this route, pharmaceuticals may go through chemical reactions and transformation, forming by-products, which are sometimes more harmful and persistent than their parent compounds. Most of these compounds are non-biodegradable in conventional treatment methods; as a result, they remain and are discharged through wastewater treatment plants' effluents into water bodies such as lakes, rivers, and estuaries [56]. In veterinary products, pharmaceuticals reach aquatic systems through subsequent outflow and manure and direct application in aquaculture [57]. Microorganisms may convert the metabolic compounds to their parental form in surface and groundwater [58]. The ecological concern related to pharmaceuticals in water resources is not directly related to their quantity but to their availability and persistence, which directly affects aquatic life through their toxicity and their potential effect on endocrine function [54,59].

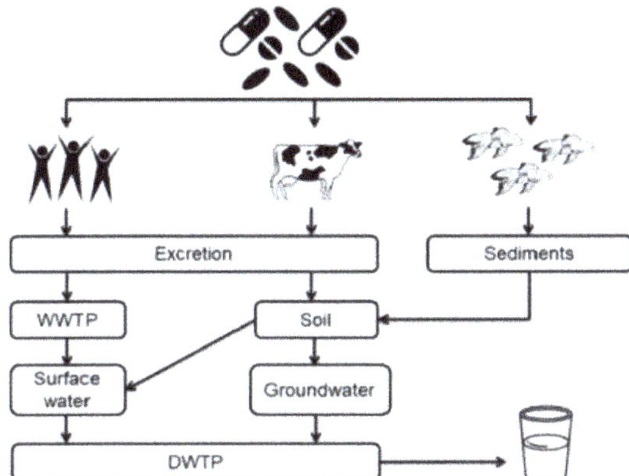

Figure 3. Pharmaceuticals' fates and environmental pathways, copied with permission from Ref. [54], Copyright, 2019, Elsevier.

5.2. Technologies for Pharmaceutical Wastewater Treatment

Pharmaceutical removal from water and wastewater is challenging due to their low concentration and resistance to degradation. Many technologies have been investigated for pharmaceutical removal from water and wastewater [60]. In this section, the pharmaceutical removal methods are discussed.

Activated sludge systems have been used for domestic and industrial wastewater treatments for a long time. Recently, the efficiency of this conventional treatment method for pharmaceuticals removal was investigated. Ren et al. [61] studied the removal of 21 parimutuels by an activated sludge treatment system. The result show that 14 compounds were biodegradable, whereas seven were non-biodegradable. Thus, activated sludge treatment methods are not efficient in completely removing pharmaceuticals from wastewater, as it is not designed for this type of pollutant. Electrocoagulation is more efficient and effective than chemical coagulation. In electrocoagulation, anodes are used to treat contaminants, and the formed coagulants are used for their degradation. Many studies investigated the use of electrocoagulation treatment methods to remove pharmaceuticals such as dexamethasone, doxycycline hyclate, hydrolyzed peptone, caffeine, sulfamethazine, and cephalexin from wastewater. The results show a high removal efficiency (generally above 90%), indicating that these systems efficiently remove pharmaceuticals [62]. The main advantages of using electrocoagulation treatment are its easy chemical maintenance and high efficiency for colloidal particle removal. However, electricity and sacrificial electrodes are the main drawbacks of using this method, as they need to be replaced [63].

Advanced oxidation processes are effective in removing pharmaceuticals that conventional biological methods cannot remove. Among these methods, the Fenton reaction represents hydroxyl radical formation by a reaction between hydrogen peroxide (H_2O_2) and Fe (II). The hydroxyl radical is considered among the strongest oxidants that can oxidize a wide range of organic matters with low selectivity [64]. Therefore, Fenton-based reactions are commonly used for degradation emergency contaminants such as pharmaceuticals. pH control is essential for the Fenton reaction; thus, this treatment technology is usually performed at an acidic pH (3–5). The Fenton reaction method is found to be an effective method for a wide range of pharmaceuticals removal such as hydroxylamine, cyclohexanone, pyridine, toluene [65]. Many reports revealed that membrane bioreactor technologies can remove more micropollutants than conventional activated sludge systems due to their high MLSS (Mixed Liquor Suspended Solids) concentration and high sludge retention time, which allow the growth of low growth bacteria [66]. These bacteria can degrade complex organic compounds. The removal of acetaminophen, carbamazepine, mefenamic acid, ibuprofen, diazepam, naproxen, and ketoprofen by a membrane bioreactor was studied. Overall, more than 85% removal efficiency was obtained [67]. The major drawback of membrane bioreactors is the fouling of membranes that need frequent cleaning and sometimes replacement [68].

Photocatalysis is a reaction in which the presence of a catalyst accelerates the photoreaction [69–71]. The main advantage of photocatalysis reactions is the need for temperature or pressure or chemical agents such as hydrogen peroxide. However, this method is costly. Titanium oxide is the most studied catalyst for photocatalysis reactions due to its biological and chemical stability, inertness, and low cost compared to highly photoactive semiconductor materials. Titanium oxide can be used many times without losing its photocatalyst activity [72,73]. Nevertheless, the separation of titanium oxide from the reaction matrix is complicated. In addition, the transformation of organic matter is incomplete, and by-products, sometimes with a higher toxicity than parent compounds, are produced. Figure 4 shows the problems of using titanium oxides as a catalyst in photocatalysis reactions for pharmaceutical removal [74]. Ozonation has been used as an advanced treatment in many wastewater treatment plants worldwide to enhance contaminant removal. Ozone is a colorless, unstable gas used as a disinfectant for organic and inorganic pollutants. Two mechanisms are used to degrade organic matter by ozonation; (1) an indirect attack by hydroxyl radicals produced by ozone decomposition, and (2) a direct electrophilic attack by ozone [75]. The main drawbacks of using ozone are the high operational cost; the by-products may be toxic; and that ozone is less soluble in water [68].

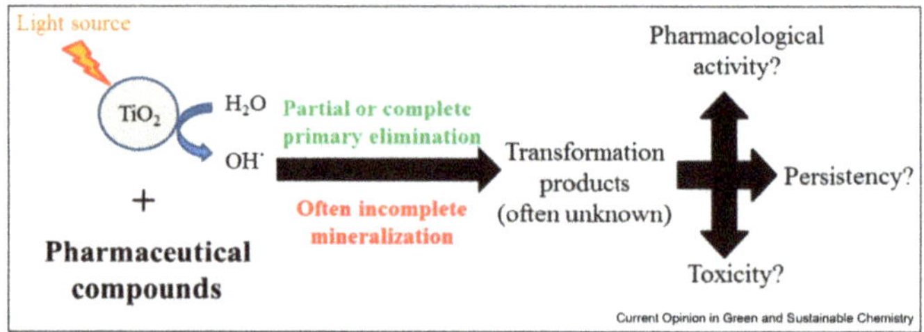

Figure 4. TiO$_2$-related problems, copied with permission from Ref. [62], Copyright, 2021, Elsevier.

6. Application of Natural Coagulants for Pharmaceutical Removal

Recently, green water and wastewater treatment technologies have gained more attention. Among these technologies, natural coagulants are a promising method for wastewater treatment [76]. In this section, the recent advancements made in using natural-based coagulants are presented.

In a recent study, Nonfodji et al. [77] prepared a natural coagulant from *Moringa oleifera* seeds, and they studied its performance for hospital wastewater treatment. The results indicate that the removal efficacy of turbidity and COD was 64 and 38%, respectively. In a subsequent study, Thirugnanasambandham and Karri [78] compared the COD, turbidity, and color removal by two types of coagulants; a natural coagulant (Azadirachta indica A. Juss) and a chemical coagulant (aluminum sulfate). Remarkably, the results indicate that natural-based coagulants may not only be effective for COD, turbidity, and color removal, but may also be economically competitive, as the operating costs were USD 0.56/m^3 and USD 1.73/m^3 for natural coagulants and the chemical ones.

In another study, Maharani et al. [79] investigated the removal of COD and BOD from pharmaceutical waste using moringa seed coagulant and tapioca starch coagulant. The results point to high BOD (Biochemical Oxygen Demand) and COD removal for both natural coagulants. For moringa, the BOD and COD removals were 90 and 71%, respectively, whereas for tapioca, they were 95 and 94% for BOD and COD, respectively. These results indicate that natural coagulants might be a promising treatment technology for pharmaceutical waste treatment. Oliva et al. [80] studied the use of rice husk ash functionalized by *Moringa oleifera* protein for amoxicillin removal from water solutions. They also investigated the effect of operating parameters such as coagulant dosage, initial amoxicillin concentration, and contact time. The results indicate that the used biomaterials are feasible for pharmaceutical removal from water. Olivera [81] examined the potential of using biomaterial extracted from *Moringa oleifera* for the extraction of diclofenac and oxytetracycline from wastewater. The results show the high potential for pharmaceutical removal from wastewater using biomaterial.

The removal percentages were 88% for diclofenac and 50% for oxytetracycline. Santos et al. [82] examined tetracycline removal from river water by using *Moringa oleifera* seeds. The results show 50% tetracycline removal efficiency at 0.5 g/L *Moringa oleifera* dosage. Iloamaeke and Chizaram [83] examined the removal of pharmaceuticals by Phoenix dactylifera seeds-based coagulants. The results show that a maximum removal efficiency of 99.86% was achieved at a 100 mg/L coagulant dosage, a 50 min settling time, and a pH of 2.

Sibartie and Ismail [84] studied the performance of H. Sabdariffa and J. Curcas as a neutral coagulant for pharmaceutical wastewater treatment. The results demonstrate that at a coagulant dosage of 190 mg/L and pH 4, the maximum removal efficiency was achieved for turbidity (5.8%) and COD (30%) by H. Sabdariffa, while J. Curcas works best at pH 3 and a coagulant dosage of 200 mg/L to remove 51% of turbidity and 32% of COD. Table 2 presents the application of natural coagulants to remove different types of pharmaceuticals.

Table 2. Application of natural coagulant for the removal of different types of pharmaceuticals.

Coagulants	Properties	Contaminants	Conditions	Main Results	Reference
Moringa oleifera seeds	Plant-based	COD (hospital wastewater)	Initial COD 238 mg/L; pH 6, 8; Coagulant dosage 0–4000 mg/L; Rapid mixing: 200 rpm for 3 min; Gentle mixing: 45 rpm for 30 min; Settling time 60 min.	*Moringa oleifera* seed polymers are promising bio-coagulants for hospital wastewater treatments.	[77]
Azadirachta indica A. Juss.	Plant-based	COD (urban sewage)	Initial COD 3030 mg/L; pH 4.5;Coagulant dosage 2000–6000 g/L; Rapid mixing: 100 rpm for 1 min; Gentle mixing: 40 rpm for 30 min; Settling time 60 min.	Natural coagulants effectively reduced COD, turbidity, and color at optimum conditions compared to chemical coagulants	[78]
Moringa seed coagulant, tapioca starch coagulants	Plant-based	COD (Pharmaceutical waste)	pH 6–8; Coagulant dosage 3780 mg/L; Rapid mixing: 100 rpm for 10 min; Gentle mixing: 60 rpm for 15 min;	The high removal efficiency observed with the use of tapioca flour coagulant is due to an amide group that contains a high positive charge	[79]
Rice husk ash functionalized by *Moringa* oleifera protein	Plant-based	amoxicillin	Dosage 500, 1000, 1500 mg/L; Contact time 30, 60, 90 min; Mixing speed 150 rpm; Initial amoxicillin concentration (100, 200, 300) mg/L	Rice husk ash functionalized by *Moringa oleifera* protein can be an effective treatment method for an antibiotic from water	[80]
Moringa oleifera adsorpant	Plant-based	Diclofenac and Oxytetracycline	pH 3–10; Dosage 2000 mg/L; Initial diclofenac and oxytetracycline concentration 0.2–1 mg/L; Stirring speed 150 rpm.	The removal efficiency is highly pH-dependent; diclofenac removal efficiency was 4.8% at pH 8 and 87.3% at pH 2, while the removal efficiency of oxytetracycline at pH 3 and 10 was 31 and 50%, respectively	[81]
Moringa oleifera seed	Plant-based	Tetracycline antibiotic	Tetracycline initial concentration 5 mg/L; Coagulant dosage 250–2500 mg/L; pH 5–8; Rapid mixing: 120 rpm for 1 min; Gentle mixing: 30 rpm for 15 min; Settling time 30 min	*Moringa oleifera* seed is a natural, simple, and environmentally friendly technology for antibiotic removal from contaminated water	[82]
Phoenix dactylifera	Plant-based	Pharmaceutical effluent	pH 4–10; Coagulant dosage 200–400 mg/L; Rapid mixing: 100 rpm for 2 min; Gentle mixing: 40 rpm for 20 min; Settling time 50 min	SEM analysis indicated that phoenix dactylifera adsorbed pharmaceutical particles on the surface; thus, phoenix dactylifera can be an effective green coagulant for emergency pollutant removal	[83]
Hibiscus Sabdariffa and Jatropha Curcas	Plant-based	Pharmaceutical Wastewater	Contaminant initial concentration 660 mg/L; pH 2–12; Coagulant dosage 40–200 mg/L; Rapid mixing: 100 rpm for 10 min; Gentle mixing: 40 rpm for 25 min; Settling time 50 min	Compared to chemical coagulants (Alum), natural coagulants such as J. Curcas have better performance in terms of pharmaceutical wastewater treatments	[84]

7. The Transition from Chemical to Natural Coagulant: Comparative Evaluation on Performance

The transition from chemical coagulation to natural coagulation can be an important step towards increasing green water treatment technology, reducing health risks and environmental pollution [23]. Natural coagulants can be obtained from plant or animal sources. Natural coagulants were discovered years ago, before chemical coagulant; over the years, the application of natural coagulants decreased due to the development of chemical coagulants. Recently, the rise of green water treatment technology, besides the environmental problems related to chemical coagulants, has motivated the consideration of natural coagulants again. This section presents a comparative discussion of natural and chemical coagulants.

Many studies evaluated the performance of natural coagulants for removing pollutants from water and wastewater; they concluded that natural coagulants can be competitive in terms of removal efficiency [26]. Table 3 presents the comparison performance of natural and chemical coagulants. The combination of chemical and natural coagulants may increase the performance of the coagulation process. In a study, the combination of alum and banana peels removed 94% of turbidity, whereas the use of alum and banana peels alone resulted in turbidity removal efficiency of 73.1 and 65.6%, respectively [85].

The advantages of using natural-based coagulants over chemical ones are: (1) natural coagulants may produce less sludge than chemical coagulant; thus, the environmental sustainability increases, while the sludge handling cost decreases; (2) the natural coagulant dosage is less than that of chemical coagulants; thus, the cost and sludge production is lower; (3) the toxicity of natural coagulants is lower than that of chemical coagulants [86]; and (4) the use of natural coagulants does not require skilled workers, as they have a low health impact and do not represent such as potential environmental hazard [23].

However, natural coagulants have some disadvantages that hinder their widespread use: (1) rapid mixing during the coagulation process induces cell rupture; thus, the organic matter load may increase and react with disinfectants in the following treatment process, resulting in disinfectant by-products [87,88]; (2) the vast majority of natural coagulants are extracted from plants, so the supply of these coagulants may be affected by seasonal production [89]; (3) natural coagulants are bio-based materials; thus, this material can decompose during long-term storage [9]; and (4) some natural coagulants are used as medicines; the high consumption of these materials in water treatment could affect their supply to the medicine sector [10].

Table 3. Comparison performance of natural and chemical coagulants.

Type of Wastewater	Chemical Coagulant	Removal Performance	Natural Coagulant	Removal Performance	Reference
Arsenic-contaminated surface water	Ferric chloride	Maximum arsenic removal of 69.3% at 40 mg/L coagulant dosage	Cellulose and chitosan	Maximum arsenic removal of 84.62% at a 1 mg/L cellulose dosage and 75.87% at a 25 mg/L chitosan dosage.	[90].
Turbidity in Surface water	Alum	Turbidity removal of 78.72% at a dosage of 100 mg/L	Sago and chitin	Turbidity removal of 69.15% at a sago dosage of 300 mg/L, and 67.73% at a chitin dosage of 300 mg/L	[91]
Paper mill industry	Alum	Turbidity removal of 97.1%, COD removal of 92.7%	*Moringa oleifera* seed	Turbidity removal of 96%, COD removal of 97.3%	[92]
Paint industry	Ferric chloride	Color removal of 89.4%, COD removal of 83.4%	Cactus	Color removal of 88.4%, COD removal of 78.2%	[93]
Concreate plant	Ferric chloride and Alum	Turbidity removal of 99.9%	*Moringa oleifera* seed	Turbidity removal of 99.9%	[94]
Confectionary	PAM	TSS removal of 93.5% COD removal of 95.9%	Cactus	TSS removal of 92.2% COD removal of 95.6%	[95]
Paper and mill	Alum	Color removal of 80% TOC removal of 40%	Chitosan	Color removal of 90% TOC removal of 70%	[96]
Dam water	Alum	Turbidity removal of 98.5%, color removal of 98.5%	Watermelon seed	Turbidity removal of 89.3%, color removal of 93.9%	[97]

8. Recommendation and Future Prospective

All the mentioned disadvantages of implementing natural coagulants for water and wastewater create challenges for future research. The current extraction methods of coagulants from plants and animals are complex; thus, a new reliable and straightforward extraction method should be developed for the easily accessible use of natural coagulants. Some studies reported a higher removal efficiency of chemical coagulants than natural coagulants. However, optimizing the natural coagulant extraction methods can increase the performance of these green coagulants; thus, intensive research is needed in this domain. The utilization of sources for natural coagulant production is a great challenge, as the water and wastewater industries consume many of these coagulants. More research needs to search for new sources, such as inedible plants or/and new medicine plants for natural coagulant production. Further investigations are required to determine the optimum conditions for a green coagulation-flocculation process for various wastewater types. More studies should be conducted to investigate the efficiency of natural coagulants for micropollutants' removal from water and wastewater.

9. Conclusions

The removal of pharmaceuticals from water and wastewater is challenging due to their low concentration and their resistance to biodegradation. Several studies reported the feasibility of using natural-based coagulants for water and wastewater treatments. The main mechanisms that natural coagulants use for pharmaceutical removal from water and wastewater are charge neutralization and polymer bridging. Plant-based natural coagulants are more affordable than animal-based ones. Although the application of natural coagulants for emergency pollutants, especially pharmaceuticals, is limited in the literature, the available data demonstrate a promising future for these bio-coagulants in this domain. A natural coagulant has advantages over a chemical coagulant as a low dosage is required, less sludge is produced, and low/no toxicity is presented. For the complete transition from chemical coagulants to natural coagulants, further research is required in areas such as developing reliable extraction methods, searching for new natural sources, determining the optimal conditions for pharmaceutical removal, and evaluating the effect of environmental parameters on the process' performance.

Author Contributions: Conceptualization and Funding, M.Y.D.A.; Writing Original Draft Preparation, A.A. and M.Y.D.A.; Writing—Review & Editing, G.A.M.A., M.J.K.B., S.S.A.A., D.E.N., M.S.S.A. and T.A.M.; validation, M.J.K.B. All authors have read and agreed to the published version of the manuscript.

Funding: The research leading to these results received funding from the Ministry of Higher Education, Research, and Innovation (MoHERI) of the Sultanate of Oman under the Block Funding Program, MoHERI Block Funding Agreement No. MoHERI/BFP/ASU/01/2021. In addition, the authors would like to thank A'Sharqiyah University for partially funding this research through the ASU Seed Fund Project No: ASU-FSFR/COE/01/2020.

Conflicts of Interest: The authors declare no conflict of interest.

References

1. Kooijman, G.; de Kreuk, M.K.; Houtman, C.; van Lier, J.B. Perspectives of coagulation/flocculation for the removal of pharmaceuticals from domestic wastewater: A critical view at experimental procedures. *J. Water Process Eng.* **2020**, *34*, 101161. [CrossRef]
2. Aziz, H.A.; Zainal, S.F.F.; Alazaiza, M.Y.D. Optimization of Coagulation-Flocculation Process of Landfill Leachate by Tin (IV) Chloride Using Response Surface Methodology. *Avicenna J. Environ. Health Eng.* **2019**, *6*, 41–48.
3. Ahmad, A.; Abdullah, S.R.S.; Hasan, H.A.; Othman, A.R.; Ismail, N.I. Plant-based versus metal-based coagulants in aquaculture wastewater treatment: Effect of mass ratio and settling time. *J. Water Process Eng.* **2021**, *43*, 102269. [CrossRef]
4. Alazaiza, M.Y.D.; Albahnasawi, A.; Ali, G.A.M.; Bashir, M.J.K.; Copty, N.K.; Amr, S.S.A.; Abushammala, M.F.M.; Maskari, T. Al Recent Advances of Nanoremediation Technologies for Soil and Groundwater Remediation: A Review. *Water* **2021**, *13*, 2186. [CrossRef]
5. Mathuram, M.; Meera, R.; Vijayaraghavan, G. Application of Locally Sourced Plants as Natural Coagulants for Dye Removal from Wastewater: A Review. *J. Mater. Environ. Sci.* **2018**, *2508*, 2058–2070.
6. Mohd-Salleh, S.N.A.; Mohd-Zin, N.S.; Othman, N. A review of wastewater treatment using natural material and its potential as aid and composite coagulant. *Sains Malays.* **2019**, *48*, 155–164. [CrossRef]
7. Madiraju, S.V.H.; Kumar, A.; Kuruppuarachchi, L.N. Examination of plant-based coagulants to replace lime and alum for surface water treatment. In Proceedings of the A&WMA's 112th Annual Conference & Exhibition, Quebec City, QC, Canada, 25–28 June 2019; pp. 1–9.
8. Kristianto, H. Recent advances on magnetic natural coagulant: A mini review. *Environ. Technol. Rev.* **2021**, *10*, 254–269. [CrossRef]
9. Saleem, M.; Bachmann, R.T. A contemporary review on plant-based coagulants for applications in water treatment. *J. Ind. Eng. Chem.* **2019**, *72*, 281–297. [CrossRef]
10. Owodunni, A.A.; Ismail, S. Revolutionary technique for sustainable plant-based green coagulants in industrial wastewater treatment—A review. *J. Water Process Eng.* **2021**, *42*, 102096. [CrossRef]
11. Rekha, B.; Sumithra, S. Natural Plant Seeds as an Alternative Coagulant in the Treatment of Mining Effluent. *Rekha Sumithra* **2020**, *6*, 15–23. [CrossRef]
12. Zaid, A.Q.; Ghazali, S.; Ahmad Mutamim, N.S.; Abayomi, O.O.; Abdurahman, N.H. Assessment of Moringa Oleifera Cake Residues (Mocr) as Eco-Friendly Bio-Coagulant. *J. Chem. Eng. Ind. Biotechnol.* **2019**, *5*, 29–38. [CrossRef]
13. Neerajasree, V.R.; Varsha Ashokan, V. Treatment of Automobile Waste Water using Plant-Based Coagulants. *Int. Res. J. Eng. Technol.* **2008**, *6*, 161–164.
14. Jones, A.N. Investigating the Potential of Hibiscus Seed Species as Alternative Water Treatment Material to the Traditional Chemicals. Ph.D. Thesis, University of Birmingham, Birmingham, UK, 2016.
15. Iqbal, A.; Hussain, G.; Haydar, S.; Zahara, N. Use of new local plant-based coagulants for turbid water treatment. *Int. J. Environ. Sci. Technol.* **2019**, *16*, 6167–6174. [CrossRef]

16. Ang, W.L.; Mohammad, A.W. State of the art and sustainability of natural coagulants in water and wastewater treatment. *J. Clean. Prod.* **2020**, *262*, 121267. [CrossRef]

17. Okoro, B.U.; Sharifi, S.; Jesson, M.A.; Bridgeman, J. Natural organic matter (NOM) and turbidity removal by plant-based coagulants: A review. *J. Environ. Chem. Eng.* **2021**, *9*, 106588. [CrossRef]

18. Pal, P.; Pal, A.; Nakashima, K.; Yadav, B.K. Applications of chitosan in environmental remediation: A review. *Chemosphere* **2021**, *266*, 128934. [CrossRef]

19. Jayalakshmi, G.; Saritha, V.; Dwarapureddi, B.K. A Review on Native Plant Based Coagulants for Water Purification. *Int. J. Appl. Environ. Sci.* **2017**, *12*, 469–487.

20. Deshmukh, S.O.; Hedaoo, M.N. Wastewater Treatment using Bio-Coagulant as Cactus Opuntia Ficus Indica—A Review. *Int. J. Sci. Res. Dev.* **2018**, *6*, 711–717.

21. Deng, X.; Jiang, W. Evaluating Green Supply Chain Management Practices under Fuzzy Environment: A Novel Method Based on D Number Theory. *Int. J. Fuzzy Syst.* **2019**, *21*, 1389–1402. [CrossRef]

22. Yusoff, M.S.; Aziz, H.A.; Alazaiza, M.Y.D.; Rui, L.M. Potential use of oil palm trunk starch as coagulant and coagulant aid in semi-aerobic landfill leachate treatment. *Water Qual. Res. J.* **2019**, *54*, 203–219. [CrossRef]

23. Kurniawan, S.B.; Abdullah, S.R.S.; Imron, M.F.; Said, N.S.M.; Ismail, N.I.; Hasan, H.A.; Othman, A.R.; Purwanti, I.F. Challenges and opportunities of biocoagulant/bioflocculant application for drinking water and wastewater treatment and its potential for sludge recovery. *Int. J. Environ. Res. Public Health* **2020**, *17*, 9312. [CrossRef] [PubMed]

24. Tawakkoly, B.; Alizadehdakhel, A.; Dorosti, F. Evaluation of COD and turbidity removal from compost leachate wastewater using *Salvia hispanica* as a natural coagulant. *Ind. Crops Prod.* **2019**, *137*, 323–331. [CrossRef]

25. Aziz, H.A.; Rahim, N.A.; Ramli, S.F.; Alazaiza, M.Y.D.; Omar, F.M.; Hung, Y.T. Potential use of *Dimocarpus longan* seeds as a flocculant in landfill leachate treatment. *Water* **2018**, *10*, 1672. [CrossRef]

26. Bahrodin, M.B.; Zaidi, N.S.; Hussein, N.; Sillanpää, M.; Prasetyo, D.D.; Syafiuddin, A. Recent Advances on Coagulation-Based Treatment of Wastewater: Transition from Chemical to Natural Coagulant. *Curr. Pollut. Rep.* **2021**, *7*, 379–391. [CrossRef]

27. Sulaiman, M.; Zhigila, D.A.; Umar, D.M.; Babale, A.; Andrawus Zhigila, D.; Mohammed, K.; Mohammed Umar, D.; Aliyu, B.; Manan, F.A. *Moringa oleifera* seed as alternative natural coagulant for potential application in water treatment: A review. *J. Adv. Rev. Sci. Res. J.* **2017**, *30*, 1–11.

28. Vishali, S.; Karthikeyan, R. Application of green coagulants on paint industry effluent—A coagulatioflocculation kinetic study. *Desalin. Water Treat.* **2018**, *122*, 112–123. [CrossRef]

29. Gautam, S.; Saini, G. Use of natural coagulants for industrial wastewater treatment. *Glob. J. Environ. Sci. Manag.* **2020**, *6*, 553–578. [CrossRef]

30. Yadav, S.; Yadav, A.; Bagotia, N.; Sharma, A.K.; Kumar, S. Adsorptive potential of modified plant-based adsorbents for sequestration of dyes and heavy metals from wastewater—A review. *J. Water Process Eng.* **2021**, *42*, 102148. [CrossRef]

31. Zainal, S.F.F.S.; Aziz, H.A.; Omar, F.M.; Alazaiza, M.Y.D. Influence of *Jatropha curcas* seeds as a natural flocculant on reducing Tin (IV) tetrachloride in the treatment of concentrated stabilised landfill leachate. *Chemosphere* **2021**, *285*, 131484. [CrossRef]

32. Shayegan, H.; Ali, G.A.M.; Safarifard, V. Amide-Functionalized Metal–Organic Framework for High Efficiency and Fast Removal of Pb(II) from Aqueous Solution. *J. Inorg. Organomet. Polym. Mater.* **2020**, *30*, 3170–3178. [CrossRef]

33. Shayegan, H.; Ali, G.A.M.; Safarifard, V. Recent Progress in the Removal of Heavy Metal Ions from Water Using Metal-Organic Frameworks. *ChemistrySelect* **2020**, *5*, 124–146. [CrossRef]

34. Wang, J.; Chen, X. Removal of antibiotic resistance genes (ARGs) in various wastewater treatment processes: An overview. *Crit. Rev. Environ. Sci. Technol.* **2020**, *52*, 571–630. [CrossRef]

35. Shewa, W.A.; Dagnew, M. Revisiting chemically enhanced primary treatment of wastewater: A review. *Sustainability* **2020**, *12*, 5928. [CrossRef]

36. Othmani, B.; Rasteiro, M.G.; Khadhraoui, M. Toward green technology: A review on some efficient model plant-based coagulants/flocculants for freshwater and wastewater remediation. *Clean Technol. Environ. Policy* **2020**, *22*, 1025–1040. [CrossRef]

37. Zainal, S.F.F.S.; Aziz, H.A.; Omar, F.; Alazaiza, M.Y.D. Sludge performance in coagulation-flocculation treatment for suspended solids removal from landfill leachate using Tin (IV) chloride and *Jatropha curcas*. *Int. J. Environ. Anal. Chem.* **2021**, 1–15. Available online: https://www.researchgate.net/profile/Motasem_Alazaiza/publication/351883646_Sludge_performance_in_coagulation-flocculation_treatment_for_suspended_solids_removal_from_landfill_leachate_using_Tin_IV_chloride_and_Jatropha_curcas/links/60eab81e1c28af34585effa5/Sludge-performance-in-coagulation-flocculation-treatment-for-suspended-solids-removal-from-landfill-leachate-using-Tin-IV-chloride-and-Jatropha-curcas.pdf (accessed on 4 December 2021). [CrossRef]

38. Jagaba, A.H.; Kutty, S.R.M.; Hayder, G.; Latiff, A.A.A.; Aziz, N.A.A.; Umaru, I.; Ghaleb, A.A.S.; Abubakar, S.; Lawal, I.M.; Nasara, M.A. *Sustainable Use of Natural and Chemical Coagulants for Contaminants Removal from Palm Oil Mill Effluent: A Comparative Analysis*; Elsevier: Amsterdam, The Netherlands, 2020.

39. Bhatia, S.; Othman, Z.; Journal, A.A.-C.E. *Coagulation–Flocculation Process for POME Treatment Using Moringa oleifera Seeds Extract: Optimization Studies*; Elsevier: Amsterdam, The Netherlands, 2007.

40. Ramesh, S.; Mekala, L. Treatment of Textile Wastewater Using *Moringa oleifera* and *Tamarindus indica*. *Int. Res. J. Eng. Technol.* **2018**, *6*, 3891–3895.

41. Chakrabarti, P.; Kale, V.; Sarkar, B.; Chakrabarti, P.P.; Vijaykumar, A. *Wastewater Treatment in Dairy Industries—Possibility of Reuse*; Elsevier: Amsterdam, The Netherlands, 2006; Volume 195, pp. 141–152. [CrossRef]

42. Yunos, F.H.M.; Nasir, N.M.; Jusoh, H.H.W.; Khatoon, H.; Lam, S.S.; Jusoh, A. *Harvesting of Microalgae (Chlorella sp.) from Aquaculture Bioflocs Using an Environmental-Friendly Chitosan-Based Bio-Coagulant*; Elsevier: Amsterdam, The Netherlands, 2017.

43. Ahmad, A.; Sumathi, S.; Journal, B.H.-C.E. *Coagulation of Residue Oil and Suspended Solid in Palm Oil Mill Effluent by Chitosan, Alum and PAC*; Elsevier: Amsterdam, The Netherlands, 2006.

44. Shak, K.; Journal, T.W.-C.E. *Coagulation–Flocculation Treatment of High-Strength Agro-Industrial Wastewater Using Natural Cassia obtusifolia Seed Gum: Treatment Efficiencies and Flocs*; Elsevier: Amsterdam, The Netherlands, 2014.

45. Teh, C.Y.; Wu, T.Y.; Juan, J.C. *Potential Use of Rice Starch in Coagulation–Flocculation Process of Agro-Industrial Wastewater: Treatment Performance and Flocs Characterization*; Elsevier: Amsterdam, The Netherlands, 2014.

46. Abidin, Z.; Ismail, N.; Yunus, R.; Ahamad, I.S.; Idris, A. A preliminary study on *Jatropha curcas* as coagulant in wastewater treatment. *Environ. Technol.* **2011**, *32*, 971–977. [CrossRef]

47. Ernest, E.; Onyeka, O.; David, N.; Blessing, O. Effects of pH, dosage, temperature and mixing speed on the efficiency of water melon seed in removing the turbidity and colour of Atabong River, Awka-Ibom. *Int. J. Adv. Eng. Manag. Sci.* **2017**, *3*, 239833. [CrossRef]

48. Kakoi, B.; Kaluli, J.W.; Ndiba, P.; Thiong'o, G. *Banana Pith as a Natural Coagulant for Polluted River Water*; Elsevier: Amsterdam, The Netherlands, 2016.

49. Shamsnejati, S.; Chaibakhsh, N.; Pendashteh, A.R.; Hayeripour, S. *Mucilaginous Seed of Ocimum basilicum as a Natural Coagulant for Textile Wastewater Treatment*; Elsevier: Amsterdam, The Netherlands, 2015.

50. González Peña, O.I.; López Zavala, M.Á.; Cabral Ruelas, H. Pharmaceuticals market, consumption trends and disease incidence are not driving the pharmaceutical research on water and wastewater. *Int. J. Environ. Res. Public Health* **2021**, *18*, 2532. [CrossRef]

51. Massima Mouele, E.S.; Tijani, J.O.; Badmus, K.O.; Pereao, O.; Babajide, O.; Zhang, C.; Shao, T.; Sosnin, E.; Tarasenko, V.; Fatoba, O.O.; et al. Removal of pharmaceutical residues from water and wastewater using dielectric barrier discharge methods—A review. *Int. J. Environ. Res. Public Health* **2021**, *18*, 1683. [CrossRef]

52. Gogoi, A.; Mazumder, P.; Tyagi, V.K.; Tushara Chaminda, G.G.; An, A.K.; Kumar, M. Occurrence and fate of emerging contaminants in water environment: A review. *Groundw. Sustain. Dev.* **2018**, *6*, 169–180. [CrossRef]

53. Kharel, S.; Stapf, M.; Miehe, U.; Ekblad, M.; Cimbritz, M.; Falås, P.; Nilsson, J.; Sehlén, R.; Bregendahl, J.; Bester, K. Removal of pharmaceutical metabolites in wastewater ozonation including their fate in different post-treatments. *Sci. Total Environ.* **2021**, *759*, 143989. [CrossRef]

54. Quesada, H.B.; Baptista, A.T.A.; Cusioli, L.F.; Seibert, D.; de Oliveira Bezerra, C.; Bergamasco, R. Surface water pollution by pharmaceuticals and an alternative of removal by low-cost adsorbents: A review. *Chemosphere* **2019**, *222*, 766–780. [CrossRef]

55. Yang, Y.; Ok, Y.S.; Kim, K.H.; Kwon, E.E.; Tsang, Y.F. Occurrences and removal of pharmaceuticals and personal care products (PPCPs) in drinking water and water/sewage treatment plants: A review. *Sci. Total Environ.* **2017**, *596–597*, 303–320. [CrossRef]

56. Couto, C.F.; Lange, L.C.; Amaral, M.C.S. Occurrence, fate and removal of pharmaceutically active compounds (PhACs) in water and wastewater treatment plants—A review. *J. Water Process Eng.* **2019**, *32*, 100927. [CrossRef]

57. Karimi-Maleh, H.; Ayati, A.; Davoodi, R.; Tanhaei, B.; Karimi, F.; Malekmohammadi, S.; Orooji, Y.; Fu, L.; Sillanpää, M. Recent advances in using of chitosan-based adsorbents for removal of pharmaceutical contaminants: A review. *J. Clean. Prod.* **2021**, *291*, 125880. [CrossRef]

58. Krishnan, R.Y.; Manikandan, S.; Subbaiya, R.; Biruntha, M.; Govarthanan, M.; Karmegam, N. Removal of emerging micropollutants originating from pharmaceuticals and personal care products (PPCPs) in water and wastewater by advanced oxidation processes: A review. *Environ. Technol. Innov.* **2021**, *23*, 101757. [CrossRef]

59. Tiwari, B.; Sellamuthu, B.; Ouarda, Y.; Drogui, P.; Tyagi, R.D.; Buelna, G. Review on fate and mechanism of removal of pharmaceutical pollutants from wastewater using biological approach. *Bioresour. Technol.* **2017**, *224*, 1–12. [CrossRef]

60. Li, Z.; Yang, P. Review on Physicochemical, Chemical, and Biological Processes for Pharmaceutical Wastewater. *IOP Conf. Ser. Earth Environ. Sci.* **2018**, *113*, 012185. [CrossRef]

61. Ren, Y.; Yang, J.; Chen, S. *The Fate of a Nitrobenzene-Degrading Bacterium in Pharmaceutical Wastewater Treatment Sludge*; Elsevier: Amsterdam, The Netherlands, 2015.

62. Alam, R.; Sheob, M.; Saeed, B.; Khan, S.U.; Shirinkar, M.; Frontistis, Z.; Basheer, F.; Farooqi, I.H. Use of electrocoagulation for treatment of pharmaceutical compounds in water/wastewater: A review exploring opportunities and challenges. *Water* **2021**, *13*, 2105. [CrossRef]

63. Ahmad, A.; Priyadarshini, M.; Das, S.; Ghangrekar, M.M. Electrocoagulation as an efficacious technology for the treatment of wastewater containing active pharmaceutical compounds: A review. *Sep. Sci. Technol.* **2021**, 1–23. [CrossRef]

64. Li, C.; Mei, Y.; Qi, G.; Xu, W.; Zhou, Y.; Shen, Y. Degradation characteristics of four major pollutants in chemical pharmaceutical wastewater by Fenton process. *J. Environ. Chem. Eng.* **2021**, *9*, 104564. [CrossRef]

65. Olvera-Vargas, H.; Gore-Datar, N.; Garcia-Rodriguez, O.; Mutnuri, S.; Lefebvre, O. Electro-Fenton treatment of real pharmaceutical wastewater paired with a BDD anode: Reaction mechanisms and respective contribution of homogeneous and heterogeneous OH. *Chem. Eng. J.* **2021**, *404*, 126524. [CrossRef]

66. Albahnasawi, A.; Yüksel, E.; Gürbulak, E.; Duyum, F. Fate of aromatic amines through decolorization of real textile wastewater under anoxic-aerobic membrane bioreactor. *J. Environ. Chem. Eng.* **2020**, *8*, 104226. [CrossRef]

67. Akkoyunlu, B.; Daly, S.; Casey, E. Membrane bioreactors for the production of value-added products: Recent developments, challenges and perspectives. *Bioresour. Technol.* **2021**, *341*, 125793. [CrossRef]

68. Gebreyohannes, A.Y.; Giorno, L.; Bekele, D.N.; Besha, A.T.; Curcio, E.; Tufa, R.A. Removal of emerging micropollutants by activated sludge process and membrane bioreactors and the effects of micropollutants on membrane fouling: A review. *J. Environ. Chem. Eng.* **2017**, *5*, 2395–2414. [CrossRef]

69. Ethiraj, A.S.; Uttam, P.; Varunkumar, K.; Chong, K.F.; Ali, G.A.M. Photocatalytic performance of a novel semiconductor nanocatalyst: Copper doped nickel oxide for phenol degradation. *Mater. Chem. Phys.* **2020**, *242*, 122520. [CrossRef]

70. Sharifi, A.; Montazerghaem, L.; Naeimi, A.; Abhari, A.R.; Vafaee, M.; Ali, G.A.M.; Sadegh, H. Investigation of photocatalytic behavior of modified ZnS:Mn/MWCNTs nanocomposite for organic pollutants effective photodegradation. *J. Environ. Manag.* **2019**, *247*, 624–632. [CrossRef]

71. Solehudin, M.; Sirimahachai, U.; Ali, G.A.M.; Chong, K.F.; Wongnawa, S. One-pot synthesis of isotype heterojunction g-C3N4-MU photocatalyst for effective tetracycline hydrochloride antibiotic and reactive orange 16 dye removal. *Adv. Powder Technol.* **2020**, *31*, 1891–1902. [CrossRef]

72. Ethiraj, A.S.; Rhen, D.S.; Soldatov, A.V.; Ali, G.A.M.; Bakr, Z.H. Efficient and recyclable Cu incorporated TiO_2 nanoparticle catalyst for organic dye photodegradation. *Int. J. Thin Film Sci. Technol.* **2021**, *10*, 169–182. [CrossRef]

73. Giahi, M.; Pathania, D.; Agarwal, S.; Ali, G.A.M.; Chonge, K.F.; Gupta, V.K. Preparation of Mg-doped TiO_2 nanoparticles for photocatalytic degradation of some organic pollutants. *Stud. UBB Chem.* **2019**, *64*, 7–18. [CrossRef]

74. Mahmoud, W.M.M.; Rastogi, T.; Kümmerer, K. Application of titanium dioxide nanoparticles as a photocatalyst for the removal of micropollutants such as pharmaceuticals from water. *Curr. Opin. Green Sustain. Chem.* **2017**, *6*, 1–10. [CrossRef]

75. Rekhate, C.V.; Srivastava, J.K. Recent advances in ozone-based advanced oxidation processes for treatment of wastewater—A review. *Chem. Eng. J. Adv.* **2020**, *3*, 100031. [CrossRef]

76. Araujo, L.A.; Bezerra, C.O.; Cusioli, L.F.; Silva, M.F.; Nishi, L.; Gomes, R.G.; Bergamasco, R. Moringa oleifera biomass residue for the removal of pharmaceuticals from water. *J. Environ. Chem. Eng.* **2021**, *6*, 7192–7199. [CrossRef]

77. Nonfodji, O.M.; Fatombi, J.K.; Ahoyo, T.A.; Osseni, S.A.; Aminou, T. Performance of *Moringa oleifera* seeds protein and *Moringa oleifera* seeds protein-polyaluminum chloride composite coagulant in removing organic matter and antibiotic resistant bacteria from hospital wastewater. *J. Water Process Eng.* **2020**, *33*, 101103. [CrossRef]

78. Thirugnanasambandham, K.; Karri, R.R. Preparation and characterization of *Azadirachta indica* A. Juss. plant based natural coagulant for the application of urban sewage treatment: Modelling and cost assessment. *Environ. Technol. Innov.* **2021**, *23*, 101733. [CrossRef]

79. Maharani, Z.; Setiawan, D.; Ningsih, E. Comparison of the Effectiveness of Natural Coagulant Performance on % BOD Removal and % COD Removal in Pharmaceutical Industry Waste. *Tibuana* **2021**, *4*, 55–60. [CrossRef]

80. Oliva, M.P.; Corral, C.; Jesoro, M.; Barajas, J.R. Moringa-functionalized rice husk ash adsorbent for the removal of amoxicillin in aqueous solution. *MATEC Web Conf.* **2019**, *268*, 01005. [CrossRef]

81. Olivera, A.R. Biosorption of Pharmaceuticals from Wastewater Using *Moringa oleifera* as Biosorbent. Ph.D. Thesis, Instituto Politécnico de Bragança, Buenos Aires, Argentina, 2020. Available online: https://bibliotecadigital.ipb.pt/bitstream/10198/22123/1/Olivera_Agustina.pdf (accessed on 4 December 2021).

82. Santos, A.F.S.; Matos, M.; Sousa, Â.; Costa, C.; Nogueira, R.; Teixeira, J.A.; Paiva, P.M.G.; Parpot, P.; Coelho, L.C.B.B.; Brito, A.G. Removal of tetracycline from contaminated water by Moringa oleifera seed preparations. *Environ. Technol.* **2016**, *37*, 744–751. [CrossRef]

83. Iloamaeke, I.M.; Julius, C. Treatment of Pharmaceutical Effluent Using seed of Phoenix Dactylifera as a Natural Coagulant. *J. Basic Phys. Res.* **2019**, *9*, 92–100.

84. Sibartie, S.; Ismail, N. Potential of Hibiscus Sabdariffa and Jatropha Curcas as Natural Coagulants in the Treatment of Pharmaceutical Wastewater. *MATEC Web Conf.* **2018**, *152*, 01009. [CrossRef]

85. Chong, K.H.; Kiew, P.L. Potential of Banana Peels as Bio-Flocculant for Water Clarification. *Prog. Energy Environ.* **2017**, *1*, 47–56.

86. Awolola, G.; Oluwaniyi, O.; Solanke, A.; Dosumu, O.; Shuiab, A. Toxicity assessment of natural and chemical coagulants using brine shrimp (*Artemia salina*) bioassay. *Int. J. Biol. Chem. Sci.* **2010**, *4*, 633–641. [CrossRef]

87. Šćiban, M.; Klašnja, M.; Antov, M.; Škrbić, B. Removal of water turbidity by natural coagulants obtained from chestnut and acorn. *Bioresour. Technol.* **2009**, *100*, 6639–6643. [CrossRef]

88. Oladoja, N.A. Headway on natural polymeric coagulants in water and wastewater treatment operations. *J. Water Process Eng.* **2015**, *6*, 174–192. [CrossRef]

89. Formicoli, T.K.; Freitas, S.; Almeida, A.; Domingos Manholer, D.; Cesar, H.; Geraldino, L.; Ferreira De Souza, M.T.; Garcia, J.C.; Freitas, T.K.F.S.; Almeida, C.A.; et al. Review of Utilization Plant-Based Coagulants as Alternatives to Textile Wastewater Treatment. In *Detox Fashion*; Springer: Singapore, 2018; pp. 27–79. [CrossRef]

90. Kumar, I.; Quaff, A.R. Comparative study on the effectiveness of natural coagulant aids and commercial coagulant: Removal of arsenic from water. *Int. J. Environ. Sci. Technol.* **2019**, *16*, 5989–5994. [CrossRef]

91. Saritha, V.; Karnena, M.K.; Dwarapureddi, B.K. "Exploring natural coagulants as impending alternatives towards sustainable water clarification"—A comparative studies of natural coagulants with alum. *J. Water Process Eng.* **2019**, *32*, 100982. [CrossRef]

92. Boulaadjoul, S.; Zemmouri, H.; Bendjama, Z.; Drouiche, N. A novel use of Moringa oleifera seed powder in enhancing the primary treatment of paper mill effluent. *Chemosphere* **2018**, *206*, 142–149. [CrossRef] [PubMed]

93. Vishali, S.; Karthikeyan, R. Cactus opuntia (ficus-indica): An eco-friendly alternative coagulant in the treatment of paint effluent. *Desalin. Water Treat.* **2015**, *56*, 1489–1497. [CrossRef]

94. de Paula, H.M.; de Oliveira Ilha, M.S.; Sarmento, A.P.; Andrade, L.S. *Dosage Optimization of Moringa oleifera Seed and Traditional Chemical Coagulants Solutions for Concrete Plant Wastewater Treatment*; Elsevier: Amsterdam, The Netherlands, 2018.
95. Sellami, M.; Zarai, Z.; Khadhraoui, M.; Jdidi, N.; Leduc, R.; Ben Rebah, F. Cactus juice as bioflocculant in the coagulation–flocculation process for industrial wastewater treatment: A comparative study with polyacrylamide. *Water Sci. Technol.* **2014**, *70*, 1175–1181. [CrossRef] [PubMed]
96. Ganjidoust, H.; Tatsumi, K.; Yamagishi, T.; Gholian, R.N. *Effect of Synthetic and Natural Coagulant on Lignin Removal from Pulp and Paper Wastewater*; Elsevier: Amsterdam, The Netherlands, 1997.
97. Muhammad, I.M.; Abdulsalam, S.; Abdulkarim, A.; Bello, A.A. Water melon seed as a potential coagulant for water treatment. *Glob. J. Res. Eng.* **2015**, *15*, 17–23.

Article

Effective Adsorption of Reactive Black 5 onto Hybrid Hexadecylamine Impregnated Chitosan-Powdered Activated Carbon Beads

Mohammadtaghi Vakili [1], Haider M. Zwain [2], Amin Mojiri [3,*], Wei Wang [4], Fatemeh Gholami [5], Zahra Gholami [6], Abdulmoseen S. Giwa [1], Baozhen Wang [1], Giovanni Cagnetta [7] and Babak Salamatinia [8]

[1] Green Intelligence Environmental School, Yangtze Normal University, Chongqing 408100, China; mvakili1981@yahoo.com (M.V.); giwasegun25@yahoo.com (A.S.G.); 20170076@yznu.cn (B.W.)

[2] College of Water Resources Engineering, Al-Qasim Green University, Al-Qasim Province, Babylon 51013, Iraq; haider.zwain@wrec.uoqasim.edu.iq

[3] Department of Civil and Environmental Engineering, Graduate School of Advanced Science and Engineering, Hiroshima University, Higashihiroshima 739-8527, Japan

[4] State Key Laboratory of Plateau Ecology and Agriculture, Qinghai University, Xi'ning 810016, Qinghai Province, China; weiwang@qhu.edu.cn

[5] New Technologies-Research Centre, University of West Bohemia, 30100 Plzeň, Czech Republic; gholami@ntc.zcu.cz

[6] Unipetrol Centre of Research and Education, a.s, Areál Chempark 2838, Záluží 1, 43670 Litvínov, Czech Republic; zgholami@gmail.com

[7] State Key Joint Laboratory of Environment Simulation and Pollution Control, Beijing Key Laboratory for Emerging Organic Contaminants Control, School of Environment, Tsinghua University, Beijing 100084, China; gcagnetta@mail.tsinghua.edu.cn

[8] Discipline of Chemical Engineering, School of Engineering, Monash University, Jalan Lagoon Selatan, Bandar Sunway, Selangor 47500, Malaysia; babak.salamatinia@monash.edu

* Correspondence: amin.mojiri@gmail.com

Received: 13 July 2020; Accepted: 6 August 2020; Published: 9 August 2020

Abstract: In this study, hexadecylamine (HDA) impregnated chitosan-powder activated carbon (Ct-PAC) composite beads were successfully prepared and applied to adsorption of the anionic dye reactive black 5 (RB5) in aqueous solution. The Ct-PAC-HDA beads synthesized with 0.2 g powdered activated carbon (PAC) and 0.04 g HDA showed the highest dye removal efficiency. The prepared beads were characterized using Fourier-transform infrared spectroscopy (FTIR) and scanning electron microscopy (SEM). Various adsorption parameters, i.e., adsorbent dosage, pH, and contact time, which affect the adsorption performance, were studied in a series of batch experiments. The obtained adsorption data were found to be better represented by Freundlich (R^2 = 0.994) and pseudo-second-order (R^2 = 0.994) models. Moreover, it was ascertained that the adsorption of RB5 onto Ct-PAC-HDA beads is pH-dependent, and the maximum Langmuir adsorption capacity (666.97 mg/g) was observed at pH 4. It was also proved that Ct-PAC-HDA beads were regenerable for repeated use in the adsorption process.

Keywords: chitosan; powder activated carbon; hexadecylamine; hybrid adsorbent; regeneration

---

1. Introduction

In recent years, the unprecedented development of industrial and urban activities has led to a significant increase in wastewater discharge into the environment, often contaminated with harmful organic pollutants (e.g., dyestuffs). Therefore, the separation/elimination of these pollutants from water sources is a goal that must be accomplished to ensure human and environmental safety [1]. Due to their high toxicity, chemical stability, and low biodegradability, dyes are a class of pollutants that are raising increasing concern, since they cause severe problems to aquatic life and human beings [2]. In particular, reactive dyes are being widely used in dyeing processes because of their notable properties such as ease of application, high color fastness, bright colors, and a wide shade gamut from black to vibrant, brilliant shades [3–5]. However, a low degree of fixation to the fabrics, high water solubility, and poor adsorption ability are the main reasons for the intense color of the effluents that contain such kind of dyes [6]. Discharge of such type of wastewater into the environment is known to threaten the ecosystem due to high toxicity and reduction of sunlight penetration into the water, thus affecting aquatic biota living functions [7]. Hence, the elimination of reactive dyes from aqueous effluents is a necessary action that must be taken to prevent further spread in the environment and adverse effects. To date, different treatment techniques have been developed for this aim. Among them, adsorption processes have received much attention from researchers due to its cost-effectiveness, simplicity, and high efficiency [8]. The properties of materials that are used as adsorbents are key factors for adsorption efficiency [9]. In recent years, the development of adsorbents obtained from non-toxic, sustainable, and renewable natural resources (e.g., biopolymers) has intensified [10].

Chitosan (Ct) is a suitable material in this area because of its non-toxicity, abundance, availability, biodegradability, and ability to adsorb organic pollutants (including reactive dyes). This biopolymer is a multifunctional cationic biodegradable polysaccharide produced by deacetylation of chitin. Because of the presence of several functional groups (in particular, NH_2 and OH) on the Ct backbone, this material has a high potential affinity to dyes. However, low porosity, small surface area, scarce acid stability, and limited adsorption capacity preclude its application at full-scale for wastewater treatment [11]. Also, raw Ct is available in powder form, thus it is not easy to separate from aqueous solutions after the adsorption process. Its crystallinity and hydrophobicity are features that reduce the liquid-to-solid mass transfer rate and induce column clogging and high pressure drops, thereby resulting in high operation costs [12]. Modification of Ct is a feasible way to overcome these limitations and improve its adsorption performance [13].

Among such modification methods, preparation of Ct-based hybrid adsorbents has received significant attention and is considered as one of the most efficient ways to improve Ct properties. The use of hybrid adsorbents is not only cost-effective, environmentally friendly, and safe, but also can reduce shortcomings of constituent materials and consequently increase their value for practical applications [14]. Hybridized adsorbents obtained from Ct and carbonaceous materials can fulfill these criteria and, therefore, they have attracted the attention of researchers as potential materials for green technology development. The combination of Ct and carbonaceous materials is an efficient method to enhance its thermochemical and mechanical properties. Additionally, carbonaceous materials can enhance functionality and pore properties of Ct and thus improve its adsorption capability [15]. It was proved by Yadaei et al. [16] that Ct-activated carbon hybrid adsorbent possesses favorable strength and porous structure. The development of hybrid adsorbents obtained from Ct and carbonaceous materials has been reported in some studies [17,18]. Carbonaceous materials are extensively applied in the purification of water as efficient adsorbents owing to their high functional group's number, high porosity, and large surface area [19]. Among them, powdered activated carbon (PAC) has been widely used in water treatment to control odor, color, and taste because of its remarkable adsorption potentiality, fast adsorption kinetic, availability, and low cost [20]. Despite such noteworthy characteristics, PAC suffers from some critical drawbacks that limit its application: small particle size and powder form, as well as difficulties in regeneration and separation from aqueous solutions [21].

The immobilization of PAC particles in some kind of matrix is one way to overcome the problems mentioned above. Preparation of Ct-PAC hybrid adsorbent can combine the advantages of Ct and PAC, showing strong adsorption capacity for different kinds of pollutants. This study combines PAC with Ct in a bead shape, thus achieving its stabilization, prevention of carbon leaching, and improved separability of the prepared adsorbent from the solution. However, the PAC would cover the surface of chitosan and occupy its active functional groups. It may negatively affect the dye removal efficiency of the prepared hybrid adsorbent. Therefore, increasing the number of functional groups (mainly amino groups) through chemical modification using different types of modification agents will improve the adsorption properties of the prepared hybrid adsorbent. Impregnation with cationic surfactants is an appropriate strategy leading to an increase in the functional groups and positive charge density of the adsorbent.

The present study aims to further enhance the adsorption capability of the prepared Ct-PAC hybrid beads using a cationic surfactant. Hexadecylamine (HDA) is a cationic surfactant possessing a positively charged hydrophilic group (NH_2) that can be applied to increase the number of amino groups already present in Ct, as well as to augment the positive charge of the synthesized adsorbent [22]. Therefore, the main objectives of this study were the preparation of Ct-PAC hybrid beads, as well as using HDA for increasing the functional groups and cationicity of the material for adsorption of reactive dye from water. To the best of our knowledge, such modification has not been reported in the literature. Thus, there is a necessity in understanding the behavior of this combination as an effective adsorbent for improved reactive dye removal. In the present research, the anionic dye reactive black 5 (RB5) is utilized as a model pollutant molecule to assess the adsorption performance of the prepared hybrid adsorbent. The effect of preparation conditions (i.e., PAC and HDA concentration) and adsorption parameters (i.e., pH of dye solution, adsorbent dosage, and contact time) on RB5 adsorption behavior is investigated. Kinetic and isotherm studies also are conducted to reveal the adsorption behavior of RB5 onto the prepared adsorbent. Moreover, the regeneration of the Ct-PAC-HDA adsorbent is assessed.

2. Materials and Methods

2.1. Chemicals and Materials

Chitosan (medium-molecular-weight 75–85% deacetylation) and hexadecylamine ($C_{16}H_{35}N$, 98%) were purchased from Sigma-Aldrich (Beijing, China). RB5, sodium hydroxide (NaOH), hydrochloric acid (HCl) and acetic acid (CH_3COOH) were all supplied by Beijing Chemical Works. A commercial PAC was obtained from Sinopharm Chemical Reagent Co., Ltd. (Shanghai, China). All chemicals in the present study were analytical grade. The chemicals were used directly without further purification.

2.2. Preparation of Adsorbent

For the preparation of the Ct-PAC solution, Ct solution was mixed with PAC, as described previously by Vakili et al. [10]. Briefly, Ct flakes (2 g) were dissolved into 100 mL acetic acid solution (3% v/v) under continuous stirring (400 rpm) at room temperature (25 °C) for 5 h. Then, the desired concentration of PAC (0.10, 0.2, 0.30, 0.40, and 0.50 g) was poured into the Ct solution and stirred for 3 h to study the effect of PAC concentration. Preparation of the HDA impregnated Ct-PAC beads was conducted following the conditions reported by Vakili et al. [22]. Concisely, the desired amount of HDA was mixed with the Ct-PAC solution at 400 rpm and 50 °C for 6 h. In the Ct-PAC solution, the concentration of HDA was varied from 0.02 to 0.1 g. Then the prepared Ct-PAC-HDA solution was poured dropwise into a 500 mL precipitation solution (NaOH, 2 M), followed by gentle stirring overnight. For removing the residual NaOH, the prepared Ct-PAC-HDA beads were washed several times using deionized water. Finally, all the beads were oven-dried at 60 °C for 12 h.

2.3. Characterization

For studying the functional groups of the prepared adsorbent, Fourier-transform infrared spectroscopy (FTIR) spectra of the Ct-PAC-HDA beads were obtained at 500–4000 cm^{-1} using a Thermo Nicolet NEXUS (Thermofisher, Waltham, MA, USA). The surface morphological features of the adsorbent were assessed employing scanning electron microscopy (SEM) using a Phenom Prox, Phenom-world, Holland.

2.4. Adsorption and Desorption Experiments

Batch adsorption experiments were designed to study RB5 adsorption onto Ct-PAC-HDA beads. A desired mass of the adsorbent was mixed with an RB5 solution (200 mL) in a 250 mL Erlenmeyer flask. Then, the mixture was shaken using an orbital shaker at 150 rpm at room temperature (25 °C) for 24 h. The optimum pH was determined by adding the prepared beads (20 mg) into 200 mL (50 mg/L) RB5 solution at different pH values ranging from 2 to 10. The solution pH was adjusted to the desired values using NaOH (0.1 M) and HCl (0.1 M) solutions. The effect of adsorbent dosage on the removal of RB5 was studied in the RB5 concentration range between 10 and 50 g/L at 100 mg/L. All of the experiments in this study were conducted in triplicate. For quantifying the residual RB5 concentration, after the adsorption process, the spectrophotometric technique was applied using a UV–Vis spectrophotometer (Hach DR 5000, Germany) at a wavelength of 597 nm. The adsorption capacity and removal efficiency of the Ct-PAC-HDA beads were calculated using the following equations:

$$q_e = ((C_0 - C_e) \times V) / W \tag{1}$$

$$RE = (C_0 - C_e/C_0) \times 100 \tag{2}$$

q_e = Adsorption capacity (mg/g)
RE = Removal efficiency (%)
C_0 = Initial RB5 concentration (mg/L)
C_e = Equilibrium RB5 concentration (mg/L)
V = RB5 solution volume (L)
W = Adsorbent mass (g)

The effect of time and the adsorption rates of the dye were evaluated through adsorption kinetic experiments, performed at an RB5 concentration of 200 mg/L. The obtained results were assessed based on pseudo-first-order (PFO; Equation (3)) and pseudo-second-order (PSO; Equation (4)) kinetic models.

$$q_t = q_e (1 - e^{-kt}) \tag{3}$$

$$q_t = (q_e v_0 t) / (q_e + v_0 t) \tag{4}$$

q_t = RB5 adsorption at time t (mg/g)
q_e = RB5 adsorption at equilibrium (mg/g)
k = PFO rate constant (1/min)
v_0 = PFO/PSO rate constant (g/mg/min)
t = time (min)

To assess the transmission of adsorbate from solution phase to the adsorbent phase at equilibrium condition, adsorption isotherm experiments were performed at six RB5 concentrations (20, 40, 80, 120, 160 and 200 mg/L) at 25 °C and with contact time of 24 h. Two adsorption isotherms, i.e., the Langmuir (Equation (5)) and Freundlich (Equation (6)), were applied to express the isotherm data.

$$q_e = (q_m C_e b) / (1 + b C_e) \tag{5}$$

$$q_e = K_F C_e^{(1/n)} \tag{6}$$

q_t = Maximum adsorption capacity of the Ct-PAC-HDA beads (mg/g)
q_m = Maximum adsorption capacity of the adsorbent (mg/g)
C_e = RB5 equilibrium concentration (mg/L)
B = Affinity of Ct-PAC-HDA beads towards RB5 (L/g)
K_F = Freundlich constant (mg/g)

N = Adsorption intensity

The reusability of the adsorbent was evaluated by mixing 30 mg of Ct-PAC-HDA beads with 200 mL of RB5 solution (100 mg/L) at pH 4 for 24 h. Afterward, saturated beads were separated and regenerated using 0.1 M of NaOH solution. Then, the regenerated adsorbents were applied again in the next adsorption experiment. The adsorption and desorption cycle was repeated until the RB5 adsorption capacity of the adsorbent dropped significantly, to study the regeneration performance of the adsorbent.

3. Results and Discussion

3.1. Effects of the Reaction Conditions

Figure 1a presents the impact of PAC concentration on the adsorption of RB5 on Ct-PAC beads. PAC had a positive effect on the RB5 adsorption performance of the Ct-PAC beads. Increasing the amount of PAC into the Ct solution up to 0.20 g increased RB5 removal from a solution with RB5 concentration of 20 mg/L, to 78.91%. This enhancement could be caused by the increase in the beads' surface area by the addition of PAC, which resulted in a broader availability of functional groups for adsorption of RB5 molecules [23]. However, a further increase in PAC concentration (up to 0.40 g) led to a decrease in the removal of RB5, to 67.29%. The presence of a high amount of PAC likely caused the collapse of the structure and, consequently, a decrease of accessibility to functional groups in the Ct-PAC beads [24]. Moreover, at higher PAC concentration (>0.40 g), the added dropwise Ct-PAC solution into the precipitation solution could not solidify to form beads. This might be due to the low entanglement rate and low polymerization of Ct in the Ct-PAC solution, caused by the existence of a high amount of PAC in Ct solution [25].

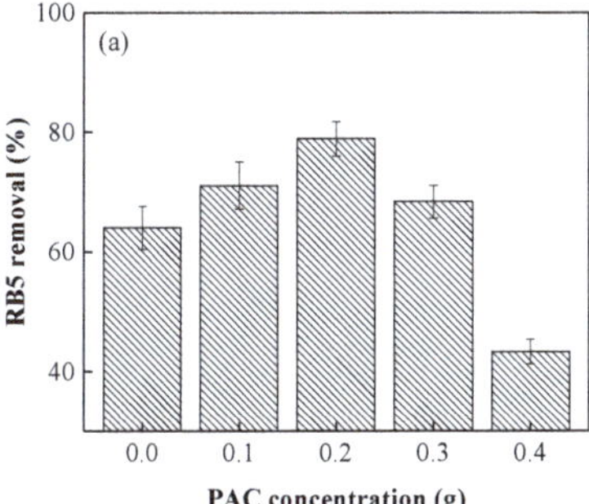

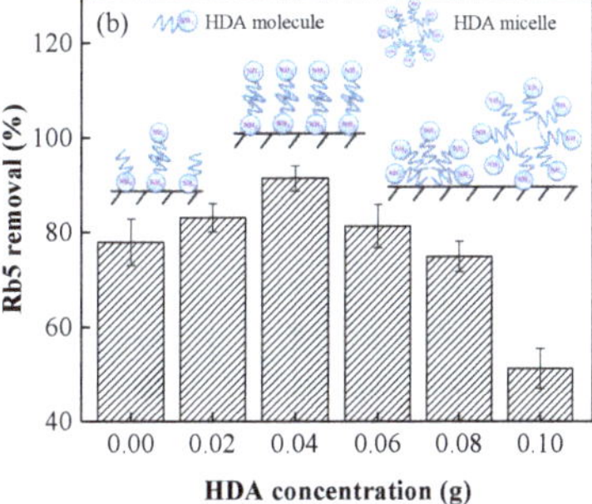

Figure 1. Effect of (**a**) powdered activated carbon (PAC) and (**b**) hexadecylamine (HDA) concentrations on the adsorption efficiency of reactive black 5 (RB5) onto the adsorbents (20 mg of beads, 200 mL of 20 mg/L of RB5, pH = 6, 25 °C, 24 h).

The effect of HDA concentration on the RB5 adsorption performance of the Ct-PAC-HDA beads was investigated, and results are displayed in Figure 1b. As noticed, the RB5 removal efficiency of the beads is initially enhanced by an increase in the amount of HDA (up to a concentration of 0.04 g). Then, a significant adsorption capacity decrease, along with the rise in the HDA concentration, is observed. Specifically, the optimal RB5 removal percentage was 91.32%. This phenomenon is likely due to the increase in the number of amine groups, as well as the positive charge of the beads attributed to the presence of HDA molecules. This led to a rise in the adsorbate–adsorbent electrostatic interactions and, thus, the higher RB5 adsorption capacity of the beads [26]. However, at higher HDA concentrations (>0.04 g), the decrease in the RB5 adsorption onto the Ct-PAC-HDA beads can be attributed to the self-aggregation of HDA molecules, by forming micelles at high concentration. These micelles can block the pores on the beads and decrease the accessibility of functional groups to RB5 molecules (Figure 1b) [22].

3.2. Adsorbent Characterization

Relevant surface functional groups on Ct-PAC beads (before and after impregnation) were verified by FTIR analysis, and the spectral data are presented in Figure 2. In the Ct-PAC beads, the main overlapping area of stretching vibrations of amine and hydroxyl groups appears as a strong and broadband in the region of 3070–3800 cm^{-1}. The smaller peak at 2913 cm^{-1} is attributed to the stretching vibration of the CH$_2$ groups [27]. Moreover, peaks at 1635, 1378, and 1012 cm^{-1} could be assigned to amide II band, N–H bending, alcoholic C–O, and C–N stretching, respectively. Compared to Ct-PAC, the FTIR spectrum of Ct-PAC-HDA presents wavenumber shift of several peaks, as well as changes in intensity. After HDA impregnation, the peaks at 3266, 2913, 1635, 1373, 985 cm^{-1} were shifted to higher frequencies: 3415, 2915, 1646, 1380, 1000 cm^{-1}, respectively. These changes might be owed to the overlap of stretching bands of amine and hydroxyl groups of HDA, CH, and PAC. These findings suggest that Ct-PAC beads chemically adsorb HDA through interaction between the amine group of the HDA molecule and the hydroxyl group on the Ct-PAC beads [22].

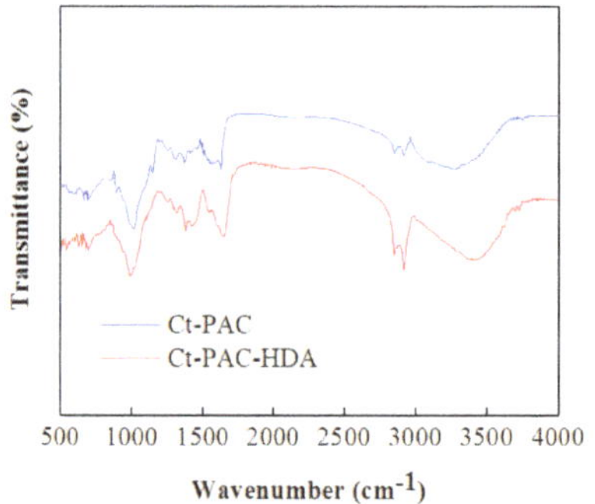

Figure 2. Fourier-transform infrared spectroscopy (FTIR) spectra of Ct-PAC and Ct-PAC-HDA beads.

The surface morphology of the prepared beads is shown in Figure 3. In both before and after impregnation, the adsorbents display a heterogeneous and uniform surface, which is favorable for adsorption of RB5. However, the Ct-PAC-HDA bead (Figure 3b) shows a more uneven, rough, and heterogeneous surface than that of the Ct-PAC bead (Figure 3a). These properties could be attributed to the presence of HDA in the structure of the Ct-PAC beads, which in the end may increase the contact area, number/availability of functional groups, and adsorption capacity of the beads.

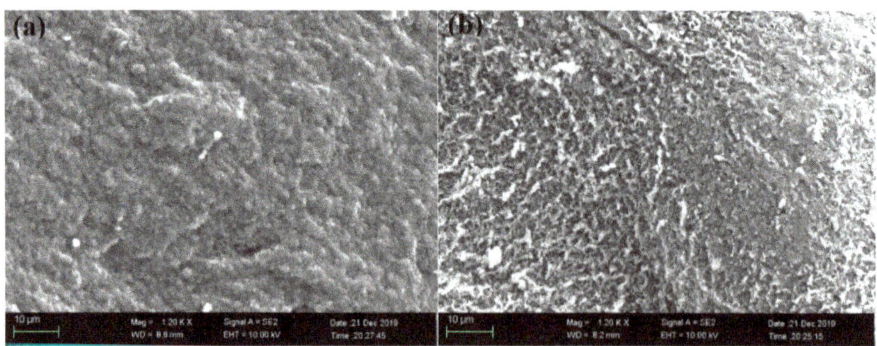

Figure 3. Scanning electron microscopy (SEM) images of (**a**) Ct-PAC and (**b**) Ct-PAC-HDA beads.

3.3. RB5 Adsorption Experiments

3.3.1. Effect of Adsorbent Dosage on RB5 Removal

The impact of the quantity of Ct-PAC-HDA beads on the removal of RB5 was analyzed, and the results are illustrated in Figure 4. The adsorption capacity increased from 105.78 mg/g to 140.9 mg/g for an increase in the amount of Ct-PAC-HDA beads from 10 to 30 g/L. Obviously, at higher adsorbent doses, increasing the surface area, as well as augmenting the number of accessible, functional groups, leads to a higher RB5 adsorption rate [28]. However, a further increase in the adsorbent quantity (>30 g/L) reduced the RB5 adsorption capacity of Ct-PAC-HDA beads. The reduction in adsorption capacity could be due to the conglomeration and interaction of adsorbent particles resulting from a high concentration of adsorbent. This phenomenon likely reduces the total surface area and increases the diffusional path length, leading to the unsaturation of the functional groups through the adsorption process [29,30]. Thus, based on the maximum dye removal efficiency and minimum adsorbent mass, the amount of Ct-PAC-HDA beads was fixed at 30 g/L.

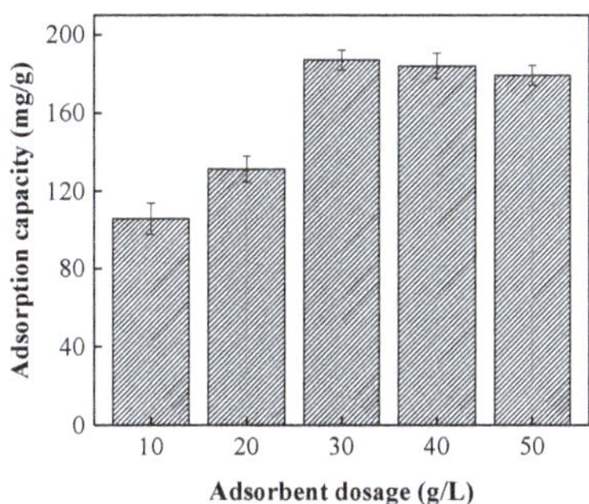

Figure 4. Effect of adsorbent dosage on the adsorption of RB5.

3.3.2. Effect of pH on RB5 Removal

The adsorption of an anionic dye onto a cationic adsorbent is generally governed by the number of charged functional groups on the adsorbent surface, which is highly dependent on its environment's pH [31]. Therefore, the effect of the solution pH on the removal of RB5 using Ct-PAC-HDA was studied, and the results are presented in Figure 5a. It was observed that RB5 adsorption of Ct-PAC-HDA beads was significantly dependent on this parameter. RB5 adsorption was higher at acidic conditions, with the maximum RB5 uptake (187.08 mg/g) obtained at pH = 4. Around this pH value, the presence of appropriate quantities of protons (H^+) induces protonation of amine groups on the Ct-PAC-HDA beads (NH_3^+) and changes the charge of the beads to markedly positive values [32]. On the other hand, in the aqueous phase the sulfonic groups ($-SO_3H$) in the RB5 molecules are converted to their anionic form (i.e., sulfonate group, ($-SO_3^-$)) because the first dissociation acidity constant of sulfonic groups in reactive dyes is very low (pKa $\approx$ 2) [33]. These phenomena result in higher RB5 uptake due to the enhanced electrostatic interactions between positively charged beads and negatively charged anionic dye molecules (Figure 5b).

$$RB5-SO_3H + H_2O \rightarrow RB5-SO_3^- \tag{7}$$

$$Adsorbent-NH_2 + H^+ \rightarrow Adsorbent-NH_3^+ \tag{8}$$

$$Adsorbent-NH_3^+ + RB5-SO_3^- \rightarrow Adsorbent-NH_3^+ -_3SO-RB5 \tag{9}$$

Nevertheless, at very low pH values (pH < 4), the uptake of RB5 decreased. This is attributed to the protonation of higher amounts of amine groups, and thus inducing repulsion among Ct-PAC-HDA components, and possibly dissolution of Ct because of glycosidic bond hydrolyzation [34]. Moreover, at very acidic pH values, which are closer to $-SO_3H$ pK_a, the sulfonic groups are shifted toward their protonated form, thus augmenting the positive charge of RB5 by conversion of anionic $-SO_3{}^-$ groups to $-SO_3H$. In the low-acidic/basic range (pH > 6), the Ct-PAC-HDA beads showed weak RB5 adsorption capacity. At high pH values, massive deprotonation of the amine groups on the beads results in the repulsion of anionic RB5 molecules [35].

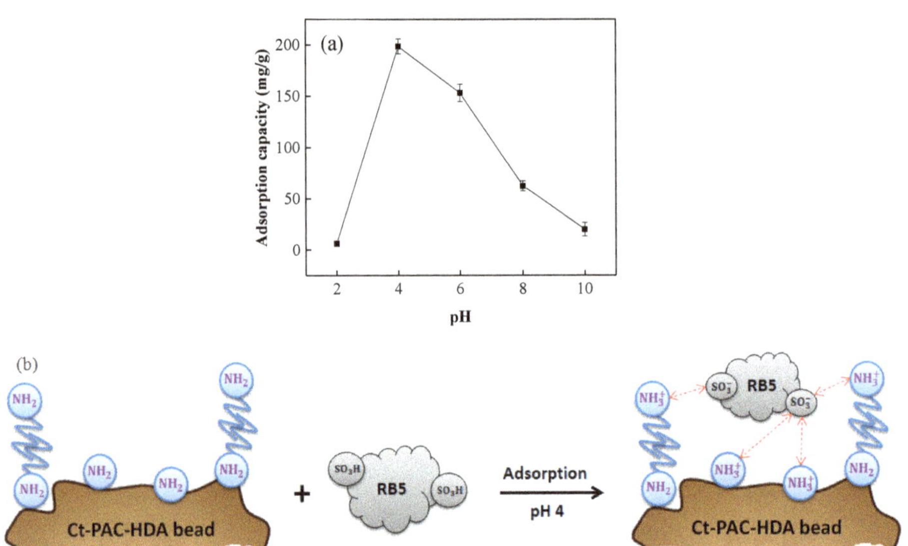

Figure 5. (**a**) Effect of pH on the adsorption of RB5 using Ct-PAC-HDA beads (30 mg of beads, 200 mL of 50 mg/L of RB5, 25 °C, 24 h) and (**b**) schematic diagram of the RB5 adsorption mechanism by Ct-PAC-HDA beads.

3.3.3. Kinetic Study of RB5 Adsorption

For estimating the removal efficiency of an adsorbent, the equilibrium time for adsorption of adsorbate is considered as one of the most critical parameters. Hence, the variation of RB5 adsorption onto the Ct-PAC-HDA beads as a function of contact time was studied (Figure 6). Results revealed that by increasing contact time, the adsorption capacity of the Ct-PAC-HDA beads increased and reached the maximum level within 10 h. Afterward, prolongation of the contact time led to a reduction of adsorption rate until it remained almost stable after 15 h (equilibrium time). The high RB5 removal rate at an initial 240 min is likely owed to a large number of unoccupied/free functional groups on the adsorbent surface. Then, gradual occupation and saturation of functional groups on the Ct-PAC-HDA beads by RB5 molecules result in a decrease in the adsorption rate until the equilibrium is reached [36].

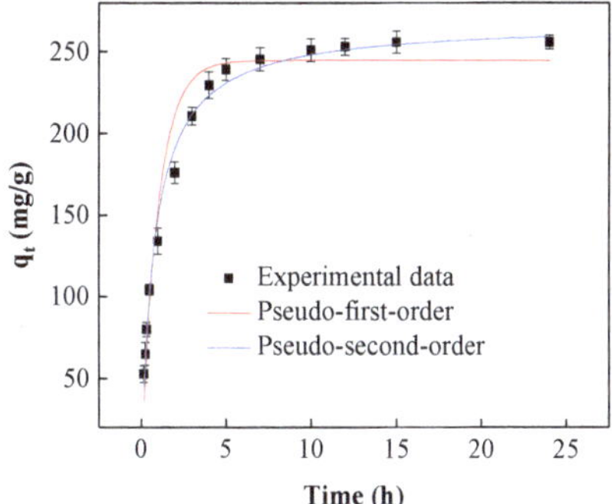

Figure 6. Adsorption kinetics of RB5 onto Ct-PAC-HDA beads (30 mg of beads, 200 mL of 200 mg/L of RB5, pH = 4, 25 °C, 24 h).

The obtained adsorption data were modeled in PFO and PSO kinetic models, and the calculated parameters are presented in Table 1. By comparing the kinetic data, it is found that RB5 adsorption is described better by the PSO kinetic model. The correlation coefficient (R^2) and the Chi-square (χ^2) values of the PSO are higher and lower, respectively, than those of the PFO model. Furthermore, the calculated adsorption capacity by the PSO model (260.12 mg/g) fits well with the experimental adsorption capacity (256.01 mg/g). These results suggest that adsorption of RB5 onto the Ct-PAC-HDA beads is mainly controlled by chemisorption [37].

Table 1. The constants obtained from the kinetic parameters for RB5 adsorption onto Ct-PAC-HDA beads.

Kinetic parameters	Values
Pseudo-first-order	
C_0 (mg/L)	200
q_e (mg/g)	256.01
q_{cal} (mg/g)	244.81 ± 5.939
k (1/min)	1.009 ± 0.0930
χ^2	7.57
R^2	0.968
Slope	−0.366
Intercept	5.123
Pseudo-second-order	
C_0 (mg/L)	200
q_e (mg/g)	256.01
q_{cal} (mg/g)	260.13 ± 3.851
V_0 (g/mg/min)	335.24 ± 18.29
χ^2	1.26
R^2	0.994
Slope	0.004
Intercept	0.003

3.3.4. Isotherm Study of RB5 Adsorption

For investigating suitable conditions for the optimized application of the prepared adsorbent, as well as the study of the nature of RB5 adsorption on Ct-PAC-HDA beads, adsorption equilibrium results were fitted by Langmuir and Freundlich isotherm models (Figure 7). Calculated model parameters (Table 2) reveal that both isotherm models show a good fit with experimental data (R^2 > 0.99). However, the Freundlich model represents a better fit for the experimental results (according to the higher R^2 and lower χ^2 estimated for the fitting of the Freundlich model, with respect to those of the Langmuir model). These findings suggest that the adsorption of RB5 onto the Ct-PAC-HDA beads was controlled by multilayer adsorption, where the functional groups on the external particle layers have a heterogeneous nature [38]. The calculated adsorption intensity (n = 1.3) indicates a strong interaction between RB5 and Ct-PAC-HDA beads (i.e., a favorable adsorption process) [39].

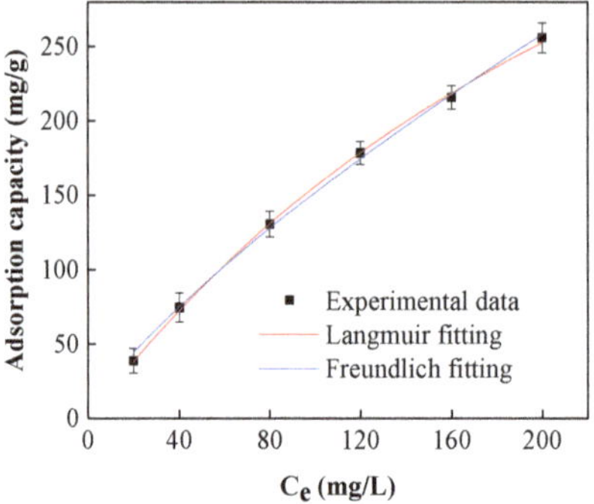

Figure 7. Adsorption isotherms of RB5 onto Ct-PAC-HDA beads (30 mg of beads, pH = 4, 25 °C, 24 h).

Table 2. The constants obtained from the isotherm parameters for adsorption of RB5 onto Ct-PAC-HDA beads.

Isotherm parameters	Values
Langmuir	
q_m (mg/g)	666.97 ± 38.14
b (l/mg)	327.82 ± 27.19
χ^2	0.225
R^2	0.997
Slope	0.002
Intercept	0.201
Freundlich	
K_F (mg/g)	4.48 ± 0.495
n	1.3 ± 0.038
χ^2	0.074
R^2	0.999
Slope	0.686
Intercept	2.279

Moreover, a comparison of the maximum Langmuir adsorption capacity value of the adsorbent prepared in the present study with those of other adsorbing materials described in the literature suggests that Ct-PAC-HDA has relatively high RB5 adsorption capacity (Table 3). The adsorption capacity of Ct-PAC-HDA beads was found to be almost 1112 times higher than that achieved by using an *Eichhornia crassipes*/chitosan composite [39]. It showed 585 times higher capacity than macadamia seed husks [40] while performing 12 times better than the peanut hull [41]. The Ct-PAC-HDA outperformed the dolomite [42], activated carbon F400 [43], and polyethyleneimine/sodium dodecyl sulphate [44] by 8.20, 3.37, and 1.61 times better removals, respectively. A 33.10% higher adsorption was observed with the developed beads as compared to the pine-fruit shell activated carbon [28]. Thereby, Ct-PAC-HDA beads might be an effective adsorbent for the elimination of reactive dyes from wastewaters with satisfactory adsorption capacity.

Table 3. Comparison of the maximum RB5 adsorption capacity of different adsorbents.

Adsorbent	pH	q_m (mg/g)	Ref.
Eichhornia crassipes/chitosan composite	3	0.60	[39]
Macadamia seed husks	3	1.14	[40]
Fly ash	7	7.18	[45]
Edible fungi activated carbon	2	19.6	[46]
Peanut hull	6.4	55.55	[41]
Pine-fruit shell	2	74.6	[28]
Dolomite	6.9	80.9	[42]
Bone char	5.2	160.0	[43]
Activated carbon F400	5.2	197.5	[43]
Chitosan/polyamide nanofibers	1	198.60	[47]
Polyacrylamide/silica nanoporous composite	2	389.58	[27]
Polyethyleneimine/sodium dodecyl sulphate	4	413.23	[44]
Bamboo activated carbon	5.2	441.7	[37]
Pine-fruit shell activated carbon	6	446.2	[28]
Ct-PAC-HDA beads	4	666.97	Present study

3.4. Desorption and Reuse of Spent Ct-PAC-HDA Beads

Reuse evaluation of adsorbent is very fundamental for potential full-scale application because it permits the assessment of the adsorbent capability to recover after a cycle of utilization (i.e., adsorption–desorption). The regeneration minimizes the need for new absorbents, recovers resources, reduces the secondary waste, and decreases the process costs [48]. For assessing the reusability of the Ct-PAC-HDA beads, regeneration experiments were performed by repeating several adsorption–desorption cycles. According to Vakili et al. [49], the regeneration of chitosan-based adsorbents saturated with reactive dyes was successfully conducted by NaOH solution; therefore, it was selected as an eluent for the regeneration of RB5 loaded Ct-PAC-HDA beads. As can be seen in Figure 8, during the first five subsequent regeneration cycles, the RB5 adsorption capacity showed only a 6% loss. Such a reduction is possibly due to incomplete desorption of RB5 molecules. Afterward, adsorption capacity dropped to 46.59 at the tenth cycle (75% loss), which might be attributed to the saturation of functional groups on adsorbent [50].

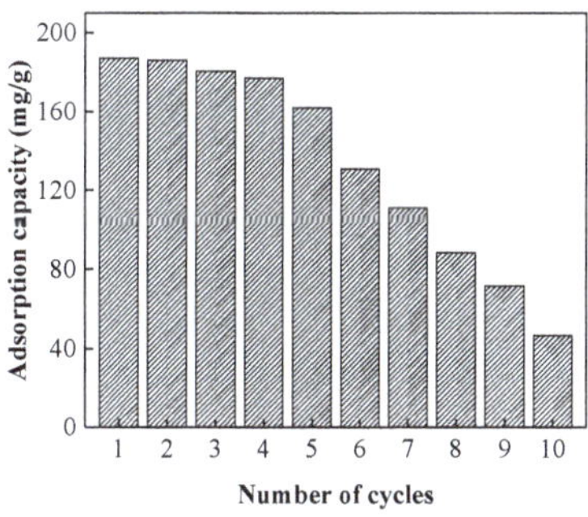

Figure 8. Effect of regeneration cycles on the RB5 adsorption capacity of Ct-PAC-HDA beads.

4. Conclusions

Ct-PAC-HDA beads were successfully prepared for the elimination of RB5 from aqueous solutions. The maximum enhancement in the adsorption performance of adsorbent was achieved using 0.04 g HDA and 0.2 g PAC. The removal of RB5 increased with decreasing pH. The isotherm and kinetic study concluded that the adsorption data fitted better to the Freundlich and PSO models than the Langmuir and PFO models. The results suggested that the adsorption of RB5 onto the Ct-PAC-HDA beads was a chemisorption process that occurred on multilayer heterogeneous surfaces. The maximum

RB5 adsorption capacity of Ct-PAC-HDA beads (666.97 mg/g) was obtained with 30 g/L of adsorbent, at acidic condition (pH 4), 30 g/L of adsorbent mass, and during 10 h (equilibrium time). Moreover, regeneration experiments demonstrated that the prepared beads have good reusability and can be regenerated at least 5 cycles without significant adsorption capacity loss. Generally, the results revealed the great potential of Ct-PAC-HDA as a promising adsorbent for the elimination of reactive dyes.

Author Contributions: Formal analysis, W.W. and B.W.; investigation, M.V.; methodology, F.G. and G.C.; software, Z.G.; supervision, A.M.; validation, A.S.G.; writing—review and editing, H.M.Z. and B.S. All authors have read and agreed to the published version of the manuscript.

Funding: The Yangtze Normal University supported this work.

Conflicts of Interest: The authors declare no conflict of interest.

References

1. Farraji, H.; Robinson, B.; Mohajeri, P.; Abedi, T. Phytoremediation: Green technology for improving aquatic and terrestrial environments. *Nippon. J. Environ. Sci.* **2020**, *1*, 1–30.
2. Vakili, M.; Rafatullah, M.; Salamatinia, B.; Abdullah, A.Z.; Ibrahim, M.H.; Tan, K.B.; Gholami, Z.; Amouzgar, P. Application of chitosan and its derivatives as adsorbents for dye removal from water and wastewater: A review. *Carbohyd. Polym.* **2014**, *113*, 115–130. [CrossRef] [PubMed]
3. Vakili, M.; Rafatullah, M.; Salamatinia, B.; Ibrahim, M.H.; Abdullah, A.Z. Elimination of reactive blue 4 from aqueous solutions using 3-aminopropyl triethoxysilane modified chitosan beads. *Carbohyd. Polym.* **2015**, *132*, 89–96. [CrossRef] [PubMed]
4. Homaeigohar, S. The nanosized dye adsorbents for water treatment. *Nanomaterials* **2020**, *10*, 1–19. [CrossRef] [PubMed]
5. Lewis, D.M. 9—The chemistry of reactive dyes and their application processes. In *Handbook of Textile and Industrial Dyeing*; Clark, M., Ed.; Woodhead Publishing: Sawston, UK, 2011; Volume 1, pp. 303–364.
6. Nabil, G.M.; El-Mallah, N.M.; Mahmoud, M.E. Enhanced decolorization of reactive black 5 dye by active carbon sorbent-immobilized-cationic surfactant (AC-CS). *J. Ind. Eng. Chem.* **2014**, *20*, 994–1002. [CrossRef]
7. Vakili, M.; Rafatullah, M.; Gholami, Z.; Farraji, H. Treatment of reactive dyes from water and wastewater through chitosan and its derivatives. *Smart Mater. Waste Water Appl.* **2016**, 347–377. [CrossRef]
8. Vakili, M.; Rafatullah, M.; Ibrahim, M.H.; Abdullah, A.Z.; Gholami, Z.; Salamatinia, B. Enhancing reactive blue 4 adsorption through chemical modification of chitosan with hexadecylamine and 3-aminopropyl triethoxysilane. *J. Water Process. Eng.* **2017**, *15*, 49–54. [CrossRef]
9. Vakili, M.; Rafatullah, M.; Ibrahim, M.H.; Abdullah, A.Z.; Salamatinia, B.; Gholami, Z. Oil palm biomass as an adsorbent for heavy metals. *Rev. Environ. Contam. T* **2014**, *232*, 61–88.
10. Vakili, M.; Mojiri, A.; Kindaichi, T.; Cagnetta, G.; Yuan, J.; Wang, B.; Giwa, A.S. Cross-linked chitosan/zeolite as a fixed-bed column for organic micropollutants removal from aqueous solution, optimization with RSM and artificial neural network. *J. Environ.* **2019**, *250*, 109434. [CrossRef]
11. Vakili, M.; Amouzgar, P.; Cagnetta, G.; Wang, B.; Guo, X.; Mojiri, A.; Zeimaran, E.; Salamatinia, B. Ultrasound-Assisted Preparation of Chitosan/Nano-Activated Carbon Composite Beads Aminated with (3-Aminopropyl) Triethoxysilane for Adsorption of Acetaminophen from Aqueous Solutions. *Polymers* **2019**, *11*, 1701. [CrossRef]
12. Vakili, M.; Mojiri, A.; Zwain, H.M.; Yuan, J.; Giwa, A.S.; Wang, W.; Gholami, F.; Guo, X.; Cagnetta, G.; Yu, G. Effect of beading parameters on cross-linked chitosan adsorptive properties. *React. Funct. Polym.* **2019**, *144*, 104354. [CrossRef]
13. Qiu, W.; Vakili, M.; Cagnetta, G.; Huang, J.; Yu, G. Effect of high energy ball milling on organic pollutant adsorption properties of chitosan. *Int. J. Biol. Macromol.* **2020**, *148*, 543–549. [CrossRef] [PubMed]
14. Lan, W.; He, L.; Liu, Y. Preparation and Properties of Sodium Carboxymethyl Cellulose/Sodium Alginate/Chitosan Composite Film. *Coatings* **2018**, *8*, 1–17. [CrossRef]
15. Ahmed, M.J.; Hameed, B.H.; Hummadi, E.H. Review on recent progress in chitosan/chitin-carbonaceous material composites for the adsorption of water pollutants. *Carbohyd. Polym.* **2020**, *247*, 116690. [CrossRef]

16. Yadaei, H.; Beyki, M.H.; Shemirani, F.; Nouroozi, S. Ferrofluid mediated chitosan@mesoporous carbon nanohybrid for green adsorption/preconcentration of toxic Cd(II): Modeling, kinetic and isotherm study. *React. Funct. Polym.* **2018**, *122*, 85–97. [CrossRef]

17. Arumugam, T.K.; Krishnamoorthy, P.; Rajagopalan, N.R.; Nanthini, S.; Vasudevan, D. Removal of malachite green from aqueous solutions using a modified chitosan composite. *Int. J. Biol. Macromol.* **2019**, *128*, 655–664. [CrossRef]

18. Yan, M.; Huang, W.; Li, Z. Chitosan cross-linked graphene oxide/lignosulfonate composite aerogel for enhanced adsorption of methylene blue in water. *Int. J. Biol. Macromol.* **2019**, *136*, 927–935. [CrossRef]

19. Vakili, M.; Rafatullah, M.; Salamatinia, B.; Ibrahim, M.H.; Ismail, N.; Abdullah, A.Z. Adsorption Studies of Methyl Tert-butyl Ether from Environment. *Sep. Purif. Rev.* **2017**, *46*, 273–290. [CrossRef]

20. Lv, S.; Zhou, Z.; Xue, M.; Zhang, X.; Yang, Z. Adsorption characteristics of reactive blue 81 by powdered activated carbon: Role of the calcium content. *J. Water Process. Eng.* **2020**, *36*, 101247. [CrossRef]

21. Zhou, J.; Ma, F.; Guo, H. Adsorption behavior of tetracycline from aqueous solution on ferroferric oxide nanoparticles assisted powdered activated carbon. *Chem. Eng. J.* **2020**, *384*, 123290. [CrossRef]

22. Vakili, M.; Rafatullah, M.; Ibrahim, M.H.; Abdullah, A.Z.; Salamatinia, B.; Gholami, Z. Chitosan hydrogel beads impregnated with hexadecylamine for improved reactive blue 4 adsorption. *Carbohyd. Polym.* **2016**, *137*, 139–146. [CrossRef] [PubMed]

23. Amouzgar, P.; Vakili, M.; Chan, E.-S.; Salamatinia, B. Effects of beading parameters for development of chitosan-nano-activated carbon biocomposite for acetaminophen elimination from aqueous sources. *Environ. Eng. Sci.* **2017**, *34*, 805–815. [CrossRef]

24. Mahaninia, M.H.; Wilson, L.D. Phosphate uptake studies of cross-linked chitosan bead materials. *J. Colloid Interface Sci.* **2017**, *485*, 201–212. [CrossRef] [PubMed]

25. Chatterjee, S.; Lee, D.S.; Lee, M.W.; Woo, S.H. Enhanced adsorption of congo red from aqueous solutions by chitosan hydrogel beads impregnated with cetyl trimethyl ammonium bromide. *Bioresour. Technol.* **2009**, *100*, 2803–2809. [CrossRef]

26. Sutirman, Z.A.; Sanagi, M.M.; Abd Karim, K.J.; Ibrahim, W.A.W.; Jume, B.H. Equilibrium, kinetic and mechanism studies of Cu (II) and Cd (II) ions adsorption by modified chitosan beads. *Int. J. Biol. Macromol.* **2018**, *116*, 255–263. [CrossRef] [PubMed]

27. Nematollahzadeh, A.; Shojaei, A.; Karimi, M. Chemically modified organic/inorganic nanoporous composite particles for the adsorption of reactive black 5 from aqueous solution. *React. Funct. Polym.* **2015**, *86*, 7–15. [CrossRef]

28. Cardoso, N.F.; Pinto, R.B.; Lima, E.C.; Calvete, T.; Amavisca, C.V.; Royer, B.; Cunha, M.L.; Fernandes, T.H.M.; Pinto, I.S. Removal of remazol black B textile dye from aqueous solution by adsorption. *Desalination* **2011**, *269*, 92–103. [CrossRef]

29. Banerjee, S.; Chattopadhyaya, M.C. Adsorption characteristics for the removal of a toxic dye, tartrazine from aqueous solutions by a low cost agricultural by-product. *Arab. J. Chem.* **2017**, *10*, S1629–S1638. [CrossRef]

30. Subramaniam, R.; Kumar Ponnusamy, S. Novel adsorbent from agricultural waste (cashew NUT shell) for methylene blue dye removal: Optimization by response surface methodology. *Water Resour. Ind.* **2015**, *11*, 64–70. [CrossRef]

31. Huang, R.; Liu, Q.; Huo, J.; Yang, B. Adsorption of methyl orange onto protonated cross-linked chitosan. *Arab. J. Chem.* **2017**, *10*, 24–32. [CrossRef]

32. El-Bindary, M.; El-Deen, I.; Shoair, A. Removal of anionic dye from aqueous solution using magnetic sodium alginate beads. *J. Mater. Environ. Sci.* **2019**, *10*, 604–617.

33. Wong, S.; Ghafar, N.A.; Ngadi, N.; Razmi, F.A.; Inuwa, I.M.; Mat, R.; Amin, N.A.S. Effective removal of anionic textile dyes using adsorbent synthesized from coffee waste. *Sci. Rep.* **2020**, *10*, 2928. [CrossRef] [PubMed]

34. Kasaai, M.R.; Arul, J.; Charlet, G. Fragmentation of Chitosan by Acids. *Sci. World J.* **2013**, *2013*, 508540. [CrossRef] [PubMed]

35. Batool, F.; Akbar, J.; Iqbal, S.; Noreen, S.; Bukhari, S.N.A. Study of Isothermal, Kinetic, and Thermodynamic Parameters for Adsorption of Cadmium: An Overview of Linear and Nonlinear Approach and Error Analysis. *Bioinorg. Chem. Appl.* **2018**, *2018*, 3463724. [CrossRef] [PubMed]

36. Dehmani, Y.; Alrashdi, A.A.; Lgaz, H.; Lamhasni, T.; Abouarnadasse, S.; Chung, I.-M. Removal of phenol from aqueous solution by adsorption onto hematite (α-Fe2O3): Mechanism exploration from both experimental and theoretical studies. *Arab. J. Chem.* **2020**, *13*, 5474–5486. [CrossRef]

37. Mondal, M.; Manoli, K.; Ray, A.K. Removal of arsenic (III) from aqueous solution by concrete-based adsorbents. *Can. J. Chem. Eng.* **2020**, *98*, 353–359. [CrossRef]

38. Al-Senani, G.M.; Al-Kadhi, N.S. Studies on Adsorption of Fluorescein Dye from Aqueous Solutions Using Wild Herbs. *Int. J. Anal. Chem.* **2020**, *2020*, 8019274. [CrossRef]

39. El-Zawahry, M.M.; Abdelghaffar, F.; Abdelghaffar, R.A.; Hassabo, A.G. Equilibrium and kinetic models on the adsorption of Reactive Black 5 from aqueous solution using Eichhornia crassipes/chitosan composite. *Carbohyd. Polym.* **2016**, *136*, 507–515. [CrossRef]

40. Felista, M.M.; Wanyonyi, W.C.; Ongera, G. Adsorption of anionic dye (Reactive black 5) using macadamia seed Husks: Kinetics and equilibrium studies. *Sci. Afr.* **2020**, *7*, e00283. [CrossRef]

41. Tanyildizi, M.Ş. Modeling of adsorption isotherms and kinetics of reactive dye from aqueous solution by peanut hull. *Chem. Eng. J.* **2011**, *168*, 1234–1240. [CrossRef]

42. Ziane, S.; Bessaha, F.; Marouf-Khelifa, K.; Khelifa, A. Single and binary adsorption of reactive black 5 and Congo red on modified dolomite: Performance and mechanism. *J. Mol. Liq.* **2018**, *249*, 1245–1253. [CrossRef]

43. Ip, A.W.M.; Barford, J.P.; McKay, G. A comparative study on the kinetics and mechanisms of removal of Reactive Black 5 by adsorption onto activated carbons and bone char. *Chem. Eng. J.* **2010**, *157*, 434–442. [CrossRef]

44. Chatterjee, S.; Chatterjee, T.; Woo, S.H. Influence of the polyethyleneimine grafting on the adsorption capacity of chitosan beads for Reactive Black 5 from aqueous solutions. *Chem. Eng. J.* **2011**, *166*, 168–175. [CrossRef]

45. Eren, Z.; Acar, F.N. Equilibrium and kinetic mechanism for Reactive Black 5 sorption onto high lime Soma fly ash. *J. Hazard. Mater.* **2007**, *143*, 226–232. [CrossRef] [PubMed]

46. Xiao, H.; Peng, H.; Deng, S.; Yang, X.; Zhang, Y.; Li, Y. Preparation of activated carbon from edible fungi residue by microwave assisted K_2CO_3 activation—Application in reactive black 5 adsorption from aqueous solution. *Bioresour. Technol.* **2012**, *111*, 127–133. [CrossRef]

47. Li, Z.; Sellaoui, L.; Dotto, G.L.; Lamine, A.B.; Bonilla-Petriciolet, A.; Hanafy, H.; Belmabrouk, H.; Netto, M.S.; Erto, A. Interpretation of the adsorption mechanism of Reactive Black 5 and Ponceau 4R dyes on chitosan/polyamide nanofibers via advanced statistical physics model. *J. Mol. Liq.* **2019**, *285*, 165–170. [CrossRef]

48. Vakili, M.; Deng, S.; Cagnetta, G.; Wang, W.; Meng, P.; Liu, D.; Yu, G. Regeneration of chitosan-based adsorbents used in heavy metal adsorption: A review. *Sep. Purif. Technol.* **2019**, *224*, 373–387. [CrossRef]

49. Vakili, M.; Deng, S.; Shen, L.; Shan, D.; Liu, D.; Yu, G. Regeneration of chitosan-based adsorbents for eliminating dyes from aqueous solutions. *Sep. Purif. Rev.* **2019**, *48*, 1–13. [CrossRef]

50. Chen, A.-H.; Huang, Y.-Y. Adsorption of Remazol Black 5 from aqueous solution by the templated crosslinked-chitosans. *J. Hazard. Mater.* **2010**, *177*, 668–675. [CrossRef]

Article

Cometabolism of the Superphylum *Patescibacteria* with Anammox Bacteria in a Long-Term Freshwater Anammox Column Reactor

Suguru Hosokawa [1], Kyohei Kuroda [2], Takashi Narihiro [2], Yoshiteru Aoi [3], Noriatsu Ozaki [1], Akiyoshi Ohashi [1] and Tomonori Kindaichi [1,*]

[1] Department of Civil and Environmental Engineering, Graduate School of Advanced Science and Engineering, Hiroshima University, 1-4-1 Kagamiyama, Higashihiroshima 739-8527, Japan; m195868@hiroshima-u.ac.jp (S.H.); ojaki@hiroshima-u.ac.jp (N.O.); ecoakiyo@hiroshima-u.ac.jp (A.O.)

[2] Bioproduction Research Institute, National Institute of Advanced Industrial Science and Technology (AIST), 2-17-2-1 Tsukisamu-Higashi, Toyohira-ku, Sapporo, Hokkaido 062-8517, Japan; k.kuroda@aist.go.jp (K.K.); t.narihiro@aist.go.jp (T.N.)

[3] Program of Biotechnology, Graduate School of Integrated Sciences for Life, Hiroshima University, 1-3-1 Kagamiyama, Higashihiroshima 739-8530, Japan; yoshiteruaoi@hiroshima-u.ac.jp

* Correspondence: tomokin@hiroshima-u.ac.jp; Tel.: +81-82-424-5718

Citation: Hosokawa, S.; Kuroda, K.; Narihiro, T.; Aoi, Y.; Ozaki, N.; Ohashi, A.; Kindaichi, T. Cometabolism of the Superphylum *Patescibacteria* with Anammox Bacteria in a Long-Term Freshwater Anammox Column Reactor. *Water* **2021**, *13*, 208. https://doi.org/10.3390/w13020208

Received: 9 December 2020
Accepted: 14 January 2021
Published: 16 January 2021

Abstract: Although the anaerobic ammonium oxidation (anammox) process has attracted attention regarding its application in ammonia wastewater treatment based on its efficiency, the physiological characteristics of anammox bacteria remain unclear because of the lack of pure-culture representatives. The coexistence of heterotrophic bacteria has often been observed in anammox reactors, even in those fed with synthetic inorganic nutrient medium. In this study, we recovered 37 draft genome bins from a long-term-operated anammox column reactor and predicted the metabolic pathway of coexisting bacteria, especially *Patescibacteria* (also known as Candidate phyla radiation). Genes related to the nitrogen cycle were not detected in Patescibacterial bins, whereas nitrite, nitrate, and nitrous oxide-related genes were identified in most of the other bacteria. The pathway predicted for *Patescibacteria* suggests the lack of nitrogen marker genes and its ability to utilize poly-N-acetylglucosamine produced by dominant anammox bacteria. Coexisting *Patescibacteria* may play an ecological role in providing lactate and formate to other coexisting bacteria, supporting growth in the anammox reactor. *Patescibacteria*-centric coexisting bacteria, which produce anammox substrates and scavenge organic compounds produced within the anammox reactor, might be essential for the anammox ecosystem.

Keywords: anaerobic ammonium oxidation (anammox); *Patescibacteria*; Candidate phyla radiation; *Candidatus* Brocadia sinica; *Candidatus* Jettenia caeni; metagenomic analysis; biological nitrogen removal; wastewater treatment

1. Introduction

Anaerobic ammonium oxidation (anammox) is a microbial process in which, under anoxic conditions, ammonia is directly oxidized to nitrogen gas with nitrite as the electron acceptor. The anammox process is mediated by a member of the phylum *Planctomycetes* [1]. Six anammox bacteria candidate genera have been proposed: *Candidatus* Brocadia, *Candidatus* Kuenenia, *Candidatus* Anammoxoglobus, *Candidatus* Jettenia, *Candidatus* Scalindua, and *Candidatus* Anammoximicrobium [2,3]. Although the physiological characteristics of several genera have been investigated [4–7], detailed physiologies remain unknown due to the lack of pure cultures [5].

Recently, most of the candidate phyla were renamed, and superphyla predicted by single-cell genomics [8] and metagenomics [9,10] were proposed. The superphylum *Patescibacteria* [8] has been proposed, which is also referred to as Candidate phyla radiation (CPR) [10]. The superphylum *Patescibacteria* has been found in various environments, such as ground water sediment, lakes, and activated sludge [9,11]. The superphylum *Patescibacteria* has also been found in anammox enrichment cultures fed with ammonia as the sole energy source and lacking an external organic carbon supply [12,13]. Speth et al. [12] reported that candidate phyla OP11 (*Microgenomates*) and WS6 (*Dojkabacteria*) supported fermentative lifestyles, and that candidate phylum OD1 (*Parcubacteria*) could have a parasitic relationship with *Bacteroidetes* in full-scale partial-nitritation/anammox reactors. However, previous studies were mostly focused on the nitrogen cycle in anammox granules; thus, the details of the carbon metabolism of *Patescibacteria* in anammox granules are still largely unknown.

The purpose of the present study was to predict the carbon metabolism of *Patescibacteria* in a freshwater anammox enrichment culture and to investigate the possibility of a cometabolic relationship between anammox bacteria and coexisting heterotrophic bacteria, especially *Patescibacteria*. The anammox culture used in this study was operated for more than 15 years, fed with ammonia as the sole energy source, and lacked an external organic carbon supply [14]; this is a model system used to elucidate cometabolism. In this study, we used metagenomic deep-sequencing analysis to assemble low-abundance members in an anammox enrichment culture, such as *Patescibacteria*. The results of this study provide insights into ecophysiological interactions and substrate/metabolite exchanges in the autotrophic anammox community.

2. Materials and Methods

2.1. Reactor Operation and Sampling

Freshwater anammox bacteria-dominated *Candidatus* Brocadia sinica was enriched using activated sludge and cultured in an up-flow column reactor for 15 years. The reactor volume was 300 or 900 mL. The temperature was maintained at 37 °C. The hydraulic retention time was set to 2.5 h. A typical freshwater anammox medium [15] was used: 3.6–5.7 mM NH_4^+, 4.3–7.1 mM NO_2^-, 1000 mg L^{-1} $KHCO_3$, 27 mg L^{-1} KH_2PO_4, 300 mg L^{-1} $MgSO_4 \cdot 7H_2O$, 180 mg L^{-1} $CaCl_2 \cdot 2H_2O$, and trace element solutions. The concentrations of NH_4^+, NO_2^-, and NO_3^- were determined following a previous report [16]. Three biomass samples were collected from the column reactor 4989, 5054, and 5073 days (Figure 1) after the start of operation (Figure S1).

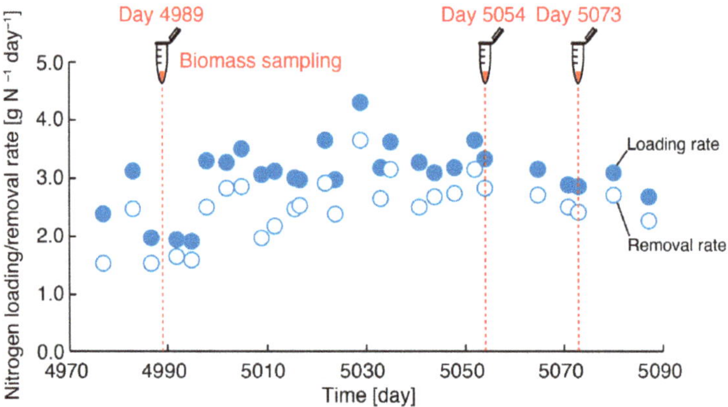

Figure 1. Nitrogen removal performance of the up-flow column reactor. Filled and open circles represent the nitrogen loading and removal rate, respectively. Biomass samples were collected on days 4989, 5054, and 5073.

2.2. DNA Extraction and DNA Sequencing

DNA was extracted using the FastDNA SPIN Kit for Soil (MP Biomedicals, Irvine, CA, USA). The extracted DNA was quantified using a Qubit 2.0 fluorometer (Thermo Fisher

Scientific, Waltham, MA, USA). Three Illumina sequencing libraries were prepared for the three samples using the TruSeq DNA PCR Free (350) kit (Illumina, San Diego, CA, USA) and paired-end-sequenced (2 × 151 bp) using shotgun sequencing on a HiSeq X instrument (Illumina, USA). A PacBio sequencing library was prepared for the sample collected on day 5054 using the SMRTbell Express Template Prep Kit (Pacific Biosciences of California Inc., Menlo Park, CA, USA) after the DNA was purified with Agencourt AMPure XP magnetic beads (Beckman Coulter Life Sciences, Danvers, MA, USA) and sequenced on a PacBio Sequel instrument (Pacific Biosciences of California, Inc., USA). A circular consensus sequence (CCS) read was generated from the Sequel data with a Phred quality score above 20 (Q20, 99%).

2.3. Bioinformatics

Raw paired-end reads from HiSeq X were trimmed using Trimmomatic v0.39 [17]. The trimmed reads from HiSeq X and CCS reads from PacBio Sequel were co-assembled with SPAdes v3.13.1 [18]. In the assembly, the draft genomes of *Candidatus* Brocadia sinica (GCA_000949635.1) and *Candidatus* Jettenia caeni (GCA_000296795.1) were used as the references (as—trusted-contigs option) because the presence of these anammox bacteria in enrichment cultures was confirmed in previous studies [4,14]. The assemblies were binned using MaxBin2 v2.2.7 [19]. The relative abundance output from MaxBin2 was also used as the abundance of each bin. The completeness and contamination of the bins were assessed using CheckM v1.1.2 [20]. For the Patescibacterial bins, the CPR marker set was used for CheckM [10]. Contamination was manually removed from the contig. Bins with high contamination (>7%) were not used for further analysis. The bins were annotated using PROKKA v1.13 [21]. Predicted amino acid sequences were annotated using KEGG BlastKOALA (KEGG Orthology and Links Annotation) [22]. The metabolic pathways obtained by the BlastKOALA annotation were visualized using KEGG (Kyoto Encyclopedia of Genes and Genomes)-Decoder v1.2 [23]. The taxonomy of each bin was estimated using a BLAST search [24]. A genome tree was constructed using PhyloPhlAn v2.0.3 [25]. The sequence data were deposited in the DDBJ database under the DDBJ/EMBL/GenBank accession number DRA011208.

3. Results and Discussion

3.1. Anammox Reactor Operation

The up-flow column reactor with the freshwater anammox medium was operated for more than 5000 d using varying nitrogen loading rates and reactor volumes (Figure S1). During the sampling period, the average nitrogen loading and removal rates were 3.1 and 2.5 g N L^{-1} d^{-1}, respectively (Figure 1). The average NH_4^+ and NO_2^- removal efficiencies were 90.5% and 93.7%, respectively. The average stoichiometric ratios of consumed NO_2^- to consumed NH_4^+ and produced NO_3^- to consumed NH_4^+ were 1.45 ± 0.18 (standard deviation) and 0.27 ± 0.05, respectively. These values were similar to previously reported ratios of 1:1 and 32:0.26 [7], respectively, indicating a stable reactor operation (stable anammox process) during the sampling period.

3.2. Genome Construction and Basic Information on Bins

In total, 0.76 billion reads were produced by metagenomic sequencing of the three samples (Table S1). After quality trimming and filtering, 0.40 billion high-quality reads (>Q20) were obtained and used for metagenomic analysis. Differences in the guanine-cytosine (GC)-contents of HiSeq X reads indicate that the composition of the microbial community of each sample differed. The combined metagenome assembly generated 5780 contigs (167.2 Mbp contigs), with an N50 value of 169,060 bp. The longest contig length was 1,758,248 bp. In total, 2460 contigs above 1000 bp were extracted from the 5780 contigs and used for binning. The reconstructed contigs were classified into 42 bins. Five of the 42 bins were excluded due to high contamination (>7%; Table 1). Two anammox bacteria, *Candidatus* Brocadia sinica and *Candidatus* Jettenia caeni, were detected. In addition, *Chloroflexi* (9 bins), *Ignavibacteriae* (2 bins), *Planctomycetes* (3 bins), *Proteobacteria* (11 bins), *Armatimonadetes* (1 bin), *Bacteroidetes* (2 bins), *Actinobacteria* (1 bin), and *Patescibacteria* (6 bins) were detected. Most of the detected bins were comparable to those reported in a previous study [26]. However, in addition, six bins belonging to the superphylum *Patescibacteria* were detected. In the present study, we focused on the metabolic analysis of *Patescibacteria*. A phylogenetic tree of the 37 bins based on protein sequences is shown in Figure 2.

Table 1. Characteristics of the bins obtained in this study.

Bin ID	Taxonomy	Complete-ness	Contamination	Bin Size (Mbp)	Number of Contigs	Relative Abundance (%)		
						Day 4989	Day 5054	Day 5073
BroJett025	*Patescibacteria, Candidatus Pacebacteria*	97.67 *	0 *	1.17	1	0.2	0.3	0.2
BroJett032	*Patescibacteria, Candidatus Pacebacteria*	100 *	0 *	1.18	5	0.1	0.2	0.1
BroJett037	*Patescibacteria,* Candidate division WS6	95.35 *	0 *	1.18	3	0.1	0.1	0.1
BroJett019	*Patescibacteria,* Candidate division WS6	95.35 *	0*	1.06	5	0.3	0.5	0.5
BroJett008	*Patescibacteria*	97.67*	2.33 *	0.57	8	2.0	0.1	0.0
BroJett034	*Patescibacteria, Berkelbacteria*	93.02 *	0 *	0.68	2	0.1	0.1	0.2
BroJett039	*Chloroflexi*	23.2	0	1.71	48	0.1	0.0	0.1
BroJett021	*Chloroflexi*	90.91	0.91	6.39	43	0.3	0.2	0.1
BroJett001	*Chloroflexi*	93.64	3.09	3.71	153	23.8	8.1	7.1
BroJett015	*Chloroflexi*	75.64	3.82	5.45	63	0.6	0.7	0.2
BroJett018	*Chloroflexi*	91.82	0.91	6.27	45	0.4	0.4	0.1
BroJett007	*Chloroflexi*	98.18	0	4.26	35	5.3	2.0	1.5
BroJett038	*Chloroflexi*	77.27	0.91	4.12	49	0.1	0.1	0.1
BroJett033	*Chloroflexi*	28.8	0	1.12	41	0.1	0.0	0.0
BroJett011	*Chloroflexi*	84.85	3.64	9.27	84	1.1	0.4	0.2
BroJett009	*Armatimonadetes*	91.76	0	2.69	23	1.4	0.4	0.4
BroJett022	*Actinobacteria*	96.98	0	2.70	5	0.3	0.2	0.0
BroJett024	*Alphaproteobacteria*	87.29	6.96	4.67	34	0.2	0.1	0.0
BroJett030	*Alphaproteobacteria*	87.95	1.2	3.49	37	0.1	0.1	0.0
BroJett013	*Alphaproteobacteria*	95.02	0.6	3.78	19	0.9	0.3	0.0
BroJett029	*Alphaproteobacteria*	75.86	6.03	4.62	51	0.2	0.1	0.0
BroJett010	*Gammaproteobacteria*	90.31	1.5	2.94	10	1.3	0.5	0.5
BroJett026	*Betaproteobacteria*	87.56	2.68	4.49	28	0.2	0.1	0.0
BroJett012	*Betaproteobacteria*	91.11	0.52	3.22	101	1.0	2.6	5.5
BroJett006	*Betaproteobacteria*	95.56	0.45	3.30	135	5.8	17.7	10.7
BroJett031	*Betaproteobacteria*	88.27	0.62	4.03	13	0.1	0.1	0.0
BroJett040	*Oligoflexia*	92.86	0	2.76	21	0.0	0.1	0.1
BroJett028	*Deltaproteobacteria*	75.91	5.38	8.63	98	0.2	0.2	0.0
BroJett017	*Ignavibacteriae*	94.97	0	3.60	22	0.5	0.1	0.1
BroJett005	*Ignavibacteriae*	96.65	0.56	3.85	120	9.9	3.7	3.2
BroJett020	*Bacteroidetes*	89.25	0	2.70	21	0.3	0.2	0.5
BroJett042	*Bacteroidetes*	93.99	0	3.74	27	0.0	0.0	0.2
BroJett014	*Planctomycetes*	81.82	4.55	4.02	32	0.9	1.0	0.6
BroJett041	*Planctomycetes* (Jettenia)	84.62	1.1	3.02	23	0.0	0.2	0.1
BroJett002	*Planctomycetes* (Brocadia)	97.8	1.65	4.10	23	14.6	39.0	58.5
BroJett003	*Planctomycetes*	95.01	2.94	3.92	437	13.1	18.0	7.6
BroJett004	*Planctomycetes*	97.66	0	3.22	20	13.0	1.8	1.0

* Calculated with the CPR marker set.

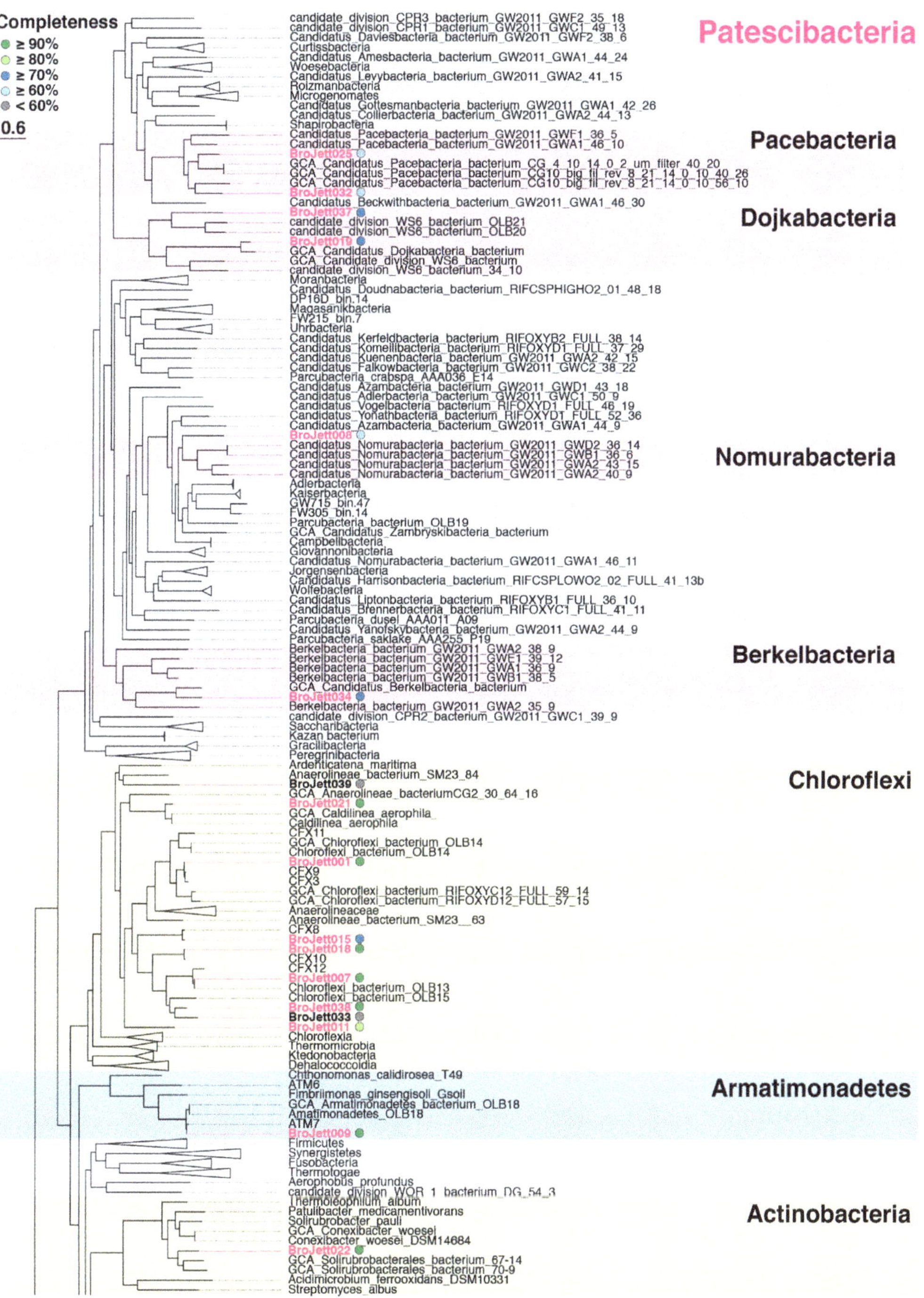

Figure 2. *Cont.*

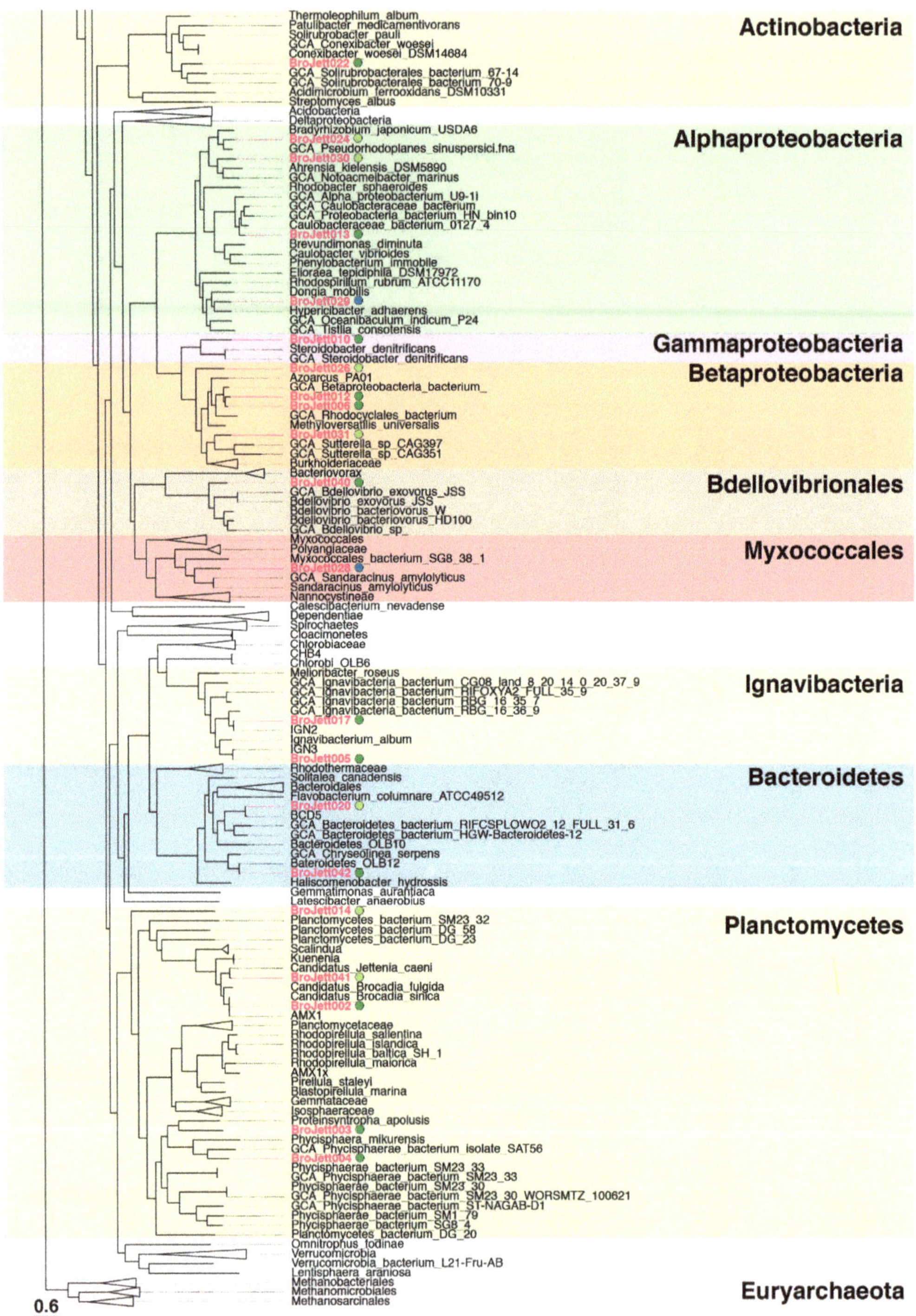

Figure 2. Phylogenetic tree of bins and related genomes. The bins found in this study are shown in bold. BroJett033 and BroJett039 had low completeness and are represented in a bold black font.

3.3. Relative Abundance

The relative abundance of each bin of the three samples was estimated from the coverage calculated using MaxBin2 (Table 1). After five bins were excluded due to high contamination, the samples collected on days 4989, 5054, and 5073 accounted for 98.6%, 99.7%, and 99.5% of the relative abundance, respectively. *Candidatus* Brocadia sinica (Bin ID: BroJett002) was the most dominant bacterium, except for the sample collected on day 4989. BroJett001, which belongs to *Chloroflexi*, was the most dominant bin of the latter sample. The relative abundance of *Candidatus* Brocadia sinica increased with increasing reactor operation. In addition, the anammox bacterium *Candidatus* Jettenia caeni (BroJett041) was detected in all samples, but its relative abundance was 0.01–0.2%. *Patescibacteria* (BroJett008), *Chloroflexi* (BroJett001 and BroJett007), *Armatimonadetes* (BroJett009), *Gammaproteobacteria* (BroJett010), *Betaproteobacteria* (BroJett006 and BroJett012), and *Planctomycetes* (BroJett002, BroJett003, BroJett004, and BroJett014) accounted for more than 1% of the relative abundance of the three samples. The relative abundance of *Patescibacteria*, except for BroJett008, was lower (0.1–0.5%).

3.4. Nitrogen Cycle

To understand the contribution of bacteria to the nitrogen cycle in the anammox reactor, we focused on the following reconstructed nitrogen maker genes: ammonia oxidation (*amoA and amoBC*), hydroxylamine oxidation (*hao*), nitrite oxidation (*nxrAB*), dissimilatory nitrate reduction (*narGH* and/or *napAB*), dissimilatory nitrate reduction to ammonium (DNRA; *nirBD* and/or *nrfAH*), nitrite reduction (*nirK* or *nirS*), nitric oxide reduction (*norBC*), nitrous oxide reduction (*nosZ*), nitrogen fixation (*nifKDH*), hydrazine dehydrogenase (*hdh*), and hydrazine synthase (*hzs*; Figure 3). The dominant anammox bacterial bin BroJett002 (*Candidatus* Brocadia sinica) and the minor anammox bacterial bin BroJett041 (*Candidatus* Jettenia caeni) contained key genes for nitrite reduction, hydrazine synthesis, and hydrazine dehydrogenation. However, bin BroJett002 lacked nitrite reduction genes (*nirS* or *nirK*), as previously reported [26,27]. Bins BroJett020 and BroJett029 contained genes for the complete denitrification of nitrate to dinitrogen gas. Lau et al. [28] reported that *Owenweeksia hongkongensis*, which is closely related to bin BroJett020 based on the 16S rRNA gene, cannot reduce nitrate. In contrast, Ali et al. [26] reported that the metagenome assembly BCD5 has key enzymes for denitrification, such as *nar*, *nir*, and *noz*. In addition, *Dongia mobilis* and *Oceanibaculum indicum*, closely related to bin BroJett029, can reduce nitrate [29,30]. Most of the bins, except for Patescibacterial bins, contained genes related to nitrate reduction and DNRA (Figure 3), and their contribution was relatively high (Figure 4). The production of NH_4^+ and NO_2^- in the anammox reactor plays an important role in supporting the anammox process. Anammox bacteria are well known to lack the N_2O production pathway [31]. Although bins BroJett010 and BroJett038 can produce N_2O and lack N_2O reduction genes, indicating the release of N_2O outside the up-flow column reactor, most of the bins can reduce N_2O to N_2 gas, except for *Patescibacteria*. The emission of N_2O from the up-flow column reactor must be investigated. Interestingly, no genes related to the nitrogen cycle were detected in bins classified as *Patescibacteria*.

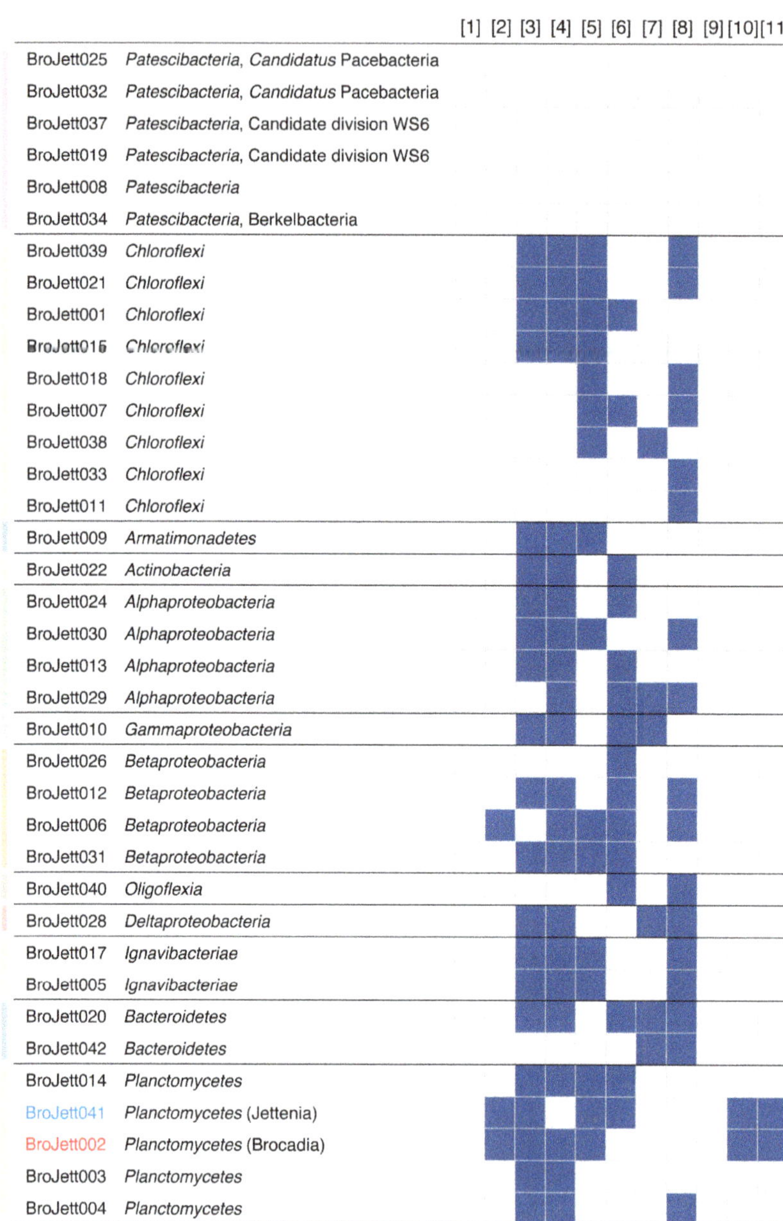

Figure 3. Presence/absence of nitrogen marker genes annotated using KEGG (Kyoto Encyclopedia of Genes and Genomes) and Blastp. The numbers in square brackets on the horizontal axis represent the following genes: [1] ammonia oxidation (amoA and amoBC, pmmo), [2] hydroxylamine oxidation (hao), [3] nitrite oxidation (nxrAB), [4] dissimilatory nitrate reduction (narGH and/or napAB), [5] DNRA (nirBD and/or nrfAH), [6] nitrite reduction (nirK or nirS), [7] nitric oxide reduction (norBC), [8] nitrous oxide reduction (nosZ), [9] nitrogen fixation (nifKDH), [10] hydrazine dehydrogenase (hdh), and hydrazine synthase (hzs). The colored lines next to the bins correspond to the phyla shown in Figure 2. The text colors of BroJett041 and BroJett002 correspond to Figure 4 and represent *Candidatus* Jettenia caeni and *Candidatus* Brocadia sinica, respectively.

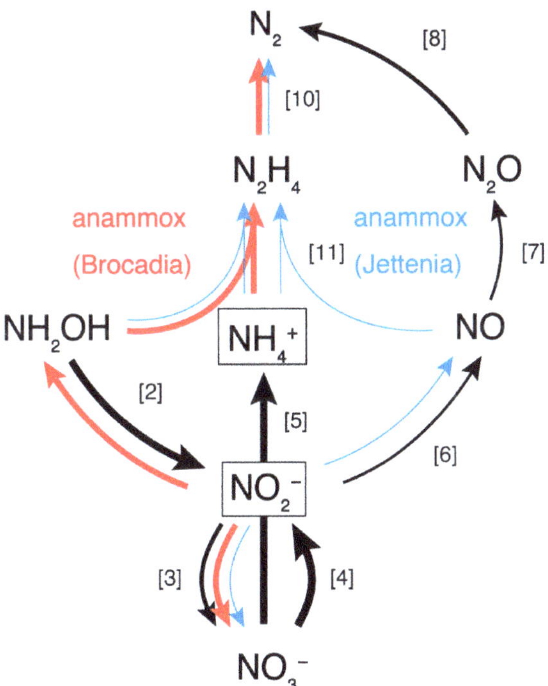

Figure 4. Nitrogen cycle in the reactor. Red and blue arrows represent the anammox processes of *Candidatus* Brocadia sinica and *Candidatus* Jettenia caeni, respectively. Black arrows represent reactions involved in the nitrogen cycle other than the anammox process. The number of each reaction corresponds to those shown in Figure 3, and represents the following genes: [2] hydroxylamine oxidation (hao), [3] nitrite oxidation (nxrAB), [4] dissimilatory nitrate reduction (narGH and/or napAB), [5] DNRA (nirBD and/or nrfAH), [6] nitrite reduction (nirK or nirS), [7] nitric oxide reduction (norBC), [8] nitrous oxide reduction (nosZ), [10] hydrazine dehydrogenase (hdh), and hydrazine synthase (hzs). The width of each arrow reflects the total abundance of the bins, except for *Candidatus* Brocadia sinica because its abundance was too high.

3.5. Genomic Features of Patescibacteria Bins

We successfully recovered six draft genome bins of *Patescibacteria* from a long-term-operated freshwater anammox column reactor (BroJett008, 019, 025, 032, 034, and 037). The taxonomic assignments of these metagenomic bins were classified as *Pacebacteria* (BroJett025 and 032), *Dojkabacteria* (BroJett019 and 037), *Patescibacteria* (BroJett008), and *Berkelbacteria* (BroJett034), based on 400 conserved protein sequences (Figure 2). The genome size and the GC-content ranged from 0.57 to 1.18 Mb and from 34.3% to 48.8%, respectively, with high completeness values of 93.0% and 100%, respectively, estimated using the CheckM software package based on the 43 CPR marker genes set [10] (Table 1). Although the incomplete Patescibacterial genomes could not conclude their whole metabolic capacities, most of the gene sets for major biosynthesis pathways, such as the tricarboxylic acid cycle, gluconeogenesis, and prerequisite electron carriers, were lacking (Figure S2). In addition, there was a lack of genomes of de novo amino acid biosynthesis pathways, except for partial biosynthesis genes for serine/glycine (BroJett008, 025, 032, 034, and 037), threonine/asparagine (BroJett037), glutamine (BroJett025), and aspartate/glutamate (BroJett032 and BroJett019; Figure S2). Similarly, there were no genes relevant to the nitrogen and sulfur cycles, suggesting that these *Patescibacteria* acquire essential nutrients from other microorganisms with symbiotic lifestyles for their growth in the reactor, which is similar to the results obtained in previous studies [32]. In contrast, a recent cell–cell association analysis based on a single amplified genome of 4829 individual cells of prokaryotes collected from subsurface field samples revealed that most of the *Patescibacteria* populations in the studied subsurface environments may not form specific physical associations with other microorganisms [33]. Instead, it was speculated that the *Patescibacteria* may rely solely on fermentation for energy conservation. In our anammox reactor, fermentative pathways for lactate (L-lactate dehydrogenase: BroJett025, 032, and 037) and formate (formate C-acetyltransferase: BroJett025) were found in Patescibacteria metagenomic bins. This suggests that *Patescibacteria* provide these fermentative by-products to bins 4 and 15, which possess lactate dehydrogenase [34] and formate dehydrogenase [35], respectively (Figure 5 and Figure S2). Although the cometabolism of these bacteria must be further studied, *Patescibacteria* may support their growth in the reactor. With respect to other possible features of the carbon cycle in the anammox reactor, we identified chitinase (BroJett037), diacetylchitobiose deacetylase (BroJett019 and BroJett032), and beta-N-acetylhexosaminidase (BroJett025 and BroJett032). Based on the function of the abovementioned chitin degradation-related genes, we speculate that chitin is converted to *N*-acetylglucosamine via chitobiose, *N*-acetylglucosamine hydrolyzes to acetate via diacetylchitobiose deacetylase [36], and acetate could be a useful carbon source for other microorganisms in the anammox reactor. Generally, chitin is a component of eukaryotic cells, such as protozoa, fungi, insects, crustaceans, and arthropods [37]. In prokaryotes, several bacteria can produce poly-N-acetylglucosamine (PGA), which is known as chitin-like polysaccharide, for biofilm formation [38]. Interestingly, metagenomic bins associated with *Candidatus* Brocadia (BroJett002), *Candidatus* Jettenia (BroJett041), and Ignavibacteria (BroJett005) encode poly-beta-1,6 N-acetyl-D-glucosamine synthase (PgaC), which is key for the biosynthesis of PGA (Table S2). This enzyme catalyzes the polymerization of uridine diphosphate-*N*-acetylglucosamine, which is synthesized from beta-D-fructose 6-phosphate generated during glycolysis, to produce PGA. In addition, an anammox bacterial bin of *Candidatus* Brocadia (BroJett002) possesses putative poly-beta-1,6 N-acetyl-D-glucosamine export porin (PgaA) and poly-beta-1,6-N-acetyl-D-glucosamine N-deacetylase (PgaB). Similar PgaABC proteins were also found in the *Candidatus* Brocadia sinica JPN1 genome. These observations imply that major microbial constituents of the reactor, including anammox bacteria, may produce PGA and that some Patescibacteria populations may utilize parts of the PGA (e.g., *N*-acetylglucosamine) for their growth. On the other hand, there were no genes of the biofilm PGA synthesis protein (PgaD) in the investigated anammox bacterial bins, which is necessary for the formation of PGA that functions as a helper protein of PgaC [38,39]. Therefore, further gene expression studies and identification of PGA materials are required for the confirmation of actual PGA production from anammox bacteria in the reactor. The utilization of organic compounds by coexisting heterotrophic bacteria has also been reported for autotrophic nitrifying biofilms, which are fed with ammonia as the sole energy source [40]. Moreover, we newly discovered that *Pacebacteria* and unclassified *Patescibacteria*, other than *Dojkabacteria* and *Microgenomates* [12], may support fermentative lifestyles in the anammox granule. Overall, *Patescibacteria* populations in the anammox reactor may play ecological roles, such as in short-chain fatty acid production and the degradation of chitin-related compounds, and they may survive depending on the PGA production by major anammox bacteria based on metagenomic information.

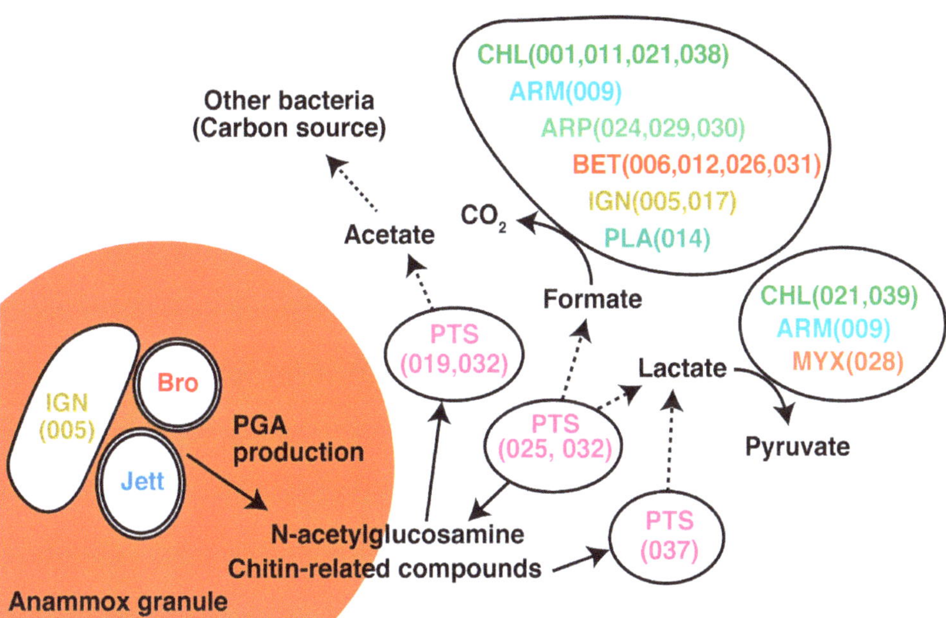

Figure 5. Putative carbon metabolic interactions among *Patescibacteria* and predominant bacteria. The numbers on the bacterial cells indicate the bin IDs shown in Table 1. PTS: *Patescibacteria*, IGN: *Ignavibacteriae*, Bro: *Candidatus* Brocadia, Jett: *Candidatus* Jettenia, CHL: *Chloroflexi*, ARM: *Armatimonadetes*, ARP: *Alphaproteobacteria*, BET: *Betaproteobacteria*, PLA: *Planctomycetes*, MYX: *Myxococcales*. Solid arrows indicate the degradation and metabolic flows of poly-*N*-acetylglucosamine/chitin-related compounds. Dashed arrows indicate the by-products from fermentation and poly-*N*-acetylglucosamine degradation. Curved solid arrows indicate metabolic reactions via formate or lactate dehydrogenases.

4. Conclusions

Six draft genome bins of *Patescibacteria* were recovered from freshwater anammox column reactors operated for more than 15 years and fed with an inorganic and synthetic nutrient medium by metagenomic deep-sequencing analysis. The metabolic capacities predicted for the six Patescibacterial bins with high completeness suggest that *Patescibacteria* can utilize chitin-related compounds and produce fermentation by-products of lactate and formate in the anammox reactor. The phylogenetically and metabolically diverse *Patescibacteria* as well as other coexisting heterotrophic bacteria ensure the effective utilization of chitin-related compounds produced by anammox bacteria, which may create a stable anammox ecosystem without other by-products. Further studies involving metatranscriptomics and metabolomics may help to elucidate the in situ ecological functions of Patescibacteria and the biological interactions with anammox bacteria in the reactor.

Supplementary Materials: The following are available online at https://www.mdpi.com/2073-4441/13/2/208/s1, Figure S1: Nitrogen loading and removal rates of the up-flow column anammox bioreactor enriched using activated sludge, Figure S2: Heat map showing the metabolic function of each bin based on KEGG and Blastp, Table S1: Summary of metagenomic data used in this study, Table S2: Summary of genes related to the Poly-beta-1,6-*N*-acetyl-D-glucosamine synthase production in bins.

Author Contributions: Conceptualization, T.K., Y.A., N.O., and A.O.; investigation, S.H.; data curation, S.H., T.K., K.K., and T.N.; original draft preparation, S.H., T.K., and K.K.; writing, review, and editing, T.K., K.K., T.N., Y.A., N.O., and A.O.; supervision, T.K.; funding acquisition, T.K. All authors have read and agreed to the published version of the manuscript.

Funding: This work was supported by the JSPS KAKENHI, Grant Number JP16H04833 and JP20H02287.

Institutional Review Board Statement: Not applicable.

Informed Consent Statement: Not applicable.

Data Availability Statement: The sequence data were deposited in the DDBJ database under the DDBJ/EMBL/GenBank accession number DRA011208.

Conflicts of Interest: The authors declare no conflict of interest.

References

1. Strous, M.; Fuerst, J.A.; Kramer, E.H.M.; Logemann, S.; Muyzer, G.; Van De Pas-Schoonen, K.T.; Webb, R.; Kuenen, J.G.; Jetten, M.S.M. Missing lithotroph identified as new planctomycete. *Nature* **1999**, *400*, 446–449. [CrossRef] [PubMed]
2. Khramenkov, S.V.; Kozlov, M.N.; Kevbrina, M.V.; Dorofeev, A.G.; Kazakova, E.A.; Grachev, V.A.; Kuznetsov, B.B.; Polyakov, D.Y.; Nikolaev, Y.A. A novel bacterium carrying out anaerobic ammonium oxidation in a reactor for biological treatment of the filtrate of wastewater fermented sludge. *Microbiology* **2013**, *82*, 628–636. [CrossRef]
3. Oshiki, M.; Satoh, H.; Okabe, S. Ecology and physiology of anaerobic ammonium-oxidizing bacteria. *Environ. Microbiol.* **2016**, *18*, 2784–2796. [CrossRef] [PubMed]
4. Ali, M.; Oshiki, M.; Awata, T.; Isobe, K.; Kimura, Z.; Yoshikawa, H.; Hira, D.; Kindaichi, T.; Satoh, H.; Fujii, T.; et al. Physiological characterization of anaerobic ammonium oxidizing bacterium "*Candidatus* Jettenia caeni". *Environ. Microbiol.* **2015**, *17*, 2172–2189. [CrossRef]
5. Awata, T.; Oshiki, M.; Kindaichi, T.; Ozaki, N.; Ohashi, A.; Okabe, S. Physiological characterization of an anaerobic ammonium-oxidizing bacterium belonging to the "*Candidatus* Scalindua"' group. *Appl. Environ. Microbiol.* **2013**, *79*, 4145–4148. [CrossRef]
6. Oshiki, M.; Shimokawa, M.; Fujii, N.; Satoh, H.; Okabe, S. Physiological characteristics of the anaerobic ammonium-oxidizing bacterium "*Candidatus* Brocadia sinica". *Microbiology* **2011**, *157*, 1706–1713. [CrossRef]
7. Strous, M.; Heijnen, J.J.; Kuenen, J.G.; Jetten, M.S.M. The sequencing batch reactor is a powerful tool for the study of slowly growing anaerobic ammonium-oxidizing microorganisms. *Appl. Microbiol. Biotechnol.* **1998**, *50*, 589–596. [CrossRef]
8. Rinke, C.; Schwientek, P.; Sczyrba, A.; Ivanova, N.N.; Anderson, I.J.; Cheng, J.F.; Darling, A.; Malfatti, S.; Swan, B.K.; Gies, E.A.; et al. Insights into the phylogeny and coding potential of microbial dark matter. *Nature* **2013**, *499*, 431–437. [CrossRef]
9. Albertsen, M.; Hugenholtz, P.; Skarshewski, A.; Nielsen, K.L.; Tyson, G.W.; Nielsen, P.H. Genome sequences of rare, uncultured bacteria obtained by differential coverage binning of multiple metagenomes. *Nat. Biotechnol.* **2013**, *31*, 533–538. [CrossRef]
10. Brown, C.T.; Hug, L.A.; Thomas, B.C.; Sharon, I.; Castelle, C.J.; Singh, A.; Wilkins, M.J.; Wrighton, K.C.; Williams, K.H.; Banfield, J.F. Unusual biology across a group comprising more than 15% of domain Bacteria. *Nature* **2015**, *523*, 208–211. [CrossRef]
11. Kindaichi, T.; Yamaoka, S.; Uehara, R.; Ozaki, N.; Ohashi, A.; Albertsen, M.; Nielsen, P.H.; Nielsen, J.L. Phylogenetic diversity and ecophysiology of Candidate phylum Saccharibacteria in activated sludge. *FEMS Microbiol. Ecol.* **2016**, *92*, fiw078. [CrossRef] [PubMed]
12. Speth, D.R.; In't Zandt, M.H.; Guerrezo-Cruz, S.; Dutilh, B.E.; Jettenm, M.S.M. Genome-based microbial ecology of anammox granules in a full-scale wastewater treatment system. *Nat. Commun.* **2016**, *7*, 11172. [CrossRef] [PubMed]
13. Kindaichi, T.; Awata, T.; Suzuki, Y.; Tanabe, K.; Hatamoto, M.; Ozaki, N.; Ohashi, A. Enrichment using an up-flow column reactor and community structure of marine anammox bacteria from coastal sediment. *Microbes Environ.* **2011**, *26*, 67–73. [CrossRef] [PubMed]
14. Tsushima, I.; Ogasawara, Y.; Kindaichi, T.; Satoh, H.; Okabe, S. Development of high-rate anaerobic ammonium-oxidizing (anammox) biofilm reactors. *Water Res.* **2007**, *41*, 1623–1634. [CrossRef]
15. Van de Graaf, A.A.; De Bruijn, P.; Robertson, L.A.; Jetten, M.S.M.; Kuenen, J.G. Autotrophic growth of anaerobic ammonium-oxidizing microorganisms in a fluidized bed reactor. *Microbiology* **1996**, *142*, 2187–2196. [CrossRef]
16. Awata, T.; Goto, Y.; Kindaichi, T.; Ozaki, N.; Ohashi, A. Nitrogen removal using an anammox membrane bioreactor at low temperature. *Water Sci. Technol.* **2015**, *72*, 2148–2153. [CrossRef]
17. Bolger, A.M.; Lohse, M.; Usadel, B. Trimmomatic: A flexible trimmer for Illumina sequence data. *Bioinformatics* **2014**, *30*, 2114–2120. [CrossRef]
18. Bankevich, A.; Nurk, S.; Antipov, D.; Gurevich, A.A.; Dvorkin, M.; Kulikov, A.S.; Lesin, V.M.; Nikolenko, S.I.; Pham, S.; Prjibelski, A.D.; et al. SPAdes: A new genome assembly algorithm and its applications to single-cell sequencing. *J. Comput. Biol.* **2012**, *19*, 455–477. [CrossRef]
19. Wu, Y.-W.; Simmons, B.A.; Singer, S.W. MaxBin 2.0: An automated binning algorithm to recover genomes from multiple metagenomic datasets. *Bioinformatics* **2016**, *32*, 605–607. [CrossRef]
20. Parks, D.H.; Imelfort, M.; Skennerton, C.T.; Hugenholtz, P.; Tyson, G.W. CheckM: Assessing of the quality of microbial genomes recovered from isolates, single cells, and metagenomes. *Genome Res.* **2015**, *25*, 1043–1055. [CrossRef]
21. Seemann, T. Prokka: Rapid prokaryotic genome annotation. *Bioinformatics* **2014**, *30*, 2068–2069. [CrossRef] [PubMed]
22. Kanehisa, M.; Sato, Y.; Morishima, K. BlastKOALA and GhostKOALA: KEGG tools for functional characterization of genome and metagenome sequences. *J. Mol. Biol.* **2016**, *428*, 726–731. [CrossRef] [PubMed]
23. Graham, E.D.; Heidelberg, J.F.; Tully, B.J. Potential for primary productivity in a globally-distributed bacterial phototroph. *ISME J.* **2018**, *12*, 1861–1866. [CrossRef] [PubMed]

24. Altschul, S.F.; Gish, W.; Miller, W.; Myers, E.W.; Lipman, D.J. Basic local alignment search tool. *J. Mol. Biol.* **1990**, *215*, 403–410. [CrossRef]

25. Asnicar, F.; Thomas, A.M.; Beghini, F.; Mengoni, C.; Manara, S.; Manghi, P.; Zhu, Q.; Bolzan, M.; Cumbo, F.; May, U.; et al. Precise phylogenetic analysis of microbial isolates and genomes from metagenomes using PhyloPhlAn 3.0. *Nat. Commun.* **2020**, *11*, 2500. [CrossRef] [PubMed]

26. Ali, M.; Shaw, D.R.; Albertsen, M.; Saikaly, P.E. Comparative genome-centric analysis of freshwater and marine anammox cultures suggests functional redundancy in nitrogen removal processes. *Front. Microbiol.* **2020**, *11*, 1637. [CrossRef]

27. Oshiki, M.; Ali, M.; Shinyako-Hata, K.; Satoh, H.; Okabe, S. Hydroxylamine-dependent anaerobic ammonium oxidation (anammox) by "*Candidatus* Brocadia sinica". *Environ. Microbiol.* **2016**, *18*, 3133–3143. [CrossRef]

28. Lau, K.W.K.; Ng, C.Y.M.; Ren, J.; Lau, S.C.L.; Qian, P.Y.; Wong, P.K.; Lau, T.C.; Wu, M. *Owenweeksia Hongkongensis* gen. nov., sp. nov., a novel marine bacterium of the phylum '*Bacteroidetes*'. *Int. J. Syst. Evol. Microbiol.* **2005**, *55*, 1051–1057. [CrossRef]

29. Liu, Y.; Jin, J.H.; Liu, Y.H.; Zhou, Y.G.; Liu, Z.P. *Dongia mobilis* gen. nov., sp. nov., a new member of the family *Rhodospirillaceae* isolated from a sequencing batch reactor for treatment of malachite green effluent. *Int. J. Syst. Evol. Microbiol.* **2010**, *60*, 2780–2785. [CrossRef]

30. Lai, Q.; Yuan, J.; Wu, C.; Shao, Z. *Oceanibaculum indicum* gen. nov., sp. nov., isolated from deep seawater of the Indian Ocean. *Int. J. Syst. Evol. Microbiol.* **2009**, *59*, 1733–1737. [CrossRef]

31. Kartal, B.; De Almeida, N.M.; Maalcke, W.J.; Op den Camp, H.J.M.; Jetten, M.S.M.; Keltjens, J.T. How to make a living from anaerobic ammonium oxidation. *FEMS Microbiol. Rev.* **2013**, *37*, 428–461. [CrossRef] [PubMed]

32. Lemos, L.N.; Medeiros, J.D.; Dini-Andreote, F.; Fernandes, G.R.; Varani, A.M.; Oliveira, G.; Pylro, V.S. Genomic signatures and co-occurrence patterns of the ultra-small Saccharimonadia (phylum CPR/Patescibacteria) suggest a symbiotic lifestyle. *Mol. Ecol.* **2019**, *28*, 4259–4271. [CrossRef] [PubMed]

33. Beam, J.P.; Becraft, E.D.; Brown, J.M.; Schulz, F.; Jarett, J.K.; Bezuidt, O.; Poulton, N.J.; Clark, K.; Dunfield, P.F.; Ravin, N.V.; et al. Ancestral absence of electron transport chains in Patescibacteria and DPANN. *Front. Microbiol.* **2020**, *11*, 1848. [CrossRef] [PubMed]

34. Chung, F.Z.; Tsujibo, H.; Bhattacharyya, U.; Sharief, F.S.; Li, S.S. Genomic organization of human lactate dehydrogenase-A gene. *Biochem. J.* **1985**, *231*, 537–541. [CrossRef] [PubMed]

35. Overkamp, K.M.; Kötter, P.; van der Hoek, R.; Schoondermark-Stolk, S.; Luttik, M.A.H.; van Dijken, J.P.; Pronk, J.T. Functional analysis of structural genes for NAD$^+$-dependent formate dehydrogenase in *Saccharomyces cerevisiae*. *Yeast* **2002**, *19*, 509–520. [CrossRef] [PubMed]

36. Tanaka, T.; Fukui, T.; Fujiwara, S.; Atomi, H.; Imanaka, T. Concerted action of diacetylchitobiose deacetylase and exo-β-D-glucosaminidase in a novel chitinolytic pathway in the hyperthermophilic archaeon *Thermococcus kodakaraensis* KOD1. *J. Biol. Chem.* **2004**, *279*, 30021–30027. [CrossRef]

37. Tharanathan, R.N.; Kittur, F.S. Chitin–the undisputed biomolecule of great potential. *Crit. Rev. Food Sci. Nutr.* **2003**, *43*, 61–87. [CrossRef]

38. Takeo, M.; Kimura, K.; Mayilraj, S.; Inoue, T.; Tada, S.; Miyamoto, K.; Kashiwa, M.; Ikemoto, K.; Baranwal, P.; Kato, D.; et al. Biosynthetic pathway and genes of chitin/chitosan-like bioflocculant in the genus *Citrobacter*. *Polymers* **2018**, *10*, 237. [CrossRef]

39. Itoh, Y.; Rice, J.D.; Goller, C.; Pannuri, A.; Taylor, J.; Meisner, J.; Beveridge, T.J.; Preston, J.F.; Romeo, T. Roles of pgaABCD genes in synthesis, modification, and export of the Escherichia coli biofilm adhesin poly-beta-1,6-N-acetyl-D-glucosamine. *J. Bacteriol.* **2008**, *190*, 3670–3680. [CrossRef]

40. Kindaichi, T.; Ito, T.; Okabe, S. Ecophysiological interaction between nitrifying bacteria and heterotrophic bacteria in autotrophic nitrifying biofilms as determined by microautoradiography-fluorescence in situ Hybridization. *Appl. Environ. Microbiol.* **2004**, *70*, 1641–1650. [CrossRef]

water

MDPI

Article

Selective Removal of Hexavalent Chromium from Wastewater by Rice Husk: Kinetic, Isotherm and Spectroscopic Investigation

Usman Khalil [1], Muhammad Bilal Shakoor [2,*], Shafaqat Ali [1,3,*], Sajid Rashid Ahmad [2], Muhammad Rizwan [1], Abdulaziz Abdullah Alsahli [4] and Mohammed Nasser Alyemeni [4]

[1] Department of Environmental Sciences and Engineering, Government College University Faisalabad, Faisalabad 38000, Pakistan; musmankhalil@yahoo.com (U.K.); mrazi1532@yahoo.com (M.R.)
[2] College of Earth and Environmental Sciences, University of the Punjab, Lahore 54000, Pakistan; sajidpu@yahoo.com
[3] Department of Biological Sciences and Technology, China Medical University, Taichung 406040, Taiwan
[4] Department of Botany and Microbiology, College of Science, King Saud University, Riyadh 11451, Saudi Arabia; aalshenaifi@ksu.edu.sa (A.A.A.); mnyemeni@ksu.edu.sa (M.N.A.)
* Correspondence: bilalshakoor88@gmail.com or bilal.cees@pu.edu.pk (M.B.S.); shafaqataligill@yahoo.com (S.A.)

Abstract: Chromium (Cr) in water bodies is considered as a major environmental issue around the world. In the present study, aqueous Cr(VI) adsorption onto rice husk was studied as a function of various environmental parameters. Equilibrium time was achieved in 2 h and maximum Cr(VI) adsorption was 78.6% at pH 5.2 and 120 mg L^{-1} initial Cr(VI) concentration. In isotherm experiments, the maximum sorption was observed as 379.63 mg g^{-1}. Among four isotherm models, Dubinin–Radushkevich and Langmuir models showed the best fitting to the adsorption data, suggesting physical and monolayer adsorption to be the dominant mechanism. The kinetic modeling showed that a pseudo-second order model was suitable to describe kinetic equilibrium data, suggesting a fast adsorption rate of Cr(VI). The results of FTIR spectroscopy indicated that mainly –OH and C–H contributed to Cr(VI) adsorption onto rice husk. This paper provided evidence that rice husk could be a cost-effective, environment-friendly and efficient adsorptive material for Cr(VI) removal from wastewater due to its high adsorption capacity.

Keywords: chromium; functional groups; isotherm; rice husk

Citation: Khalil, U.; Shakoor, M.B.; Ali, S.; Ahmad, S.R.; Rizwan, M.; Alsahli, A.A.; Alyemeni, M.N. Selective Removal of Hexavalent Chromium from Wastewater by Rice Husk: Kinetic, Isotherm and Spectroscopic Investigation. *Water* **2021**, *13*, 263. https://doi.org/10.3390/w13030263

Received: 29 December 2020
Accepted: 8 January 2021
Published: 22 January 2021

Publisher's Note: MDPI stays neutral with regard to jurisdictional claims in published maps and institutional affiliations.

1. Introduction

Aquatic systems have been contaminated by the addition of heavy metals over the past few decades. Major heavy metals of concern in water bodies are zinc (Zn), chromium (Cr), lead (Pb), silver (Ag), nickel (Ni) and copper (Cu) [1]. Industrial sources, combustion by-products, smelters and foundries, automobiles and the paint industry are major sources of heavy metals contamination in aqueous systems. In the natural environment, the release of toxic heavy metals, for example, irrigation of agricultural land with sewage, allows non-biodegradable and persistent heavy metals to enter into the food chain [2,3].

Contamination of water with Cr, specifically hexavalent Cr (Cr(VI)), is a major problem for the environment and humans, considering its toxicity and disease-causing effects. Chromium occurs as ores and in various compound forms in the earth's crust. The sources of Cr contamination of water bodies include the disposal of industrial effluents from leather tanning, electroplating, metal finishing, pigments, dyes, paints and ceramics industries [2,4]. Chromium exists in two stable oxidation forms in the natural environment: Cr(VI) and trivalent Cr (Cr(III)). Cr(III) is less toxic and can be used as a micronutrient for the human body [5]. In contrast, Cr(VI) is an extremely stable, mobile and toxic species with relatively higher solubility in water than Cr(III). The toxicity of Cr(VI) is about 500 times higher as compared to Cr(III), thus Cr(VI) has been classified as Class-I human carcinogen [6]. According to the USEPA, 0.05 mg L^{-1} is the maximum contamination level (MCL) of Cr(VI) in domestic water supplies [5]. Therefore, high Cr(VI) concentrations in water should be decreased below the MCL before discharging.

Various techniques, such as precipitation, solvent extraction, flotation, evaporation, reverse osmosis and ion exchange, have been introduced to deal with Cr(VI) originating from industrial effluents and wastewater [7,8]. However, these conventional methods have proved to be expensive and impractical due to higher capital cost and energy requirement [9–11]. In the recent past, adsorption has emerged as an inexpensive and efficient method for the removal of Cr(VI) from contaminated water. In previous studies different commercial adsorbents and agricultural residues, such as tea waste, turmeric waste, *Jatropha curcus*, citrus peel, water melon peel, coconut husk, wheat straw and bamboo, etc., were used as adsorbents for the removal of Cr(VI) and other heavy metals from contaminated water [12–14].

In Asia, including Pakistan, rice-producing countries are producing rice husk as an agricultural waste material. Approximately 500 million metric tons is the annual rice production worldwide, of which 10%–20% is rice husk [15]. About 70%–85% organic matter exists in rice husk, which contains sugars, cellulose and lignin, etc., while silica is also present in the cellular membrane [16]. Rice husk could be a good material for the removal of Cr(VI) from contaminated water because different surface functional groups, including hydroxyl and carboxyl groups, are present on the surface of rice husk [17,18]. Although literature is available in which rice husk was used for the removal of aqueous Cr(VI), more work is required to reveal the optimum environmental conditions, such as a relatively higher pH (close to natural water) for maximum efficiency. Moreover, to our knowledge, in Pakistan limited/no data exists in which rice husk was used as an adsorbent for Cr(VI) removal from wastewater, particularly in the presence of co-occurring ions such as sulfate, phosphate and nitrate. Thus, the potential aim of this study was the elimination of Cr(VI) from contaminated water by applying rice husk as an adsorbent. The effect of various environmental parameters, such as pH, sorbent dose, initial metal concentration, contact time and co-occurring ions, was studied. Finally, the mechanism involved in the adsorption of Cr(VI) from wastewater by rice husk was revealed using isotherm and kinetic modeling as well as the FTIR spectroscopy technique.

2. Materials and Methods

2.1. Experimental Chemicals and Analysis

A stock solution of Cr(VI) (up to 4000 mg L^{-1}) for a batch adsorption system was prepared by adding a defined quantity of K$_2$Cr$_2$O$_7$ to deionized water (DW). Glassware and plasticware were washed completely with DW followed by soaking in 3% HNO$_3$ solution. The study was carried out in triplicates. The Cr(VI) concentration in sorption solutions was determined using an atomic absorption spectrometer (AAS) (Model, Nova 350, Analytik Jena).

2.2. Biosorbent Material Collection and Preparation

The material (rice husk) for biosorbent preparation was collected from a rice mill. It was sun-dried and oven-dried (65 °C) for 2 and 3 days, respectively. The material was ground thoroughly after drying and sieved (size < 250 μm).

2.3. Chromium Adsorption Using Batch System

Experiments were conducted in 50 mL falcon plastic vials, and NaCl solution having 0.01 M concentration was employed as a background working solution. The shaking time and experimental temperature were applied as 2 h and 20 ± 2 °C, respectively, during adsorption studies. Specifically, to check the pH effect spanning 3 to 10, 25 mL working solution was prepared with Cr(VI) concentration of 120 mg L^{-1} and 0.6 g L^{-1} rice husk dosage. The required pH for each working solution was adjusted using either (0.01 M) HCl solution or (0.01 M) NaOH solution. The working solutions were agitated for 2 h followed by centrifugation and filtration. The remaining concentration of Cr(VI) was estimated using AAS as described above.

To conduct the isotherm study, initial concentrations of Cr(VI) were varied from 10 to 250 mg L^{-1} with 2 h of shaking time and rice husk dosage of 0.6 g L^{-1}. The pH in the experiment was maintained at 5.2 (found from pH experiments). The remaining concentration of Cr(VI) was estimated using AAS as described above.

The influence of sorbent dosage was determined using a defined amount of rice husk spanning 0.1 to 1.3 g L^{-1}; i.e., 0.1, 0.3, 0.6, 0.9, 1.1, 1.3 g L^{-1}. The adsorption solution pH was set at 5.2 while the applied Cr(VI) concentration was 120 mg L^{-1}. The mixture was agitated for 2 h followed by centrifugation and careful filtration. The remaining Cr(VI) in the samples was analyzed on AAS.

In the kinetic study, 5.2 pH level, 120 mg L^{-1} Cr(VI) concentration and 0.6 g L^{-1} rice husk dosage were used. The sorption solution mixture was agitated for contact time spanning 0.016 to 24 h. The remaining Cr(VI) in the samples was analyzed on AAS.

In order to check the influence of co-occurring ions (sulfate, nitrate and phosphate) in the water, 50 mg L^{-1} of each ion (sulfate, nitrate and phosphate) was added to the sorption working solution. The remaining Cr(VI) in the sorption solution was estimated on AAS.

The removal rate of Cr(VI) by rice husk was computed as follows (Equation (1)):

$$\% \text{ Cr removal} = \frac{C_o - C_e}{C_o} \times 100 \tag{1}$$

The q_e, Cr(VI) adsorbed (mg g^{-1}) was obtained as shown in Equation (2):

$$qe = \frac{(C_o - C_e)V}{m} \tag{2}$$

In both the equations, equilibrium Cr(VI) concentration (mg L^{-1}), initial Cr(VI) concentration (mg L^{-1}), volume of working solution (L), and biomass of rice husk (g) are illustrated by C_e, C_o, V and m, respectively.

2.4. Modeling and Statistical Analysis

The modeling of data obtained from kinetic and isotherm experiments was carried out using Microsoft$^\circledR$ Excel 2010 and Sigma Plot version 10 [12].

2.5. Analyses of Rice Husk Using FTIR Spectroscopy

The functional groups present on the rice husk surface were identified by collecting the FTIR spectra (TENSOR-II, Bruker, Ettlingen, Germany) spanning 400–4000 cm^{-1} wavenumber.

3. Results and Discussion

3.1. Influence of Varying pH Levels

The pH of the sorption working solution is a very important parameter during the adsorption process. The solution pH can have a direct effect by influencing the physicochemical features (diffusion process, surface charge, surface binding and speciation of metals) of both adsorbate and adsorbent [19,20]. Variations in pH could alter speciation of heavy metals and the surface charge of adsorbents.

Rice husk as sorbent was analyzed with respect to different values of pH levels during Cr(VI) sorption. The results showed that the adsorbent's (rice husk's) capacity to remove the specific quantity of Cr(VI) was dependent upon the pH value of the solution. Figure 1a shows that from pH value 3 to 5.2, the Cr(VI) sorption percentage was increased and reached at maximum value of 78.6% (Table S1). At a relatively low pH value (5.2), the hydroxyl (OH$^-$) ions having negative charge were neutralized with excess H$^+$ ions (positive charges) and the adsorption onto rice husk succeeded by enabling the diffusion of dichromate ions.

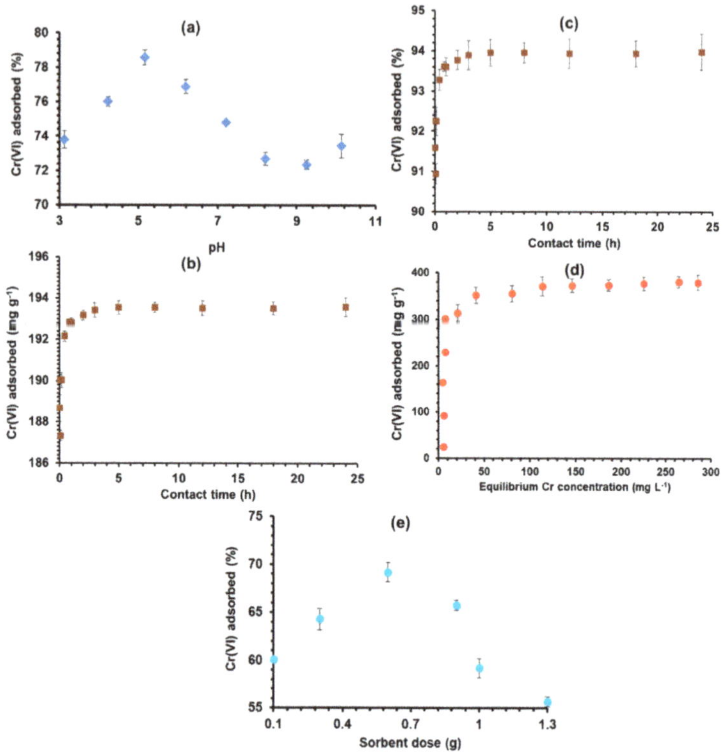

Figure 1. The Cr(VI) adsorption as a function of pH (**a**), contact time (**b,c**), initial Cr concentration (**d**) and sorbent dose (**e**) by rice husk from Cr(VI)-contaminated water.

The species of Cr(VI) exists as $HCrO_4^-$, $HCr_2O_7^-$, H_2CrO_4 (pH < 1), $Cr_2O_7^{2-}$ and CrO_4^{2-} (pH 2–6) depending upon the pH of the solution, thus the pH could play a major role in metal ions' sorption onto biosorbents [21]. In this study, an increase in pH of the solution caused transformation of $HCrO_4^-$ to CrO_4^{2-} and $Cr_2O_7^{2-}$, which could be due to the considerable interaction of negatively charged ions of Cr(VI) with positive charges from the biosorbent. Sorption of Cr(VI) declined with the additional rise in the solution pH level (>5.2) due to the existence of hydroxyl ions (OH^-) in a higher amount at increased pH and thus biosorbent capacity decreased. Moreover, the surface of the adsorbent becomes negatively charged with the presence of a higher amount of OH^- ions, thus decreasing Cr(VI) ions' diffusion, which results in the production of negative repulsive forces between the biosorbent and Cr(VI) ions [16,19,20].

3.2. Contact Time

The optimum sorption solution agitation time was determined for better performance of rice husk in removing Cr(VI). Figure 1b,c describes the contact time study for the percentage removal and average uptake of Cr(VI) by rice husk.

Figure 1b shows the ability of rice husk in mg g^{-1} for successful removal of Cr(VI) with a range of contact times. It was observed that the rice husk's adsorption capacity was 192.82 mg g^{-1} at 1 h agitation time and improved a little (193.42 mg g^{-1}) when equilibrium was obtained at 2 h agitation time and became almost constant afterwards (Table S2). To reach equilibrium, sorption kinetics have two phases, a rapid stage and slower stage [20,22–24]. At the beginning, adsorption was very fast and removal percentage of Cr(VI) increased quickly for 1 h. The percentage of Cr(VI) removed by rice husk reached

93.60% in 1 h and, after achieving equilibrium at 2 h contact time, the removal of Cr(VI) increased slightly further (93.89%) as shown in Figure 1c.

The increasing contact time after 2 h had a negligible effect on Cr(VI) adsorption. At contact time of 2 h, the maximum Cr(VI) removal obtained showed that in the sorption process the Cr(VI) uptake capacity of rice husk was more at the beginning of the adsorption process owing to availability of abundant cavities for capturing Cr(VI) ions. Moreover, at the start, the excess of active binding sites enhanced the rate of sorption, which was decreased with increasing agitation time (<2 h) due to fraction surface lessening and strong competition among Cr(VI) ions for adsorption sites [12,25,26].

Ali et al. [27] also observed that the maximum sorption percentage removal of Cr(VI) was achieved at 100 and 120 min contact time, while no further increase in Cr(VI) removal was detected with the rise in contact time.

3.3. Initial Chromium Concentration

The initial concentration of sorbate is also a critical parameter since it produces a driving force which transfers the sorbate ions onto the biosorbent from liquid solution [19].

Different Cr(VI) ion concentrations spanning from 10 to 250 mg L^{-1} were used for this study. Figure 1d shows that the sorption capacity of rice husk was enhanced from 24.31 to 351.92 mg g^{-1} with rising Cr(VI) level from 10 to 120 mg L^{-1} (Table S3). When the initial concentration of Cr(VI) reached 40 mg L^{-1}, the rise in adsorption rate became relatively slow, even at the equilibrium level of 120 mg L^{-1} [19].

The Cr(VI) ions' sorption was at a minimum when the Cr(VI) level was low but a significant sorptive reaction among sorption sites and Cr(VI) ions was favored by the increase in Cr(VI) level until equilibrium was reached at 120 mg L^{-1} [28]. Afterwards, increasing the initial Cr(VI) level had negligible effect on Cr(VI) adsorption by rice husk.

3.4. Sorbent Dosage

The sorbent dose effect was investigated with 120 mg L^{-1} initial concentration of Cr(VI), 5.2 solution pH and 2 h contact time with 0.1 to 1.3 g L^{-1} rice husk dose. Results showed that with the rising dosage of rice husk, the removal rate of Cr(VI) was increased from 60.07% to 69.18% as shown in Figure 1e and Table S4. Due to the increase in rice husk dose (up to 0.6 g L^{-1}), higher surface area and exchangeable sites were available for Cr(VI) but the Cr(VI) percentage elimination was decreased considerably from 65.75% to 55.65% as the rice husk dosage was increased from 0.9 to 1.3 g L^{-1}. This reduction in Cr(VI) removal could be because of the overcrowding/overlapping of biosorbent particles, thus reducing the number of binding sites for Cr(VI) ions [29,30].

3.5. Co-occurring Ions

Industrial wastewater contains many other anions such as $NO_3{}^-$, $PO_4{}^{-3}$ and $SO_4{}^{-2}$ and sorption of Cr(VI) could be disturbed in the presence of these anions. Figure 2 shows the co-occurring ions' effect on sorption of Cr(VI) in water: it was found that the occurrence of $PO_4{}^{-3}$, $NO_3{}^-$ and $SO_4{}^{-2}$ seriously affected the Cr(VI) sorption by rice husk.

During the Cr adsorption study with rice husk, the greatest decline in Cr(VI) sorption percentage (60.12%) was owing to the co-existence of $SO_4{}^{-2}$ ions. $NO_3{}^-$ and $PO_4{}^{-3}$ ions' presence also showed reduction in the percentage Cr(VI) removal to 72.01% and 68.5%, respectively, but the Cr(VI) sorption in the existence of $PO_4{}^{-3}$ and $NO_3{}^-$ was higher as compared to $SO_4{}^{-2}$. From the results, it can be concluded that the sorption potential of rice husk was significantly affected when $SO_4{}^{-2}$ ions were present compared to $NO_3{}^-$ and $PO_4{}^-$ ions.

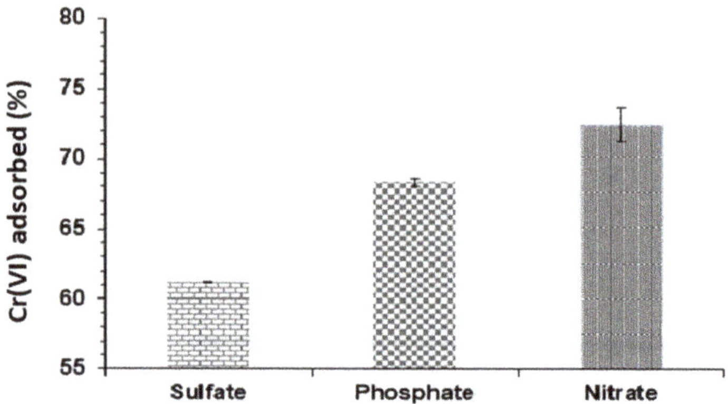

Figure 2. The Cr(VI) adsorption (%) in the occurrence of competing ions.

3.6. Sorption Kinetic Modeling

Various mathematical models could evaluate the biosorption kinetics. Under various experimental conditions, the batch sorption behavior could be explained clearly by a kinetics model [31,32]. Thus, to determine the rate of biosorption in a batch system, the kinetics models pseudo-first order (PFO) and pseudo-second order (PSO) were applied. Figure 3 demonstrates PFO and PSO for rice husk as a sorbent while Table 1 represents the constant rate (k1, k2), regression coefficients (R^2) and qe values of rice husk as a biosorbent. It was observed that the PSO model showed a better fit to experimental kinetic results compared to PFO. The results of modeling showed that the R^2 value of PSO (0.99) was higher than that of PFO (0.75) and the q_e of PSO was also close to the experimental data. Considering the sorption mechanism, Cr(VI) sorption by rice husk was due to the chemisorption process as PSO was a more appropriate fit to the kinetic equilibrium data [19].

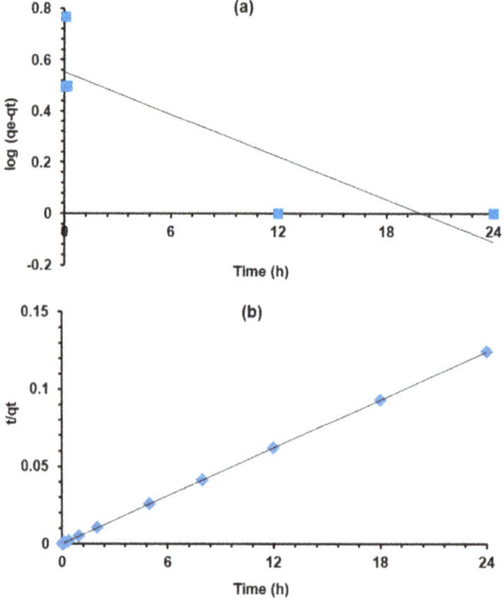

Figure 3. Kinetic modeling with (**a**) pseudo-first order (PFO) and (**b**) pseudo-second order (PSO) data for Cr(VI) adsorption onto rice husk.

Table 1. Parameters of kinetic modeling for Cr(VI) elimination by rice husk.

Pseudo-First Order	Pseudo-Second Order
k_1 (min^{-1})	k_2 $(\text{g mg}^{-1}\,\text{min}^{-1})$
0.02	0.005

3.7. Sorption Isotherm Modeling

Four non-linear isotherms were employed to delineate the Cr(VI) adsorption onto rice husk as shown in Figure 4 and Table 2.

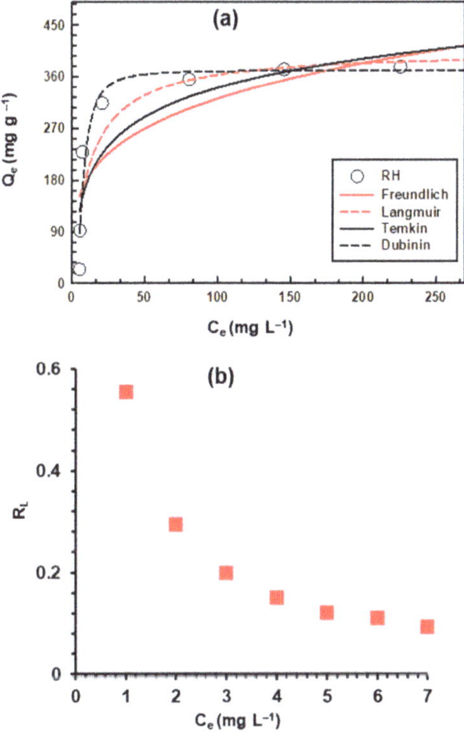

Figure 4. Isotherm models (**a**) and separation factor (**b**) for Cr(VI) adsorption onto rice husk. C_e, equilibrium concentration; C_0, initial concentration.

The Q_L value obtained from the Langmuir model was observed to be 33.68 mg g^{-1} for Cr(VI) adsorption onto rice husk. The Langmuir model fitting demonstrated that the Cr(VI) sorption by rice husk from wastewater was mainly owing to monolayer sorption, which was controlled by chemisorption [17]. The separation factor (R_L) demonstrated that Cr(VI) sorption by rice husk is a highly favorable reaction ($R_L \leq 1$), as shown in Figure 4b.

The Freundlich isotherm model showed that R^2 and Q_f values were 0.70 and 99.5 mg^{1-n} g^{-1} L^n while the Temkin model demonstrated that the R^2 value was 0.77 for Cr(VI) sorption onto rice husk. Lower heat of sorption (b) was observed with the Temkin model for Cr(VI) sorption, which indicates that a linear decrease in (b) value established a superior coverage of Cr(VI) on the rice husk surface, as shown in Table 2.

Table 2. Parameters of isotherm modeling for Cr(VI) elimination by rice husk.

Isotherm Modelr	Parameters	Obtained Value
Langmuir	Q_L (mg g^{-1})	33.68
	R^2	0.84
	K_L (L g^{-1})	0.08
Freundlich	Q_F (mg^{1-n} g^{-1} L^n)	99.5
	R^2	0.70
	n	0.25
Temkin	b	1.04
	R^2	0.77
	A	73.28
Dubinin–Redushkevich	Q_D (mg g^{-1})	371.73
	R^2	0.93
	E (kJ g^{-1})	0.02

The Dubinin–Radushkevich model fit for Cr(VI) sorption is shown in Figure 4a. The Dubinin–Radushkevich model showed a higher R^2 value for rice husk. In Dubinin–Redushkevich model bonding, energy (*E*) was calculated to clarify the Cr(VI) adsorption onto rice husk (Table 2). The sorption process is physical if $E < 8$ kJ g^{-1}; it might be illustrated by diffusion mechanism if $E > 16$ kJ g^{-1} or chemisorption when $E = 8–16$ kJ g^{-1}. In the current study, the value of *E* was 0.02 kJ g^{-1}, hence the dominant mechanism could be physical sorption [32]. It can be deduced from the isotherm results that Cr(VI) adsorption onto rice husk was mainly due to physical sorption and the monolayer sorption process. It was found that the best fitting model was the Dubinin–Radushkevich model followed by the Langmuir model compared to the other two models for Cr(VI) adsorption by rice husk, as indicated by higher R^2 values (0.93 and 0.84, respectively).

3.8. Cr (VI) Biosorption Mechanism through FTIR Spectroscopy

Surface functional groups involved in Cr(VI) sorption onto the biosorbent surface were quantitatively analyzed, and for this purpose FTIR spectra of Cr(VI) loaded and unloaded biosorbent were obtained. Figure 5 describes the FTIR spectra of rice husk (a) no Cr and (b) Cr adsorbed.

In Figure 5, the peaks at 3354 and 3377 cm^{-1} of natural rice husk and Cr(VI) adsorbed rice husk, respectively, indicated the –OH stretching vibrations with the association of macromolecules (pectin, lignin and cellulose) [33–35]. The C-H bands were recorded at 2880 and 2884 cm^{-1} of natural rice husk and Cr(VI)-adsorbed rice husk, respectively, which might be due to the methoxy, methyl and methylene functional groups [36]. The absorption peaks at 1280, 1508 and 1682 cm^{-1} of natural rice husk whereas peaks at 1295, 1541 and 1683 cm^{-1} of Cr(VI)-adsorbed rice husk showed the involvement of ionic carboxylic groups (COO–). There were bands at 706 and 777 cm^{-1} of natural rice husk while those at 738 and 835 cm^{-1} of Cr(VI) rice husk were due to the vibrations of the –NH$_2$ group [29].

The FTIR spectral peaks of natural rice husk at 3354, 2880, 1651, 1507, 1280, 777 and 706 cm^{-1} were shifted to the peaks 3377, 2884, 1685, 1541, 1295, 835 and 738 cm^{-1} after loading of Cr(VI). The changes in spectral bands' positions might be ascribed to the Cr(VI) sorption owing to involvement of functional groups on the biosorbent surface through the ion exchange mechanism [37]. Results showed that functional groups –OH and C-H were mainly involved in the process of sorption with highest peaks, while the other functional groups, such as COO– and –NH$_2$, were also involved with low peaks. The same responses with almond shell and apricot shell were recorded previously regarding the sorption of Cr(VI) [38]. Rice straw, rice bran and sawdust were also observed as a biosorbent for sorption of Cr(VI), and similar results of functional group analyses were noted [39].

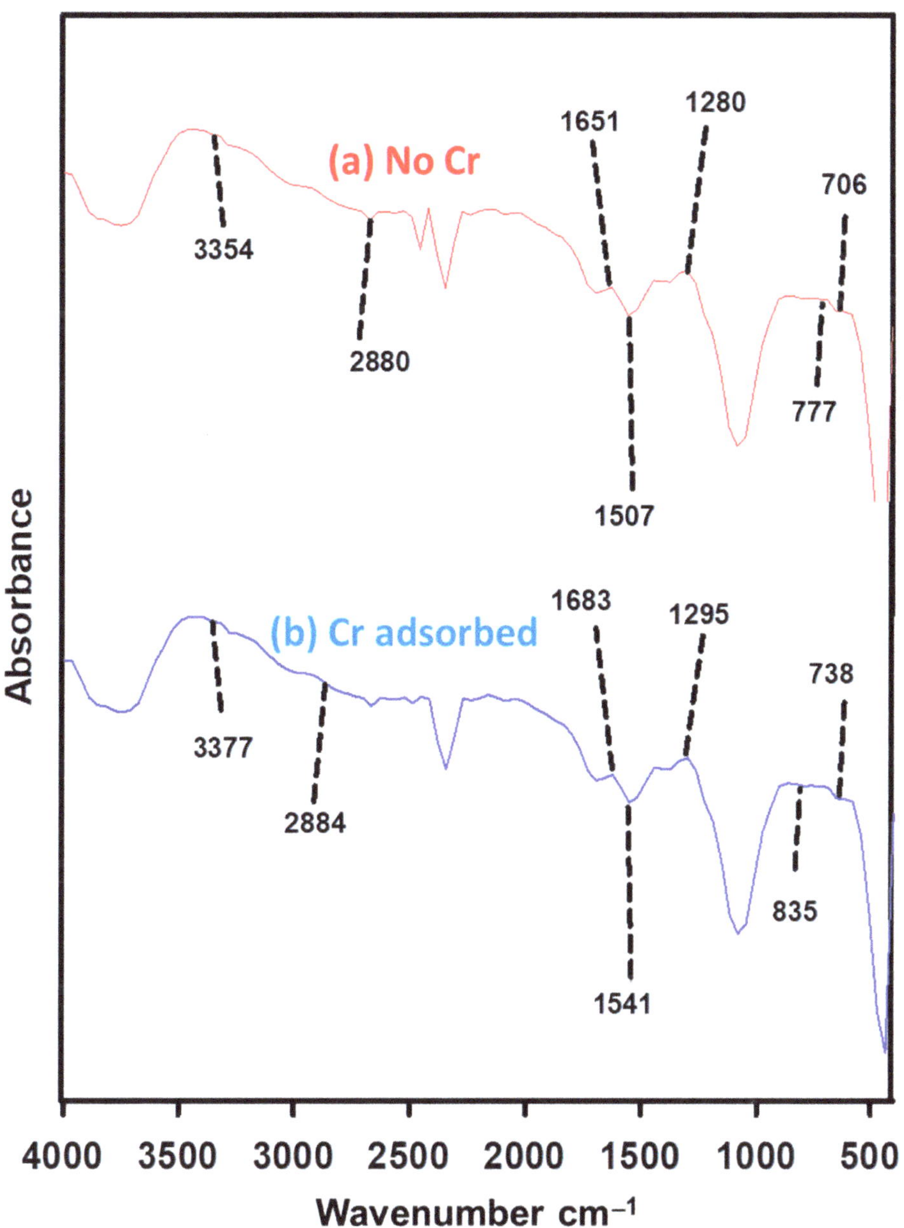

Figure 5. The FTIR spectra of rice husk (**a**) no Cr(VI); (**b**) Cr(VI)-adsorbed.

Table 3 compares the adsorption capacities of rice husk used in this study with those in other studies. It can be observed from Table 3 that rice husk showed relatively higher Cr(VI) adsorption potential compared to biosorbents previously used by several researchers.

Table 3. Comparison of various biosorbents with rice husk for Cr(VI) adsorption.

Biosorbent	Adsorption (mg g^{-1})	References
Litchi peel	7.05	Yi et al. [40]
Foxtail millet shell	11.70	Peng et al. [41]
Grapefruit peelings	39.06	Rosales et al. [42]
Freshwater snail shell	8.85	Vu et al. [43]
Pomegranate seeds	3.32	Ghaneian et al. [44]
Bamboo shoot shell	28.72	Hu et al. [45]
Rose biomass	5.26	Aman et al. [46]
Waste *Chlorella vulgaris* biomass	43.3	Xie et al. [47]
Rice husk	351.92	This study

4. Conclusions

Rice husk was used successfully for Cr(VI) biosorption from contaminated water streams. The results showed that rice husk removed 78.6% Cr(VI) from the wastewater and the highest Cr(VI) elimination was found at 5.2 pH. The data obtained from various studied parameters showed that 2 h contact time, 120 mg L^{-1} initial metal concentration and 0.6 g L^{-1} rice husk dosage were perfect for efficient adsorption of Cr(VI) from polluted water. Dubinin–Radushkevich and Langmuir models provided the best fit to the equilibrium data indication that the Cr(VI) onto rice husk was due to physical and monolayer sorption processes. The pseudo-second order model fitting proposed that adsorption of Cr(VI) was a fast process. The data from the effect of co-occurring ions showed that the presence of sulfate ions during sorption of Cr(VI) reduced the rice husk's capacity for Cr(VI) adsorption. The FTIR analysis suggested that –NH$_2$, alkyl, –OH and COO– functional groups were the main species involved in Cr(VI) removal. Overall, since rice husk used here was an agricultural waste which is inexpensive, accessible and proved to be an effective sorbent, therefore, this prepared biosorbent is expected to be economically feasible for the treatment of Cr(VI)-containing wastewater.

Supplementary Materials: The following are available online at https://www.mdpi.com/2073-4441/13/3/263/s1, Table S1: Effect of solution pH on Cr adsorption by rice husk; Table S2: Effect of contact time on Cr adsorption by rice husk; Table S3: Effect of initial Cr concentration on Cr adsorption by rice husk; Table S4: Effect of sorbent dose on Cr adsorption by rice husk.

Author Contributions: Conceptualization, S.A. and M.B.S.; methodology, U.K.; software, M.B.S. and U.K.; validation, A.A.A. and M.R.; formal analysis, S.A. and M.B.S.; investigation, U.K.; resources, S.A. and S.R.A.; data curation, U.K.; writing—original draft preparation, U.K. and M.B.S.; writing—review and editing, M.B.S. and M.R.; visualization, U.K. and M.N.A.; supervision, S.A. and M.B.S.; project administration, M.B.S., A.A.A. and M.N.A.; funding acquisition, M.B.S., A.A.A. and M.N.A. All authors have read and agreed to the published version of the manuscript.

Funding: This work was funded by Researchers Supporting Project number (RSP 2020/236), King Saud University, Riyadh, Saudi Arabia. The authors are also grateful to the Government College University, Faisalabad, Pakistan for its support. This work was also funded by the Higher Education commission, Pakistan.

Institutional Review Board Statement: Not Applicable.

Informed Consent Statement: Not Applicable.

Data Availability Statement: Not Applicable.

Acknowledgments: The authors are grateful to the University of the Punjab, Lahore, Pakistan for its support. This work was also funded by Researchers Supporting Project number (RSP 2020/236), King Saud University, Riyadh, Saudi Arabia.

Conflicts of Interest: The authors declare no conflict of interest.

References

1. Ahalya, N.; Kanamadi, R.D.; Ramachandra, T.V. Biosorption of chromium (VI) from aqueous solutions by the husk of Bengal gram (Cicer arientinum). *Electr. J. Biotechnol.* **2005**, *8*, 258–264. [CrossRef]
2. Dhanakumar, S.; Solaraj, G.; Mohanraj, R.; Pattabhi, S. Removal of Cr (VI) from aqueous solution by adsorption using cooked tea dust. *Ind. J. Sci. Technol.* **2007**, *1*, 1–6. [CrossRef]
3. Mahvi, A.H.; Naghipour, D.; Vaezi, F.; Nazmara, S. Teawaste as an adsorbent for heavy metal removal from industrial wastewaters. *Am. J. Appl. Sci.* **2005**, *2*, 372–375. [CrossRef]
4. Gopalakrishnan, S.; Kannadasan, T.; Velmurugan, S.; Muthu, S.; Vinoth Kumar, P. Biosorption of chromium (VI) from industrial effluent using neem leaf adsorbent. *Res. J. Chem. Sci.* **2013**, *3*, 48–53.
5. Monnot, A.D.; Christian, W.V.; Paustenbach, D.J.; Finley, B.L. Correlation of blood Cr(III) and adverse health effects: Application of PBPK modeling to determine non-toxic blood concentrations. *Crit. Rev. Toxicol.* **2014**, *44*, 618–637. [CrossRef]
6. Levankumar, L.; Uthukumaran, V.; Gobinath, M. Batch adsorption and kinetics of chromium(VI) removal from aqueous solutions by *Ocimum americanum* L. seed pods. *J. Hazard. Mater.* **2009**, *161*, 709–713. [CrossRef] [PubMed]
7. Dave, P.N.; Pandey, N.; Thomas, H. Adsorption of Cr (VI) from aqueous solutions on tea waste and coconut husk. *Indian J. Chem. Technol.* **2012**, *19*, 111–117.
8. Aman, T.; Kazi, A.A.; Sabri, M.U.; Bano, Q. Potato peels as solid waste for the removal of heavy metal copper (II) from waste water/industrial effluent. *Colloids Surfaces B Biointerfaces* **2008**, *63*, 116–121. [CrossRef]
9. Elmolla, E.S.; Hamdy, W.; Kassem, A.; Abdel Hady, A. Comparison of different rice straw based adsorbents for chromium removal from aqueous solutions. *Desalin. Water Treat.* **2016**, *57*, 6991–6999. [CrossRef]
10. Chakraborty, R.; Asthana, A.; Singh, A.K.; Jain, B.; Susan, A.B.H. Adsorption of heavy metal ions by various low-cost adsorbents: A review. *Int. J. Environ. Anal. Chem.* **2020**, 1–38. [CrossRef]
11. Afroze, S.; Sen, T.K. A review on heavy metal ions and dye adsorption from water by agricultural solid waste adsorbents. *Water Air Soil Pollut.* **2018**, *229*, 225. [CrossRef]
12. Shakoor, M.B.; Niazi, N.K.; Bibi, I.; Murtaza, G.; Kunhikrishnan, A.; Seshadri, B.; Shahid, M.; Ali, S.; Bolan, N.S.; Ok, Y.S.; et al. Remediation of arsenic-contaminated water using agricultural wastes as biosorbents. *Crit. Rev. Environ. Sci. Technol.* **2016**, *46*, 467–499. [CrossRef]
13. Ahmed, I.; Attar, S.J.; Parande, M.G. Removal of Hexavalent Chromium (Cr (VI)) from Industrial Wastewater by Using Biomass Adsorbent (Rice Husk Carbone). *Int. J. Adv. Eng. Stud.* **2012**, *1*, 92–94.
14. Park, D.; Yun, Y.S.; Ahn, C.K.; Park, J.M. Kinetics of the reduction of hexavalent chromium with the brown seaweed Ecklonia biomass. *Chemosphere* **2007**, *66*, 939–946. [CrossRef] [PubMed]
15. Bello, O.S.; Atoyebi, O.M.; Adegoke, K.A.; Fehintola, E.O.; Ojo, A.O. Removal of toxicant chromium (VI) from aqueous solution using different adsorbents. *J. Chem. Soc. Pak.* **2015**, *37*, 190.
16. Vempati, R.K.; Musthyala, S.C.; Mollah, M.Y.A.; Cocke, D.L. Surface analyses of pyrolysed rice husk using scanning force microscopy. *Fuel* **1995**, *74*, 1722–1725. [CrossRef]
17. Gao, W.; Li, H.; Karnowo; Song, B.; Zhang, S. Integrated Leaching and Thermochemical Technologies for Producing High-Value Products from Rice Husk: Leaching of Rice Husk with the Aqueous Phases of Bioliquids. *Energies* **2020**, *13*, 6033. [CrossRef]
18. Basri, M.S.M.; Mustapha, F.; Mazlan, N.; Ishak, M.R. Optimization of Rice Husk Ash-Based Geopolymers Coating Composite for Enhancement in Flexural Properties and Microstructure Using Response Surface Methodology. *Coatings* **2020**, *10*, 165. [CrossRef]
19. Khalil, U.; Shakoor, M.B.; Ali, S.; Rizwan, M. Tea Waste as a Potential Biowaste for Removal of Hexavalent Chromium from Wastewater: Equilibrium and kinetic studies. *Arab. J. Geosci.* **2018**, *11*, 573. [CrossRef]
20. Shakoor, M.B.; Nawaz, R.; Hussain, F.; Raza, M.; Ali, S.; Rizwan, M.; Ahmad, S. Human health implications, risk assessment and remediation of As-contaminated water: A critical review. *Sci. Total Environ.* **2017**, *601*, 756–769. [CrossRef]
21. Ucun, H.; Bayhan, Y.K.; Kaya, Y.; Cakici, A.; Algur, O.F. Biosorption of chromium (VI) from aqueous solution by cone biomass of Pinus sylvestris. *Bioresour. Technol.* **2002**, *85*, 155–158. [CrossRef]
22. ALOthman, Z.A.; Naushad, M.; Ali, R. Kinetic, equilibrium isotherm and thermodynamic studies of Cr (VI) adsorption onto low-cost adsorbent developed from peanut shell activated with phosphoric acid. *Environ. Sci. Pollut. Res.* **2013**, *20*, 3351–3365. [CrossRef] [PubMed]
23. Singanan, M. Removal of lead (II) and cadmium (II) ions from wastewater using activated biocarbon. *Sci. Asia* **2011**, *37*, 115–119. [CrossRef]
24. Sattar, M.S.; Shakoor, M.B.; Ali, S.; Rizwan, M.; Niazi, N.K.; Jilani, A. Comparative efficiency of peanut shell and peanut shell biochar for removal of arsenic from water. *Environ Sci Pollut Res.* **2019**, *26*, 18624–18635. [CrossRef] [PubMed]
25. Tajernia, H.; Ebadi, T.; Nasernejad, B.; Ghafori, M. Arsenic removal from water by sugarcane bagasse: An application of response surface methodology (RSM). *Water Air Soil Pollut.* **2014**, *225*, 2028. [CrossRef]
26. Okafor, P.C.; Okon, P.U.; Daniel, E.F.; Ebenso, E.E. Adsorption capacity of coconut (*Cocos nucifera* L.) shell for lead, copper, cadmium and arsenic from aqueous solutions. *Int. Electrochem. Sci.* **2012**, *7*, 12354–12369.
27. Ali, I.H.; Al Mesfer, M.K.; Khan, M.I.; Danish, M.; Alghamdi, M.M. Exploring Adsorption Process of Lead (II) and Chromium (VI) Ions from Aqueous Solutions on Acid Activated Carbon Prepared from Juniperus procera Leaves. *Processes* **2019**, *7*, 217. [CrossRef]
28. Razmovski, R.; Šćiban, M. Biosorption of Cr (VI) and Cu (II) by waste tea fungal biomass. *Ecol. Eng.* **2008**, *34*, 179–186. [CrossRef]

29. Abid, M.; Niazi, N.K.; Bibi, I.; Farooqi, A.; Ok, Y.S.; Kunhikrishnan, A.; Ali, F.; Ali, S.; Igalavithana, A.D.; Arshad, M. Arsenic (V) biosorption by charred orange peel in aqueous environments. *Int. J. Phytoremediation* **2016**, *18*, 442–449. [CrossRef]

30. Sahmoune, M.N.; Louhab, K.; Boukhiar, A. Advanced biosorbents materials for removal of chromium from water and wastewaters. *Environ. Prog. Sustain. Energy* **2011**, *30*, 284–293. [CrossRef]

31. Pehlivan, E.; Tran, H.T.; Ouedraogo, W.K.I.; Schmidt, C.; Zachmann, D.; Bahadir, M. Sugarcane bagasse treated with hydrous ferric oxide as a potential adsorbent for the removal of As(V) from aqueous solutions. *Food Chem.* **2013**, *138*, 133–138. [CrossRef] [PubMed]

32. Ahmad, M.; Lee, S.S.; Rajapaksha, A.U.; Vithanage, M.; Zhang, M.; Cho, J.S.; Lee, S.E.; Ok, Y.S. Trichloroethylene adsorption by pine needle biochars produced at various pyrolysis temperatures. *Bioresour. Technol.* **2013**, *143*, 615–622. [CrossRef] [PubMed]

33. Nadeem, R.; Manzoor, Q.; Iqbal, M.; Nisar, J. Biosorption of Pb (II) onto immobilized and native Mangifera indica waste biomass. *J. Ind. Eng. Chem.* **2015**, *35*, 185–195. [CrossRef]

34. Garg, U.K.; Kaur, M.P.; Garg, V.K.; Sud, D. Removal of hexavalent chromium from aqueous solution by agricultural waste biomass. *J. Hazard. Mater.* **2007**, *140*, 60–68. [CrossRef] [PubMed]

35. Ali, S.; Rizwan, M.; Shakoor, M.B.; Jilani, A.; Anjum, R. High sorption efficiency for As (III) and As (V) from aqueous solutions using novel almond shell biochar. *Chemosphere* **2020**, *243*, 125330. [CrossRef] [PubMed]

36. Owalude, S.O.; Tella, A.C. Removal of hexavalent chromium from aqueous solutions by adsorption on modified groundnut hull. *Beni-Suef Univ. J. Basic Appl. Sci.* **2016**, *5*, 377–388. [CrossRef]

37. Vinodhini, V.; Das, N. Mechanism of Cr (VI) biosorption by neem sawdust. *Am.-Eurasian J. Sci. Res.* **2009**, *4*, 324–329.

38. Khazaei, I.; Aliabadi, M.; Hamed, M.H. Use of agricultural waste for removal of Cr(VI) from aqueous solution. *Iran. J. Chem. Eng.* **2011**, *8*, 11–23.

39. Singha, B.; Naiya, T.K.; Kumar, B.A.; Das, S.K. Cr (VI) ions removal from aqueous solutions using natural adsorbents–FTIR studies. *J. Environ. Prot.* **2011**, *2*, 729. [CrossRef]

40. Yi, Y.; Lv, J.; Liu, Y.; Wu, G. Synthesis and application of modified Litchi peel for removal of hexavalent chromium from aqueous solutions. *J. Mol. Liquids* **2017**, *225*, 28–33. [CrossRef]

41. Peng, S.H.; Wang, R.; Yang, L.Z.; He, L.; He, X.; Liu, X. Biosorption of copper, zinc, cadmium and chromium ions from aqueous solution by natural foxtail millet shell. *Ecotoxicol. Environ. Saf.* **2018**, *165*, 61–69. [CrossRef] [PubMed]

42. Rosales, E.; Meijide, J.; Tavares, T.; Pazos, M.; Sanromán, M.A. Grapefruit peelings as a promising biosorbent for the removal of leather dyes and hexavalent chromium. *Proc. Saf. Environ. Prot.* **2016**, *101*, 61–71. [CrossRef]

43. Vu, X.H.; Nguyen, L.H.; Van, H.T.; Nguyen, D.V.; Nguyen, T.H.; Nguyen, Q.T.; Ha, L.T. Adsorption of Chromium (VI) onto Freshwater snail shell-derived biosorbent from aqueous solutions: Equilibrium, kinetics, and thermodynamics. *J. Chem.* **2019**. [CrossRef]

44. Ghaneian, M.T.; Bhatnagar, A.; Ehrampoush, M.H.; Amrollahi, M.; Jamshidi, B.; Dehvari, M.; Taghavi, M. Biosorption of hexavalent chromium from aqueous solution onto pomegranate seeds: Kinetic modeling studies. *Int. J. Environ. Sci. Technol.* **2017**, *14*, 331–340. [CrossRef]

45. Hu, H.; Gao, Y.; Wang, T.; Sun, L.; Zhang, Y.F.; Li, H. Removal of hexavalent chromium, an analogue of pertechnetate, from aqueous solution using bamboo (*Acidosasa edulis*) shoot shell. *J. Radioanal. Nucl. Chem.* **2019**, *321*, 427–437. [CrossRef]

46. Aman, A.; Ahmed, D.; Asad, N.; Masih, R.; Abd ur Rahman, H.M. Rose biomass as a potential biosorbent to remove chromium, mercury and zinc from contaminated waters. *Int. J. Environ. Stud.* **2018**, *75*, 774–787. [CrossRef]

47. Xie, Y.; Li, H.; Wang, X.; Ng, I.S.; Lu, Y.; Jing, K. Kinetic simulating of Cr (VI) removal by the waste Chlorella vulgaris biomass. *J. Taiwan Inst. Chem. Eng.* **2014**, *45*, 1773–1782. [CrossRef]

Article

A New Polyvinylidene Fluoride Membrane Synthesized by Integrating of Powdered Activated Carbon for Treatment of Stabilized Leachate

Salahaldin M. A. Abuabdou [1], Zeeshan Haider Jaffari [2], Choon-Aun Ng [1], Yeek-Chia Ho [3,*] and Mohammed J. K. Bashir [1,*]

[1] Department of Environmental Engineering, Engineering and Green Technology Faculty, Universiti Tunku Abdul Rahman, Kampar 31900, Malaysia; sabdou6@1utar.my (S.M.A.A.); ngca@utar.edu.my (C.-A.N.)

[2] Department of Environmental Engineering and Management, Chaoyang University of Technology, No. 168, Jifeng E. Rd, Wufeng District, Taichung 413310, Taiwan; engr.zeeshanhaiderjaffari@gmail.com

[3] Centre of Urban Resource Sustainability, Department of Civil and Environmental Engineering, Institute of Self-Sustainable Building, Universiti Teknologi PETRONAS, Seri Iskandar 32610, Malaysia

* Correspondence: yeekchia.ho@utp.edu.my (Y.-C.H.); jkbashir@utar.edu.my (M.J.K.B.); Tel.: +60-5-4688888 (ext. 4559) (M.J.K.B.); Fax: +60-5-466-7449 (M.J.K.B.)

Citation: Abuabdou, S.M.A.; Jaffari, Z.H.; Ng, C.-A.; Ho, Y.-C.; Bashir, M.J.K. A New Polyvinylidene Fluoride Membrane Synthesized by Integrating of Powdered Activated Carbon for Treatment of Stabilized Leachate. *Water* **2021**, *13*, 2282. https://doi.org/10.3390/w13162282

Academic Editor: Giovanni Esposito

Received: 11 June 2021
Accepted: 28 July 2021
Published: 20 August 2021

Abstract: Stabilized landfill leachate contains a wide variety of highly concentrated non-biodegradable organics, which are extremely toxic to the environment. Though numerous techniques have been developed for leachate treatment, advanced membrane filtration is one of the most environmentally friendly methods to purify wastewater effectively. In the current study, a novel polymeric membrane was produced by integrating powdered activated carbon (PAC) on polyvinylidene fluoride (PVDF) to synthesize a thin membrane using the phase inversion method. The membrane design was optimized using response surface methodology (RSM). The fabricated membrane was effectively applied for the filtration of stabilized leachate using a cross-flow ring (CFR) test. The findings suggested that the filtration properties of fabricated membrane were effectively enhanced through the incorporation of PAC. The optimum removal efficiencies by the fabricated membrane (14.9 wt.% PVDF, 1.0 wt.% PAC) were 35.34, 48.71, and 22.00% for COD, colour and NH$_3$-N, respectively. Water flux and transmembrane pressure were also enhanced by the incorporated PAC and recorded 61.0 L/m^2·h and 0.67 bar, respectively, under the conditions of the optimum removal efficiency. Moreover, the performance of fabricated membranes in terms of pollutant removal, pure water permeation, and different morphological characteristics were systematically analyzed. Despite the limited achievement, which might be improved by the addition of a hydrophilic additive, the study offers an efficient way to fabricate PVDF-PAC membrane and to optimize its treatability through the RSM tool.

Keywords: stabilized leachate; membrane fabrication; filtration technology; phase inversion technique; powdered activated carbon (PAC)

1. Introduction

Sanitary landfills are the widely applied technique to tackle municipal solid waste (MSW). Inappropriately, the majority of these landfills do not fulfill the normal discharged limits [1]. In developing countries such as Malaysia, more than 80% of the MSW produced was received by open duping and landfill sites [2]. This resulted in the generation of highly contaminated leachate, which is the liquid generated due to the precipitation above these solid litters and could be toxic to the surrounding environment. This leachate could contaminate the sources of fresh water if not carefully treated before discharging to the environment [3]. Stabilized leachate, which is more than ten years old, has lower BOD$_5$/COD ratio. Thus, it is almost impossible to treat this kind of leachate using some biological treatment technique [4]. To date, various purification techniques such as adsorption [5], coagulation [6], advanced oxidation [7], electro-Fenton [8], and combinations of these processes [9,10] have been successfully introduced to eliminate the organic contaminates from stabilized leachate. Among these techniques, membrane filtration could be one of the most suitable purification process [11]. The membranes acted as a selective barrier to achieve the objective of separation and purification. Nonetheless, there are still some shortcomings in membrane technology such as membrane fouling upon the higher contaminant concentration [12]. Fouling could affect the separation efficiency as well as permeability of membrane, which are the vital factors in the membrane filtration [13]. Several strategies, including pre-treatment of feed [14], optimization of operating parameters [15], selection and modification of membrane [16], hydraulic flushing [17], and applied field enhancement [18], have been performed to alleviate membrane fouling and water flux rate. Under different circumstances, the workability of membrane can be improved through the membrane characteristics and performance of treatment process. Hence, investigation of membrane characterization can be separated into four groups: membrane activity (permeability, surface wettability, average pore size, and porosity); morphological characterization (surface chemistry and roughness, and external and internal membrane texture); treatment efficiency (separation performance); and antifouling evaluation (pore size decrease and cake formation) [19].

Synthetic polymers such as polypropylene (PP), polyvinylidene fluoride (PVDF), and polysulfone (PS) are commonly applied in the membrane fabrication due to their higher flux, antifouling ability, and separation efficiency [20]. Among all these synthetic polymers, PVDF polymer proved to be an ideal membrane fabrication material due its durability [21], good thermal stability and higher chemical resistance [22]. Additionally, the PVDF polymer can also help to extend the membrane life, as well as reduce the damage caused by the concentrated pollutants [23]. However, the PVDF membranes antifouling capability could be enhanced due to its hydrophobic nature [24]. Many researchers have successfully applied dry–wet phase inversion technique to boost their membrane performance [25]. For instance, Zhou et al. [26] developed an ultrafiltration PVDF membrane using nanoparticles of titanium dioxide (TiO_2) and polyvinylpyrrolidone (PVP) as blended additives to increase the fouling resistance and water permeability. The addition of PVP-TiO_2 increases the average pore size and porosity of membrane, leading to the higher flux and hydrophilicity of membrane with more than 91.4% removal performance against sulfonamide antibiotics water. Moreover, polyethylene glycol and poly(acrylic acid) were also applied in the fabrication of membrane through chemical reaction with a key focus of enhancing hydrophilicity. Their batch filtration experiments clearly exhibit an increase in critical flux and a declined fouling rate. Similarly, various reports have presented effective ways to boost the antifouling abilities of PVA-based membranes due to their hydrophilic properties [27,28].

Recently, the incorporation of activated carbon (AC) on the surface of the membrane has proven to be an effective way to boost the membrane rejection performance [29]. The utilization of AC in membrane is a relatively new technology for the elimination of organic contaminates for wastewater, which not only enhances the adsorption capacity of AC, but also improves the particle removal capabilities of membrane [30].

To date, there are quite a number of studies which clearly demonstrate that the usage of PAC can significantly improve the filterability of membranes [13]. However, evaluation of PAC addition into PVDF flat sheet membranes with different concentrations, in terms of their treatment efficiency and productivity, has not been investigated. Therefore, the current study was performed to observe the potential of incorporating PAC, for the first time, into the PVDF polymeric membrane for stabilized landfill leachate purification. Furthermore, fabricated membrane was optimized using RSM technique, and the membrane properties and morphologies were systematically characterized.

2. Materials and Methods

2.1. Collection of Leachate

Leachate sample was taken form Sahom landfill site located in Perak, Malaysia, which is an operative landfill site with a daily production of 100 tonnes of MSW in average [31]. After collection of leachate sample, it was stored in a refrigerator at 4 °C. Initial leachate characterization was performed using standardized methods of water and wastewater [32]. All measurements, including dissolved oxygen (DO), colour, chemical oxygen demand (COD), 5-day biochemical oxygen demand (BOD_5), and ammoniacal nitrogen (NH_3-N), were undertaken in triplicate.

2.2. Materials

The PVDF polymer (Kynar®740) was purchased from Afza Maju trading (Terengganu, Malaysia), and utilized after drying for 24 h at 70 °C. 1-Methyl-2-pyrrolidone (NMP, 99.5%) was purchased from Sigma–Aldrich. Methanol, (99.8%) was supplied by Chem Soln. Ultra-pure distilled water (DI) was utilized throughout the experiments. PAC was purchased from R&M Chemicals. The AC was charcoal-based, and consists of sulfide, chloride, calcium, sulphate, iron, lead, zinc, and copper. This PAC density was 1.8–2.1 kg/m^3 with pH (4–7). Particle size analysis (PSA) and field emission scanning electron microscopy (FESEM) tests were used to investigate the distribution and the size of PAC particles, respectively. All these chemical materials were of analytical grade, and used without additional treatment.

2.3. Experiment's Design and Optimization Process

Central Composite Design (CCD) is the design method used in response surface methodology (RSM) for the membrane fabrication's experimental design [33]. Both CCD and RSM were run by version 8 from the Design Expert. For membrane dope solution design, two factors, the polymer (PVDF) weightage and the additive (PAC) weightage, were set into the CCD. Based on preliminary experiments and the extensive literature [34,35], the total mass of fabricated membrane dope was fixed at 100 g, which represents 100% of the dope weight, thus each 1 g of the dope element is equivalent to 1% weightage. The dosage of the PVDF was set within the range of 10 g to 18 g, and the amount of PAC was set within the range of 0 g to 2 g. Regarding the CCD, the alpha value was selected to be 1.0, and thus the centre points were 14.0 and 1.0 wt.% for the polymer content and additive content, respectively. The rest of the dope weight (to complete 100 g) is the NMP solvent. The total concentration of PVDF/PAC was kept at 20% (as maximum) and 10% (as minimum), as concentration higher than 20% resulted in solutions of extremely high viscosity, and was difficult to be casted on the glass plate, while clumsy, non-thick membrane was the result of using concentration less than 10%. Five responses, which are the removing efficiencies of COD, colour, and NH_3-N, as well as maximum transmembrane pressure (max. TMP), and pure water flux, were also set into the CCD to have the full design of experiments. The influence of various parameters was optimized by RSM using a combination of statistical and numerical techniques. In the current work, nine experiments were reinforced with four replications to assess the pure error [36]. The 13 different membranes were applied in double repetition and have their effluent collected. The quadratic model for every response was investigated by analysis of variances (ANOVA) to identify the results significancy, and to find the represented quadratic model after eliminating irrelevant terms. The frontal sign of each model term signifies to either antagonistic or synergistic effect on the response when it is positive or negative, respectively [4]. In RSM, it does mention that Prob > F less than 0.050 indicates model terms are significant, and Prob > F with the values greater than 0.10 indicates model term is not significant. "Not significant", in the description of lack of fit, is regarded a decent model, as it means the experimental reading is fitting the model [37]. Additionally, a good experimentally fitted data will have a higher coefficient (R^2) value. The higher the R^2 value, the closer the experimental data towards the predicted graph model by the RSM [38,39]. Selection of the best membrane takes into consideration the membrane purification performance. Desirability value closer to 1.0 used to be selected as the ideal design for the data.

2.4. PVDF-PAC Membrane Fabrication

2.4.1. Dope Preparation

To produce the polymeric membrane, PVDF and NMP were applied as polymer and solvent, respectively. Figure 1 presents the process used for the dope preparation. Initially, the polymeric PVDF was entirely dissolved in the NMP solvent at a temperature ranged between 60 and 70 °C using a heating mantle (Figure 1a). In order to achieve a better permeate flux of the synthesized membrane, the heating mantle temperature should always be maintained within the above stated range [21]. The dope solution containing dissolved PVDF polymer in the NMP solvent was then infused into a clean Schott bottle. After that, the required amount of PAC was inserted into the dope solution to generate the dope for hybrid membrane. Lastly, the Schott bottle containing the dope solution was placed into a sonicator bath (Cole-Parmer, Vernon Hills, IL, USA) for eight hours to confirm the homogeneous mixing of the additives without any air bubbles raised in the prepared dope [40].

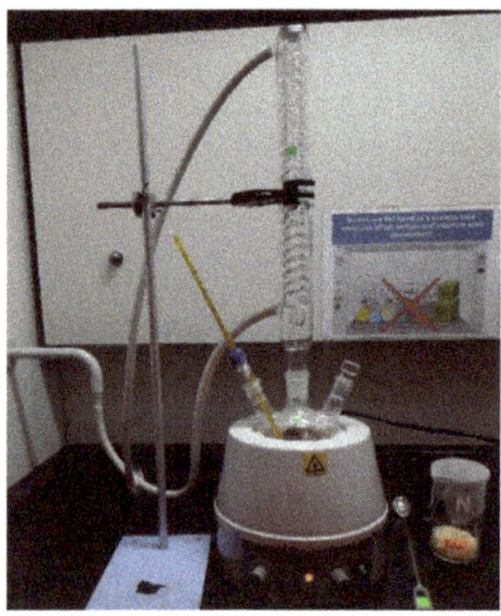

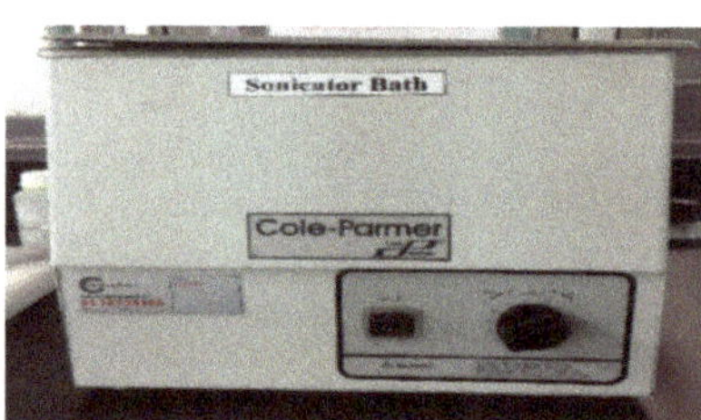

(a) Polymer dope preparation   (b) Adding of PAC   (c) Membrane dope sonicator

Figure 1. PVDF-PAC membrane dope preparation process.

2.4.2. Membrane Casting

A semi- automated membrane casting machine (TECH INC, Chennai, India) was applied to synthesize a flat sheet membrane using the dry–wet phase process, as illustrated in Figure 2a. The membrane was produced at temperature 27 °C to 30 °C with an approximate thickness of 60 μm based on literature reports [41,42]. After 60 s of membrane casting above the glass board, it was submerged into a distilled water (DW) basin for 180 s (Figure 2b). As a result, a thin layered polymeric film was generated, which separated from the glass plate. Later, the newly produced membrane was transferred into a DW coagulation bath and remained there for 24 h. Afterwards, a methanol bath was used for 8 h, as shown in Figure 2c, to perform a post-treatment to ensure the excess solvent in the membrane can be removed completely [43]. Finally, the membrane was dried 24 h at the ambient temperature with 60% humidity, as shown in Figure 2d, to be ready to use in the filtration process [13].

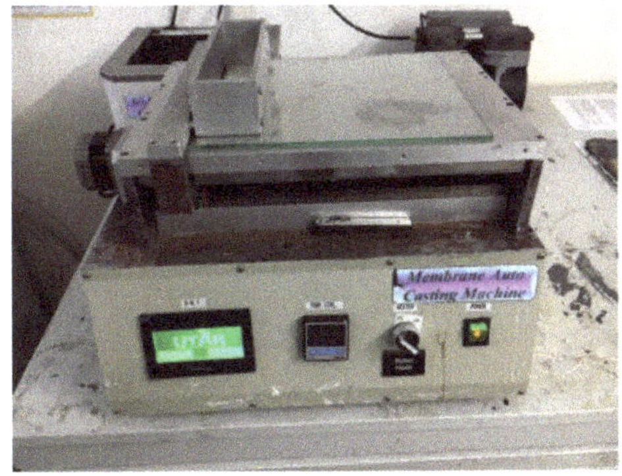

(a) Flat sheet membrane casting

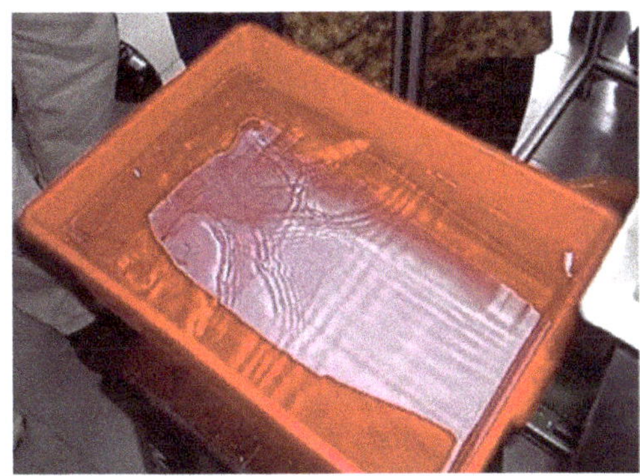

(b) Membrane solidification

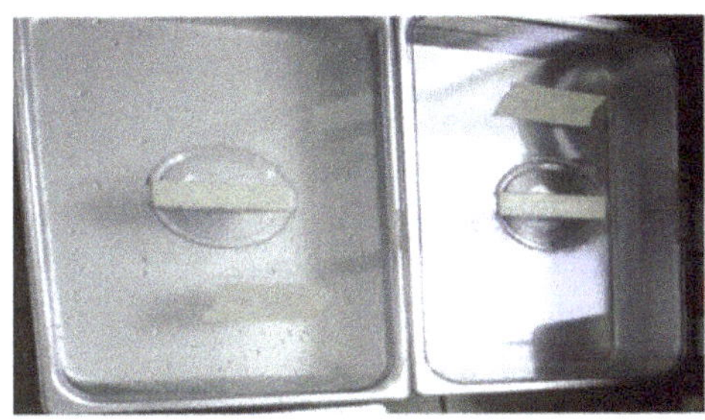

(c) Membrane neutralization

(d) Membrane drying

Figure 2. The casting process of flat sheet PVDF-PAC membrane.

2.5. Membrane Performance and Characterization

The produced membranes have been characterized to investigate their treatment efficiencies, fouling, and permeability properties and surface morphologies. To ensure the accuracy of the findings, all of the tests have been duplicated. Each time, a fresh membrane has been utilized to investigate their characteristics and performance.

2.5.1. Treatment Efficiency

The membrane filtration performance was investigated using laboratory scale cross-flow filtration setup with a 3.34 cm disc diameter, as exhibited in Figure 3. The membrane rejection capabilities were studied against the treatment of landfill leachate. Before each experiment, initial characterization of leachate was measured to eliminate the small errors which occurred due to the minor changes in organics concentration with time. The steady flux for all individual membranes was acquired by a constant (200 mL/min) flow for 120 min. The volume of permeate, along with the recorded transmembrane pressure, were noted down under the flow of 200 mL/min for different intervals of time (0.5, 1, 2, 5, 10, 20, 40, 90, and 120 min).

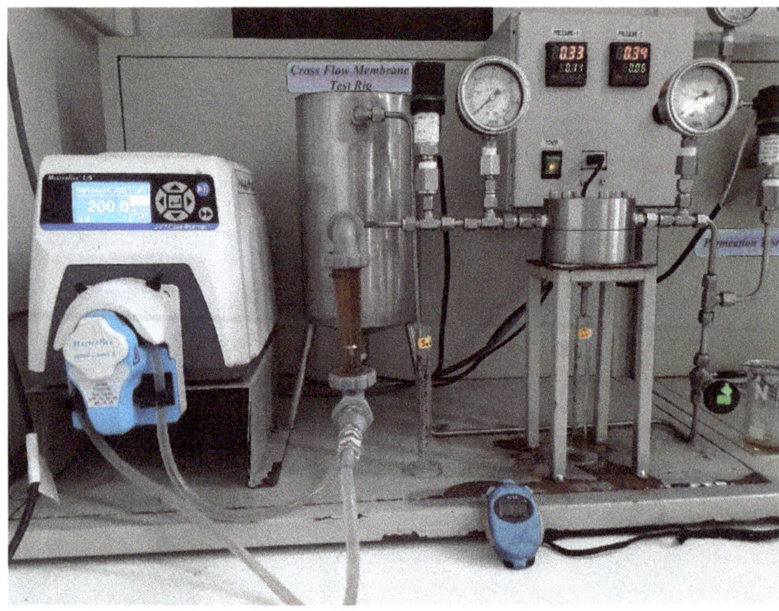

Figure 3. CFR test configuration (filtration treatment set).

Final leachate characterizations were evaluated in terms of removing efficiencies for the COD, colour, and NH_3-N pollutants using Equation (1):

$$\text{Removal eficiency } \% = \frac{(C_F - C_P)}{(C_F)} \times 100(\%) \tag{1}$$

where C_F is the contaminant concentration at the feed (mg/L) and C_P is the contaminants concentrations in the permeated solution (mg/L). All contaminants' concentrations were checked using the UV-V spectrophotometer (Hach DR6000, Loveland, CO, USA) in prior and post of filtration practice.

2.5.2. Productivity of Membrane

Pure flux plays a dynamic role in the membrane productivity evaluation. Permeability of membrane was investigated through the pure water flux, which was measured via a dead-end filtration apparatus, as illustrated in Figure 4. A metallic ring having 5 mm average pore size and 8.76 cm^2 effective permeate area was applied to support the membrane. Initially, the impurities present in the membrane were removed by submerging the membrane in DW for 30 min. Then, a stable flux was achieved by pre-compacting the membrane with N_2 gas at a pressure of 30 KPa for 2 min. After 30 min, the permeated water volume was noted at a similar pressure of 30 KPa. The pure water flux can be calculated using the Equation (2):

$$J = \frac{V}{A \times t} \ (\text{L/m}^2 \cdot \text{h}) \tag{2}$$

where V is the permeated pure water volume (L), A is the membrane effective surface area (m^2), and t is the time of permeation (h).

Figure 4. Dead-end test (pure water permeation set-up).

2.5.3. Antifouling Valuation

Throughout the membrane filtration process, the overall decrease in flux, alongside the improvement of transmembrane pressure, were mainly caused by either membrane fouling, concentration polarization, or a combination of both [44]. Both of these components can be attained from the experimental data using both of the leachate permeate flux and maximum transmembrane pressure (Max. TMP) values which are measured by the cross-flow ring test. Max. TMP was applied to indicate the antifouling ability of fabricated membranes [45].

2.5.4. Morphological Characteristics

It is a well-known fact that the membrane properties and performance are highly dependent on its morphology (pore size, surface texture, and microstructure). Therefore, investigation of membrane morphologies is considered a significant factor in the effectiveness evaluation of the produced membranes.

Fourier transform infrared spectroscopy (FTIR, Perkin Elmer Lambda 35, Waltham, MA, USA) was applied to investigate the membrane surfaces chemical compositions. The FTIR spectra ranged between 4000–400 cm^{-1}.

EDX is a chemical microanalysis method used for quantitative, qualitative, and elemental mapping examination. Octane Silicon Drift Detector (SDD, EADX Inc., Mahwah, NJ, USA) was used at high voltage of 15 kV, using Mn Kα as source of energy. The fabricated PVDF-PAC membranes with different compositions were measured by INCA Energy 400 software (Firmware INCA, Version V1.09R13), along with the image taken by the Quanta FEG 450 instrument.

FESEM (Quanta FEG 450, FEI, Hillsboro, OR, USA) was applied to record the cross-sectional and surface morphologies of the fabricated membrane. The cross-sectional morphologies were investigated by fracturing the membranes in liquid nitrogen and immediately cutting them after air drying. FESEM measurement starts by placing the sample on carbon tape, which was attached with the sample stub. The sample was also coated with the platinum nanoparticles in auto fine coater (JFC-1600, SUTD-MIT International Design Centre, Singapore) before performing the analysis.

An atomic force microscopy (AFM, Dimension 5000, Bruker AXS, Santa Barbara, CA, USA) was also applied to study the surface morphologies and roughness of the synthesized membranes. Herein, membranes were cut into small square pieces (1 × 1 cm) and pasted on a glass slide. Sample scanning were performed using a probe-optical microscope on tapping mode and images of 10 μm × 10 μm were taken by AFM. The root-mean-square roughness (R_q) and average roughness (R_a) was applied to measure the surface roughness for each membrane.

Porosity of membrane could be easily defined as the pore's volume divided by the membrane total volume. Wet membranes were weighed after carefully wiping the surface (Ww). Afterwards, these membranes were dried in an oven at 50 °C for 24 h and weighed again (Wd). The porosity of membrane ε (%), was measured by gravimetric method using Equation (3) [25]

$$\varepsilon = \frac{(Ww - Wd)/\rho w}{\frac{Ww-Wd}{\rho w} + Wd/\rho p} \times 100\%$$ (3)

where, Ww is the weight of wet membrane (kg), Wd represents the weight of dry membrane (kg), ρw is the density of water (1000 kg/m^3), and ρp, the polymer density (1770 kg/m^3 for PVDF).

Based on the measured distilled water flux, the average pore size (d) of the membrane was calculated by the Guerout–Elford–Ferry equation, Equation (4) [46].

$$d = \sqrt{\frac{(2.9 - 1.75\varepsilon)8\delta lV}{\varepsilon \, A \, \Delta P \, t}}$$ (4)

Herein, ε is membrane porosity (%), δ, the water viscosity (8.9×10^{-4} Pa s), l represents membrane thickness (60×10^{-6} m), V is the volume of the distilled water penetrating through the membrane (m^3), t is the experimental time interval (s), A, the effective membrane surface area (m^2), and ΔP is the working pressure (30 kPa).

3. Results and Discussion

3.1. Landfill Leachate Characteristics

Table 1 displays the key characteristics of the raw leachate sample of more than 10 years in age. The lower BOD$_5$ to COD ratio (0.074) was another strong indication of highly stabilized leachate sample [3]. The other quality parameters of leachate, such as COD, BOD$_5$, NH$_3$-N, colour, and pH values, were around 1188 mg/L, 89 mg/L, 313 mg/L, 1360 PtCo/L, and 8.33, respectively. These obtained values were also compared with the standard discharged limits set by the Malaysian Environmental Quality was conducted (Table 1) [47]. As shown in Table 1, the COD, colour, and NH$_3$-N concentrations were found to be far greater than the standard discharged limits.

Table 1. Raw leachate characteristics.

Parameter	Unit	Value Range	Average	Malaysia Discharge Standards
DO	mg/L	2.43–5.19	3.81	-
COD	mg/L	846–1530	1188	400
BOD$_5$	mg/L	55–122	89	20
BOD$_5$/COD	-	0.065–0.080	0.074	0.05
Colour	PtCo/L	1040–1680	1360	100
NH$_3$-N	mg/L	164–462	313	5
Suspended Solids	mg/L	75.0–80.0	77.5	50
pH	-	7.97–8.68	8.33	6.0–9.0
Turbidity	NTU	15.9–70.2	43.1	-
Electrical Conductivity	mS	13.22–22.77	18.00	-
Temperature	°C	27–30	28	40

3.2. PAC Characterization

Analysis test of the particle size was conducted to investigate the particle size distribution of fine samples in terms of volume. The particle size distribution of PAC sample is shown in Figure 5a. It can be seen from Figure 5 that PAC has small particle sizes which varied between (0.02–50 μm) in diameter. The average particle diameter of the PAC is 25 μm. It is evident from Figure 5a that the distribution curve of PAC particles could be counted a uniform-distribution curve. The percentage of adsorption is higher for those adsorbents have smaller particle size due to the availability of more surface area [48]. The surface morphology of PAC was visualized via FESEM, with a magnification of 10,000×, as shown in Figure 5b. FESEM micrographs of PAC, shows uniform size particles, which confirmed the results obtained from the particle size analysis. To some extent, the PAC surface having small cavities, pores, and more rough surfaces indicates the presence of an interconnected porous network. Increasing the particles' number of an adsorbent material by decreasing its particles size resulted in increasing the adsorption surface area, and thus the material adsorption characteristics [49].

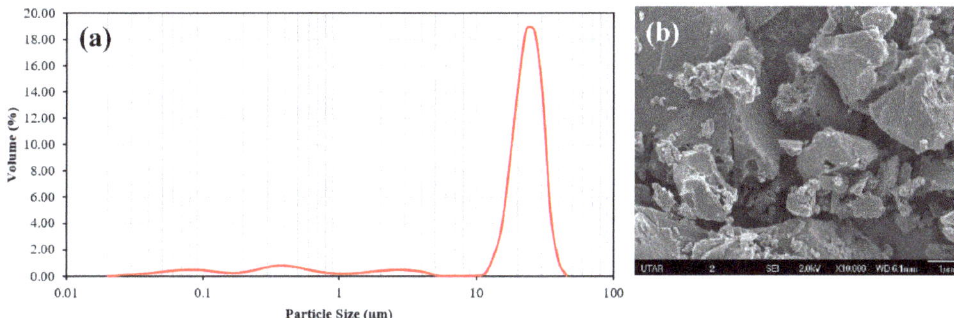

Figure 5. PAC characterization: (**a**) Particle size distribution; (**b**) FESEM image at 10,000× magnification.

3.3. Membrane Filtration and Experimental Results

Herein, the relationship among the independent factors (PVFD and PAC dosage in membrane) and responses (COD, NH_3-N, colour removal, max. TEM, and pure water flux) were thoroughly investigated. There were 13 different experiments performed on the PVDF and PAC composition based on the central RSM composite design, as shown in Table 2. CFR test was performed to investigate the pollutants removal efficiency together with the max. TEM, while dead-end test was executed to measure the pure water flux.

Table 2. Experimental results for the PVDF-PAC membranes (RSM design).

	Factors			Responses			
			Removal Efficiency (%) *				
Run Order	PVDF (wt.%)	PAC (wt.%)	COD	Colour	NH_3-N	Pure Water Flux ** ($L/m^2 \cdot h$)	Max. TMP (bar)
1	10.00	0.00	14.8	15.1	10.9	90.2	0.46
2	10.00	2.00	29.1	42.3	7.5	127.7	0.42
3	12.00	1.00	32.2	44.6	18.3	89.3	0.48
4	14.00	0.50	28.2	39.6	19.6	64.0	0.66
5	14.00	1.00	37.2	56.3	23.8	79.9	0.67
6	14.00	1.00	35.5	50.3	19.3	72.9	0.63
7	14.00	1.00	35.5	56.2	21.3	72.2	0.62
8	14.00	1.00	35.7	51.1	21.5	70.3	0.61
9	14.00	1.00	32.2	51.5	19.9	69.9	0.60
10	14.00	1.50	33.2	52.7	19.2	83.1	0.55
11	16.00	1.00	37.1	41.0	22.5	31.8	0.68
12	18.00	0.00	29.1	26.7	21.2	26.2	1.00
13	18.00	2.00	20.9	15.6	17.3	32.9	0.78

* Estimated by Equation (1). ** Estimated by Equation (2).

The COD, colour, and NH$_3$-N removal efficiencies were found to be around 14.8–37.2, 14.6–56.3, and 7.5–23.8%, respectively, while the pure flux and max. TMP were ranged between 26.2–127.7 L/m^2·h and 0.42–1.00 bar, respectively. ANOVA analysis was performed for the further investigation on the obtained experimental results.

It is observed from Table 2 that an increase in both PVDF and PAC concentrations on the membrane leads, to some extent, to an increase in the contaminants removal. When PVDF and PAC concentration are higher than 14 wt.% and 1.0 wt.%, respectively, the removal efficiency starts to decrease with increasing the amount of PVDF and PAC. This behaviour was attributed to the combination effect between polymer and additive in dope. This leads to the creation of large volume voids with increasing polymer dosage, and allows the small particles of contaminants to pass through the membrane [50].

3.3.1. Removal Efficiency of Contaminants

Table 3 depicts the empirical model using the data obtained from COD, colour, and NH$_3$-N removals. F-values of the model, together with the low probability values (P > F > 0.05), clearly suggest that the models were significant for all responses.

Table 3. ANOVA results and quadratic models of PVDF-PAC membranes for COD, colour, and NH$_3$-N removal efficiencies.

Source	COD Removal (%)		Colour Removal (%)		NH$_3$-N Removal (%)	
	F-Value	**Prob > F**	**F-Value**	**Prob > F**	**F-Value**	**Prob > F**
Model	25.62	0.0002 (S) [a]	31.93	<0.0001 (S) [a]	24.34	0.0003 (S) [a]
A-PVDF (wt.%)	4.34	0.0759	3.89	0.0840	55.81	0.0001
B-PAC (wt.%)	4.19	0.0800	5.69	0.0441	6.37	0.0396
AB	32.25	0.0008	21.10	0.0018	0.032	0.8634
A^2	0.42	0.5375	97.03	<0.0001	0.24	0.6372
B^2	12.18	0.0101	-	-	3.62	0.0988
Lack of Fit	1.39	0.3665 (NS) [b]	3.27	0.1386 (NS) [b]	0.17	0.9088 (NS) [b]
	Std. Dev.	1.98	Std. Dev.	4.28	Std. Dev.	1.40
	Mean	30.85	Mean	41.69	Mean	18.64
	R^2	0.9482	R^2	0.9411	R^2	0.9456
	Adj R^2	0.9112	Adj R^2	0.9116	Adj R^2	0.9067
	C.V. %	6.42	C.V. %	10.26	C.V. %	7.52

[a] Significant. [b] Not significant.

The significant model terms for COD removals in the ANOVA analysis were sorted in descending order depending upon the influential terms (AB, B^2, A, B, and A^2). It was clearly seen that the PVDF and PAC (AB) had the highest impact on the COD removal with an F-value of around 32.25, followed by the quadratic term of PAC concentration (B^2), PVDF concentration (A), PAC concentration (B), and finally the quadratic term of PVDF concentration (A^2) with an F-value of 12.18, 4.34, 4.19, and 0.42, respectively. The quadratic terms of PVDF concentration together with the linear terms of PAC and PVDF contents caused a positive effect on the COD removal. Nonetheless, interaction and quadratic terms of PAC exhibited negative effects. In fact, an increase in the COD removal was recorded upon the change in the liner terms of PVDF and PAC concentrations, and PVDF concentration with quadratic term from lower to higher level. Hence, this change is complemented by the outstanding COD removal using PVDF-PAC membrane. On the other hand, a decline in COD removal was recorded when the interaction term and quadratic term of PAC was in the higher level.

The quadratic term of PVDF contents (A^2) has the most significant effect towards the colour removal rate. This is due to the highest F-value (97.03), where other terms had the values of 21.10, 5.69, 3.89, respectively. The PAC content (B) had a progressive influence on the colour removing. However, the quadratic term of PVDF, PVDF concentration, and interaction among the PVDF and PAC displayed a negative effect. Thus, the removal of colour was enhanced with the enhancement of the PVDF contents in membrane fabrication until the optimum amount (>14 wt.% PVDF).

Additionally, in case of NH_3-N removal, the A, B, B^2, A^2, and AB were sorted in descending order of their effecting strength. The highest F-value of 55.81 was recorded for the linear term of PVDF concentration (A), and thus it had the huge effect in NH_3-N removal. On the other hand, the lowest F-value of 0.032 was recorded for interaction term, which regarded to have a negligible effect on the model. The PVDF linear term only offered a strong influence on removing of NH_3-N, while the remaining terms were found to be the negligible influencers. Hence, the NH_3-N removal was increased upon enhancing the PVDF contents in membranes. However, for PAC concentration after the point (PAC = 1.0 wt.%); when either the quadratic term of PVDF or PAC, or the interaction term is in the significant level, the NH_3-N removal starts to decrease.

The lack of fit F-statistic was statistically not significant, as the values of (P) were higher than 0.05. A significant lack of fit suggests that there may be some systematic variation unaccounted for the proposed models. This may be due to the exact replicate values of the independent variables in the models that provide an estimate of pure error [15]. The correlation coefficient value (R^2) resulted in the present study for COD removal (0.9482), colour removal (0.9411), and NH_3–N removal (0.9456), indicating that only 5.18, 21.09, 5.89, and 5.44% of the total dissimilarity might not be explained by the empirical models. Zielinska et al. [10] stated that the correlation coefficient should be more than 0.80 for a good fit of a model. Moreover, the C.V.% of the obtained models for COD, colour, and NH_3-N removals were 6.42%, 10.26%, and 7.52%, respectively, which designates an adequate model [51].

In the current study, all insignificant model terms which have limited effects were eliminated from the study to improve the model. Based on the findings, the response surface models for COD, colour, and NH_3-N removal efficiency were constructed to predict responses, which were considered reasonable. The final regression models, in terms of their coded factors, are expressed by the second-order polynomial equations, and are presented in Table 3.

Typically, it is vital to study the effect of the operational factors on the different responses. The effect of PVDF and PAC concentration on the responses of COD, colour, and NH_3-N removals over PVDF-PAC membranes could be evaluated using perturbation and three-dimensional (3D) response surface plots (Figure 6). Perturbation plots show the comparative effects of independent variables on the responses. For instance, in Figure 6, the different sharp curvatures in PVDF concentration (A) and PAC concentration (B) show that the three responses (COD, colour, and NH_3-N removal efficiency) were very sensitive to the fabrication variables, but with different behaviours. In other words, PVDF and PAC contents have a major function in the treatment process under the experimental conditions. This is another confirmation of the important effects of the independent variables (PVDF and PAC concentrations) on the treatment removal efficiency. Therefore, the 3D surface response and contour plots of the quadratic models were utilized to assess the interactive relationships between independent variables and responses. The 3D response surface was introduced as a function of PVDF and PAC concentrations. Figure 6a,c shows a symmetrical 3D surface response for both COD and NH_3-N removals. In the meantime, the removal of colour presents a different 3D surface (Figure 6b), which indicates that colour removal was influenced differently by experimental factors than the other responses.

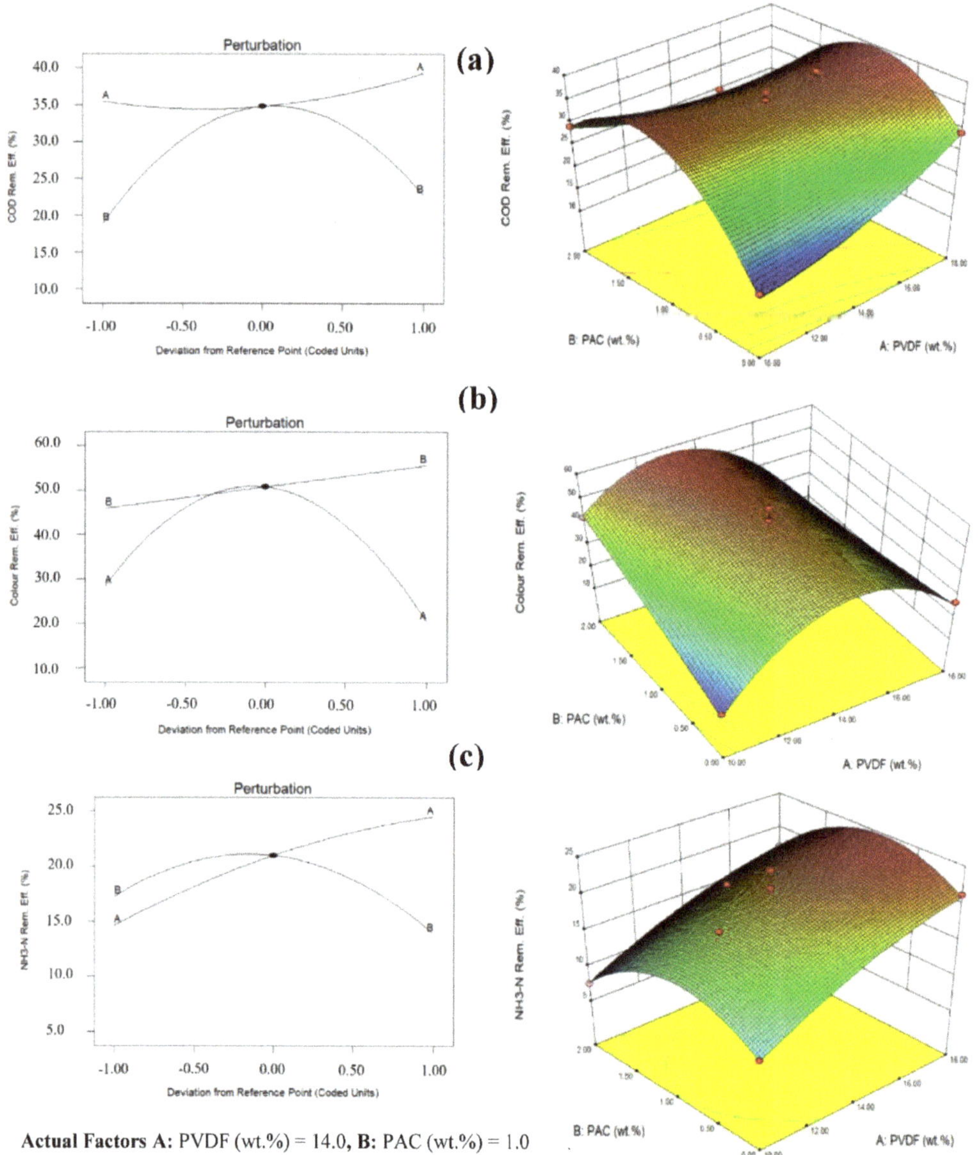

Actual Factors A: PVDF (wt.%) = 14.0, **B:** PAC (wt.%) = 1.0

Figure 6. Perturbation plots (**left**) and 3D response surface (**right**) of PVDF-PAC fabricated membrane for the removing efficiency of (**a**) COD, (**b**) colour, and (**c**) NH_3-N.

Figure 6a,c indicated that the responses for COD and NH_3-N removal rate was sufficiently enhanced upon the increase in PVDF contents in applied membranes. On the other hand, the increase in PAC contents in membrane fabrication led, to the removal of COD and NH_3-N to some extent. It was seen that, when the PAC contents in membrane were higher than 1.0 wt.%, the removal rate for COD and NH_3-N began to decline. According to Figure 6c, for the removal of NH_3-N, the effect of interaction between PVDF and PAC concentrations have a noteworthy influence on removal percent. The NH_3-N removal were gradually increased with the increasing of PAC concentration to some extent, which means that the incorporated PAC has enhanced the membrane performance in terms of NH_3-N removal, in addition to the main separation action gained by the membrane texture itself. This good result might be ascribed to the high adsorption characteristics of the used PAC, which significantly improved the fabricated membrane efficiency [5,52].

However, PVDF concentration has limited effect on COD removal efficiency compared with the PAC content. Where 35.5 and 38.5% of COD were removed at minimum and maximum PVDF concentration (10.0 and 18.0 wt.%), respectively, 18.5 and 35.5% of COD removal were removed at minimum and medium PAC concentration (0.0 and 1.0 wt.%), respectively. Likewise, the minimum NH_3-N removal was found to be 7.5% at membrane concentration of 10.0 wt.% PVDF and 2.0 wt.% PAC, while the maximum NH_3-N removal (24.5%) was observed at the PVDF and PAC concentration of 18.0 and 1.0 wt.%, respectively.

On the other hand, the 3D response surface in Figure 6b displays a different effect of interaction between the experimental factors on the colour removal rates. It was observed that an increase in the concentration of PVDF in the membrane leads to an improvement in the colour removal to some degree. When the concentration of PVDF was higher than (14 wt.%), the colour removal performance starts to decrease. This behaviour was credited to the combined effects of additive and polymer in the dope. This leads to creating large volume voids with increasing polymer dosage, and lets the fine particles from contaminants to permeate through the membrane [50]. Meanwhile, the enhancement of PAC concentration in a membrane drove a steady increase in the colour removal efficiency. As witnessed in Figure 6b, the predicted minimum and maximum efficiencies of colour removal were 15.0 and 56.5% present at fabrication concentrations of (18.0 wt.% PVDF, 0.0 wt.% PAC), and (14.0 wt.% PVDF, 1.0 wt.% PAC), respectively. This also confirms the effectiveness of PAC content in enhancing the removal performance of the filtration process using PVDF fabricated membrane.

Despite the incorporation of PAC into membrane enhancing the COD, colour, and NH_3-N removal, the filtered leachate still did not meet the Malaysian Discharge Standard (Table 1). This is due to the highly concentrated pollutants of leachate that resulted in a reduction in membrane efficiency owing to the clogging caused by influent SS component. Therefore, a pre-treatment process such as PAC adsorption is suggested to be used before the membrane treatment [33].

3.3.2. Pure Flux and Transmembrane Pressure Studies

By applying the factorial regression analysis on the experimental data related to PVDF-PAC membranes, both max. TMP and pure water flux responses were well agreed to a linear model of the second degree, as shown in the ANOVA analysis presented in Table 4. In a general linear model or a multiple regression model: $Y = \beta_0 + \sum_{i=1}^{k} ss_i \, X_i + \varepsilon$, where: Y is the response, X_i is the independent factor, k is the number of variables, β_0 is the constant term, β_i represents the coefficient of the linear, and ε is the random error or noise [53]. The final linear models obtained for each response has been expressed by the first order polynomial equation, as presented in the last raw of Table 4.

The fitted model for the pure water flux suggests a large F-value (53.56), suggesting that the model is significant. As the value of Prob > F of all terms is less than 0.050, this suggests that all the model terms are significant. Based on their F-values, the PVDF concentration term (A) has the highest influence on the model, followed by PAC concentration term, and lastly the combination term. The term of PAC concentration presents a positive effect on pure flux, while the other two terms have been found to be negative influencers. Hence, the pure water flux was raised only with enhancing PAC contents in the membrane while, in contrast, it is decreased with the increasing of the PVDF content of a membrane.

Table 4. ANOVA results and quadratic models of PVDF-PAC membranes for pure flux and max. TMP.

Source	Pure Flux (L/m^2·h)		Max. TMP (bar)	
	F-Value	Prob > F	F-Value	Prob > F
Model	53.56	<0.0001 (S) [a]	49.62	<0.0001 (S) [a]
A-PVDF (wt.%)	144.45	<0.0001	131.07	<0.0001
B-PAC (wt.%)	11.86	0.0073	13.01	0.0057
AB	4.38	0.0658	4.78	0.0566
A^2	-	-	-	-
B^2	-	-	-	-
Lack of Fit	5.18	0.0681 (NS) [b]	3.38	0.1307 (NS) [b]
	Std. Dev.	7.36	Std. Dev.	0.041
	Mean	70.03	Mean	0.63
	R^2	0.9470	R^2	0.9430
	Adj R^2	0.9293	Adj R^2	0.9240
	C.V. %	10.50	C.V. %	6.56
Model equation coded, (wt.%)	+70.03 −41.68 * A +11.94 * B −7.70 * A * B		+0.63 +0.22 * A −0.070 * B −0.045 * A * B	

[a] Significant. [b] Not significant.

On the other hand, the suggested model of max. TMP was significant with a high F-value (49.62), as can be seen from Table 4. Based on its effect on the model from the highest to the lowest, the model terms can be arranged as follows: PVDF content, PAC content, and the combination of both, with F-values of 131.07, 13.01, and 4.78, respectively. However, the PVDF concentration is the only factor which showed a positive influence on the max. TMP, due to the positive sign of its term; this indicates a worse impact on the max. TMP, as it could be increased with the increasing of PVDF content on the fabricated membrane. On the other hand, PAC concentration exhibited a better effect on the max. TMP, which showed a reduction in max. TMP occurred due to the increasing of the PAC content.

Additionally, both of the models display a non-significant lack of fit F-value, which indicates that well fitted models have been selected to present the experimental results with minor pure errors [15].

The R^2 values obtained in the present study for pure flux and max. TMP were 0.9470 and 0.9430, respectively. The high value of R^2 represents good agreement between the observed and the calculated results within the experimental ranges [37]. Moreover, C.V. % for the water flux and TMP were 10.50% and 6.56%, respectively. Where these small values indicate good fitness of the models [51].

Based on these findings, the resulted response surface models in the current work for predicting the two responses (pure flux and max. TMP) were considered reasonable.

The influence of integrated PAC and the interaction of content's concentrations on the max. TMP can be explored by the plots of perturbation and 3D response surface, as shown in Figure 7. From perturbation plots at Figure 7, it is easy to notice that pure flux and max. TMP responses are very sensitive to the experimental factors, and to conclude that both have a different (inversed) behaviour regarding the PVDF and PAC concentration values. As can be seen from Figure 7a, increasing of PVDF concentration (A) resulted in a linear decrease in pure water flux and increase in max. TMP, which attributed to the reduction in membrane porosity due to the increase in polymer concentration, which is well recognized for the system of a single polymer casting solution [50]. However, PAC concentration (B) showed a different effect, as any increase in its value causes a linear increment on the pure water flux, but a decrease in max. TMP.

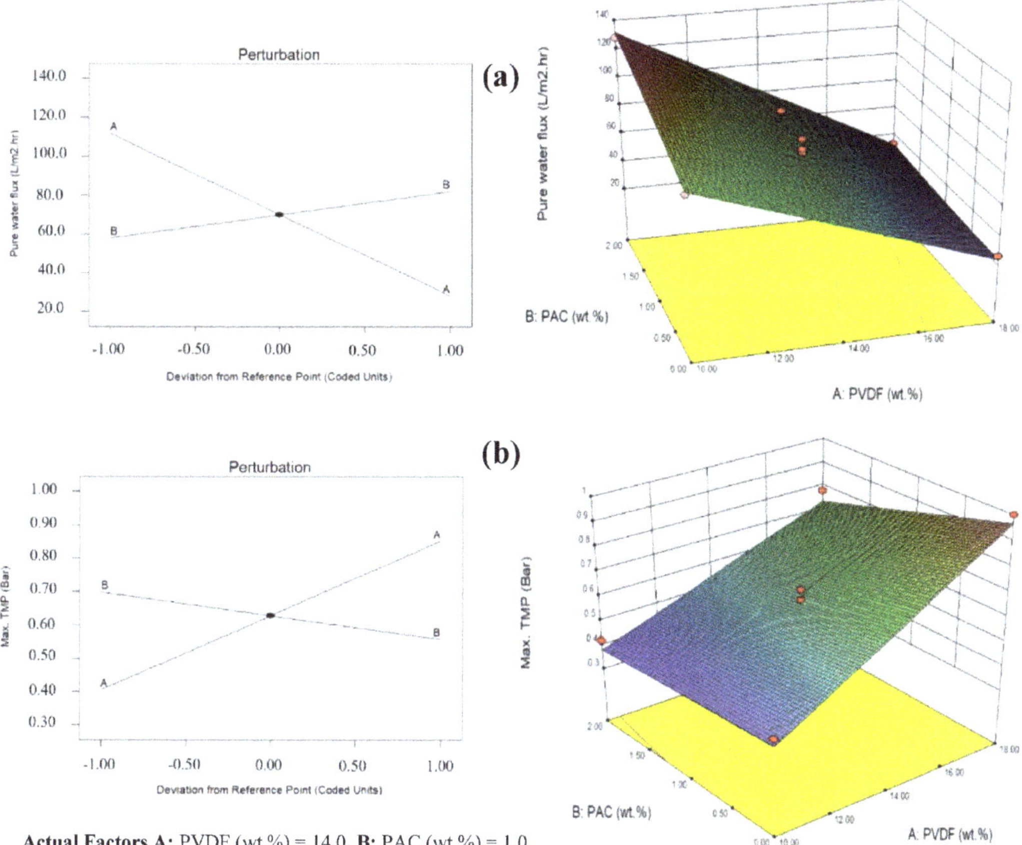

Actual Factors A: PVDF (wt.%) = 14.0, **B:** PAC (wt.%) = 1.0

Figure 7. Perturbation plots (**left**) and 3D response surface (**right**) of PVDF-PAC fabricated membrane for (**a**) pure water flux and (**b**) max. TMP.

Minimum and maximum predicted pure fluxes (26.0 and 128.5 L/m²·h) were found at the membrane compositions of 18.0 wt.% PVDF with 0.0 wt.% PAC, and 10.0 wt.% PVDF with 2.0 wt.% PAC, respectively. On the other hand, lowest and highest max. TMP according to the suggested model were found to be 0.38 and 0.98 bar at membranes of compositions (10.0 wt.%) PVDF with (2.0 wt.%) PAC, and 18.0 wt.% PVDF with 0.0 wt.% PAC, respectively. From the findings, membranes with lower PVDF concentration and high PAC concentration (10.0 wt.% PVDF and 2.0 wt.% PAC) exhibited the best water permeation and antifouling properties. Nonetheless, this membrane still falls short to produce the highest removing rates of COD, colour, and NH_3-N based on the previous discussion.

3.4. Fabricated Membrane Characterization

The morphology of produced membrane can explain the effect of dope composition on membrane performance. A collection of membranes composed from different concentrations of PVDF and PAC (wt.%) were chosen from the fabricated membranes to represent the different membrane compositions, and consequently to be investigated by the morphological studies. These membranes were: FM1 with the content of (10.0 wt.% PVDF-0.0 wt.% PAC) to represent minimum PVDF concentration with no PAC; (10.0 wt.% PVDF-2.0 wt.% PAC) to represent minimum PVDF with high PAC, denoted as FM2; (14.0 wt.% PVDF-1.0 wt.% PAC) to represent intermediate composition of both PVDF and

PAC, named FM3; and finally FM4 with 18.0 wt.% PVDF and 0.0 wt.% PAC to represent maximum concentration of PVDF without PAC.

The FTIR spectrum of PVDF-PAC fabricated membranes with the various compositions is illustrated in Figure 8. It is clearly observed from Figure 8 that membranes displayed semi-typical distinctive spectra along the range of 4000 and 400 cm^{-1}. Characteristic chemical groups are witnessed in the band of all membranes at waves with lengths 3020, 2990, 2370, 1400, 1070, 875, 590, and 490 cm^{-1} with altered vibrations of strength depends on the different membrane compositions. The spectrum shows bands at 2990 and 3020 cm^{-1} which are attributed to the symmetric and asymmetric stretching vibrations of C-H coming from ketones and carboxylic acids [54], where vibrations at 1070 and 1400 cm^{-1} presented the deformation peaks of C-F related to PVDF.

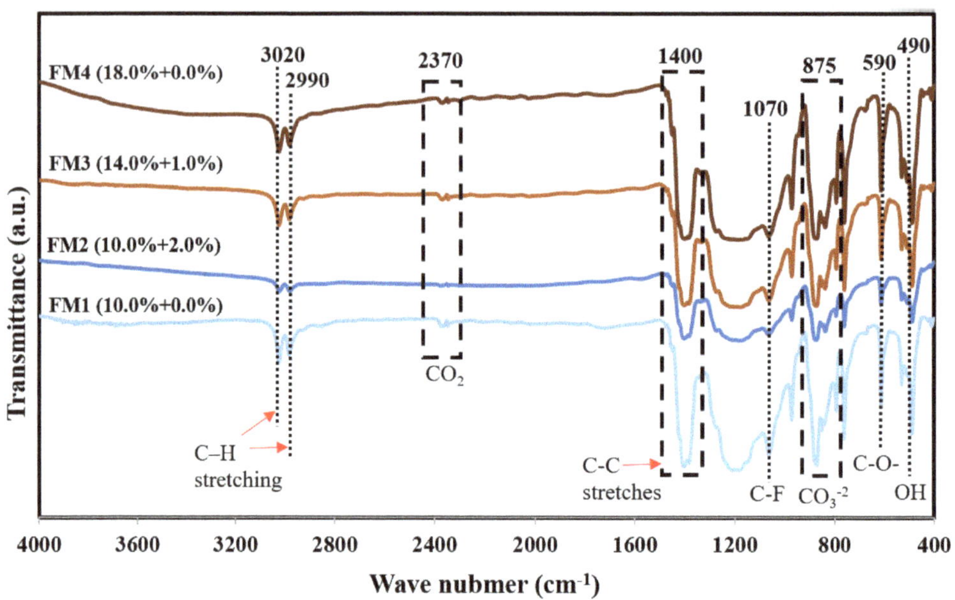

Figure 8. FTIR spectra for PVDF-PAC membranes with different concentrations.

The notable peaks of the various membranes at 2370, 875, and 590 cm^{-1}, assigned to CO_2, CO_3^{-2}, and C-O- groups, respectively, were the features distinctive of neutralization methanol, used after membrane casting [55]. Moreover, the OH group detected at 490 cm^{-1} is attributed to the DW used for membrane solidification during the casting process [56]. Figure 8 also confirmed that the recorded wave numbers in the spectrum of both membranes without PAC (FM1 and FM4) have higher frequencies in comparison with the spectrums of the other two membranes with PAC content (FM2 and FM3).

Furthermore, it could be observed that the peaks of the membrane with higher content of PAC (FM2) have lower vibrations compared with the membranes with lower PAC content (FM3). Evidently, the peaks become narrow with less strength at the increasing of PAC weight, indicating that the hydrogen bonds were constructed well between PVDF polymer chains and the hydroxyl groups from PAC, which reduces the PVDF hydrophobic tendency [57]. These outcomes confirmed that PAC was well integrated to PVDF membranes, and partially relocated on the membrane surface, which leads to membrane treatment efficiency enhancement.

To investigate the elemental composition present in the fabricated PVDF-PAC membranes with different compositions, EDX analysis was recorded in the binding energy region from 0 to 15 keV as exhibited in Figure 9. The PVDF characterized elements C and F were clearly observed in the spectra of the pure PVDF membranes (without PAC), while the AL element, which characterizes the presence of PAC, appeared only at the PVDF membranes incorporated with PAC [33]. Figure 9b,d shows the EDX analysis of 2.0 and 1.0 wt.% PAC, respectively. It is clearly witnessed that the presence of PAC was presented well.

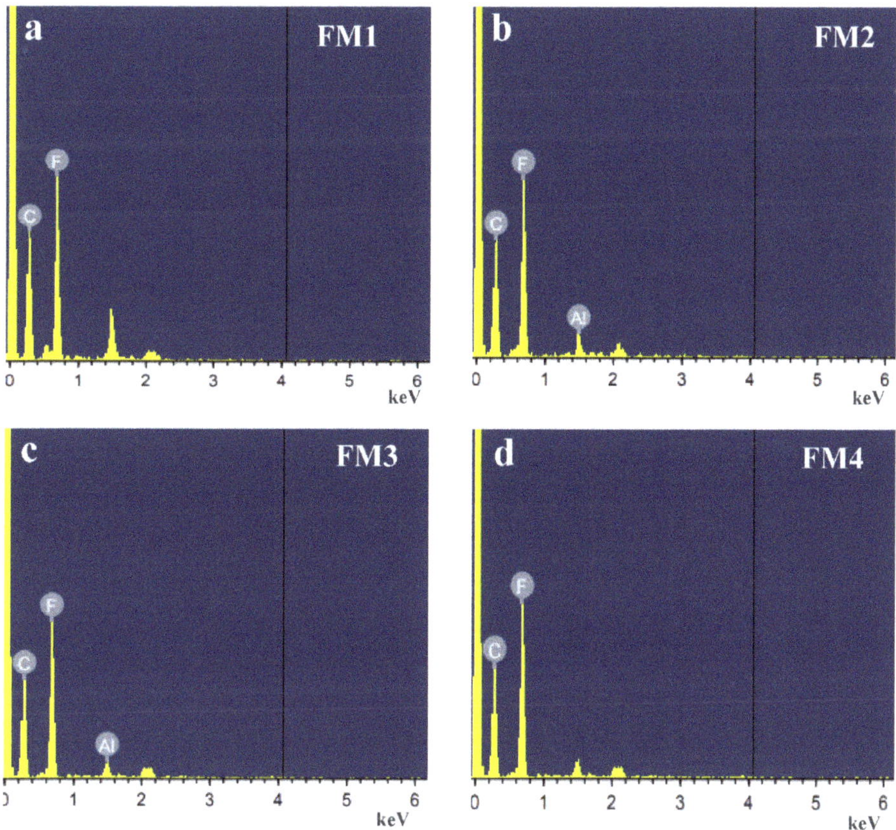

Figure 9. EDX analysis for the selected PVDF-PAC fabricated membranes with different compositions: (**a–d**) for (FM1–FM4).

Table 5 shows the atomic percentages of the different elemental compositions of the selected membranes (FM1–FM4). From the EDX findings, the weight percentages of elemental AL on the FM2 and FM3 were determined as 1.04 and 0.79, respectively, which confirmed the presence PAC with representative weights on the integrated membranes.

Table 5. Elemental compositions of selected PVDF-PAC fabricated membranes based on EDX mapping.

Sample	Composition (wt.%)		Elements Weight (%)			
	PVDF	PAC	C	F	AL	Total
FM1	10.0	0.0	61.69	38.31	0.00	100.00
FM2	10.0	2.0	60.85	38.11	1.04	100.00
FM3	14.0	1.0	61.04	38.17	0.79	100.00
FM4	18.0	0.0	60.63	39.37	0.00	100.00

Figure 10 presents the FESEM images for produced membranes with different compositions, which show the top surface morphology of membranes, along with its cross section. As can be seen from Figure 9a–d, there were many small pores available on the surface of FM1 membrane which contains the lowest PVDF polymer content (10.0 wt.%). Furthermore, the number and size of these pores start to be decreased, first on membrane FM3, with PVDF content 14.0 wt.% and PAC content 1.0 wt.%, followed by FM2 membrane with the highest PAC content (2.0 wt.%), while the membrane FM4 has a semi-impermeable surface due to its high PVDF polymer content (18.0 wt.%) with no PAC content. This was in agreement with the findings earlier discovered by Kunst and Sourirajan [58].

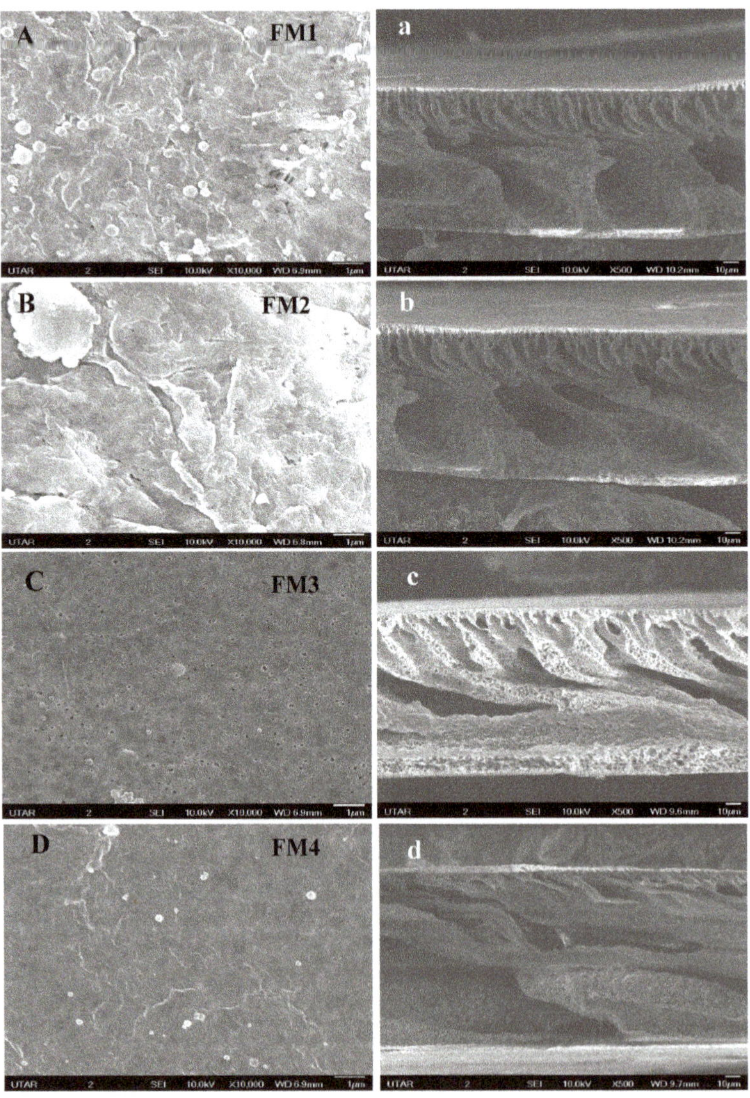

Figure 10. FESEM morphologies of PVDF-PAC membranes with different compositions (FM1 to FM4): (**a–d**) cross-sections and (**A–D**) top surfaces.

Referring to membrane cross sections on Figure 10a–d, all membranes display the formation of macrovoid with loosely packed structures. Typically, the membrane consists of two layers, which are a spongy porous support layer and a dense top finger-like layer. The establishment of these configurations can be attributed to the instantaneous demixing of polymer and solvent during the process of phase inversion.

FM1 membrane, with only PVDF and the weight of 10.0 wt.%, displayed an unimproved finger-like formation and a sponge-like support layer containing large, unconnected pores, delimited by polymer walls (see Figure 10a). The finger-like voids turn become flat, bigger, and even strained to the bottommost of the fabricated membranes with an increase in PAC concentration (i.e., in FM2 and FM3), and the spherical voids of the sponge-like structures connect more closely with themselves (Figure 10b,c). However, the FM4 membrane, containing the highest concentration of PVDF, gives thin, smaller, and non-stretched figure-like pores, with less connection to the little sponge-like pores located on the cross-section's bottom. This produces low membrane flux due to the greater amount of polymer contributing a higher membrane viscosity, which lead to a decease in the membrane porosity and pore size. The overall FESEM micrographs have proved the significant effect of the PAC presence in improving the fabricated membrane characteristic in terms of membrane rejection, and therefore removal rate of contaminants.

Furthermore, an AFM test was carried out to investigate the membrane top surface, along with its roughness, as shown in Figure 11. The FM2 membrane might contain some extra PAC particles which made its top surface rougher compared to others (Figure 11b). Having less depth of facial peaks and valleys, the FM4 membrane surface (Figure 11d) is relatively smooth due to containing only PVDF polymer which received a homogeneous mixing at the preparation phase of dope solution [59]. However, the peaks and valleys of FM1 and FM3 membranes reduced gradually compared to FM2, where FM3 has the smoothest surface compared with other membranes (see Figure 11a–d). To confirm all above observations, the values of membrane surface roughness (R_q and R_a) given in Figure 11 can be considered.

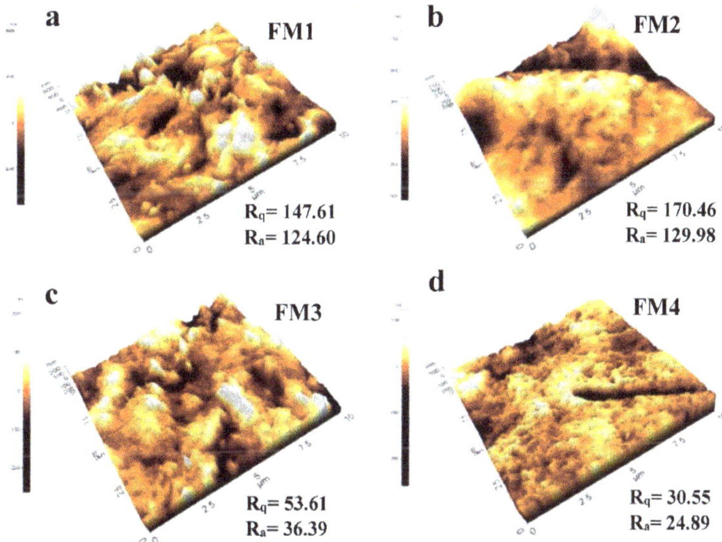

Figure 11. AFM top surface images with average membrane roughness values (nm) for different compositions of selected PVDF-PAC membranes: (**a–d**) for (FM1 to FM4).

For membrane permeability analysis, the impact of PAC addition to membrane permeability, in terms of porosity and average pore size, were evaluated for the produced PVDF membranes. As presented in Table 6, the porosity and average pore size of the fabricated PVDF membranes incorporated with PAC were higher than the other membranes without PAC. Based on Table 6, the resulted fabricated membranes were "micro-filtration", and the highest mean values of porosity and average pore size were achieved by FM2 membrane at the values 77.48% and 24.43 µm, respectively. On the other hand, the lowest values of the same corresponding permeability parameters were found using membrane FM4 at 48.38% and 12.15 µm, respectively. These findings agreed with the above morphological results.

Table 6. Permeability measurements for selected fabricated PVDF-PAC membranes.

Membrane	Composition		Porosity	Average Pore Size
	PVDF	**PAC**	(%) [a]	(µm) [a]
FM1	10.0	0.0	57.25 ± 0.18	15.34 ± 0.05
FM2	10.0	2.0	77.48 ± 0.50	24.43 ± 0.15
FM3	14.0	1.0	72.86 ± 0.20	21.27 ± 0.07
FM4	18.0	0.0	48.38 ± 0.62	12.15 ± 0.24

[a] Each parameter is expressed as average value $\pm$ standard deviation.

3.5. Membrane Treatment Optimization

The best synthesized membrane has been selected using the RSM tool, where the membrane efficiencies of COD, colour, and NH_3-N removal were optimized during this study.

Based on the DoE software, the operational conditions (PVDF weight and PAC weight) were targeted to be within the range. While the dependents of treatment performance (COD, colour, and NH_3-N removal) were chosen as "maximum" to achieve the ultimate filtration treatment. The other responses were remained "within the range". Accordingly, the optimization tool assimilates the singular desirability into a particular number, and then aims to optimize the function.

Consequently, the composition of the optimum membrane, together with respective rates of removal efficiency, were obtained. The optimum removals and their corresponding water flux and max. TMP are presented in Table 7.

Table 7. Predicted and experimental removal efficiencies of the optimum PVDF-PAC membrane with the corresponding operating condition.

Operating Conditions		Desirability	Optimum Conditions	
PVDF (wt.%)	**PAC (wt.%)**			
14.90	1.00	0.870	Selected	
Response		**Predicted Result**	**Experimental Result**	**Error (%)**
Removal of COD (%) *		35.34	36.63	3.65
Removal of colour (%) *		48.71	49.50	1.62
Removal of NH_3-N (%) *		22.00	23.84	8.36
Pure water flux (L/m^2·h)		61.00	61.10	0.16
Max. TMP (bar)		0.67	0.64	4.48

* Optimum value.

The membrane with 14.9 wt.% of PVDF and 1.0 wt.% of PAC was found to be the optimum, and thus selected as the best membrane design, having optimum removal efficiency according to its highest desirability (0.870) [60].

As shown in Table 7, 35.34, 48.71, and 22.00% removal of COD, colour and NH_3-N, respectively, was predicted by the software under optimized operational conditions. The corresponding (non-optimized) water flux and max. TMP were found at the values 61.00 L/m$^2\cdot$h and 0.67 bar, respectively. An additional experimentation was then performed to confirm the optimum findings.

As illustrated in Table 7, the error column indicates the differences between the predicted and laboratory values, which shows that the lab experiments agree well with the response values estimated by the software. However, less agreement between the predicted and the laboratory result was obtained in case of NH$_3$-N removal (8.36% error).

3.6. Membrane Performance Comparison with Other Reported Studies

The performance of the optimum fabricated membrane with other reported PVDF produced membranes is shown in Table 8. It can be noticed from Table 8 that the current study offered the smoother surface among the existing works based on the average roughness (R_a = 36.39 nm), which accordingly improves the removing performance and antifouling properties of the created membrane [61]. There exists few values of pure flux that are higher than the achieved in the current work, such as the flux of 143.24 L/m$^2\cdot$h produced by Penboon et al. [62]. The low value of pure water flux of the current work (61.00 L/m$^2\cdot$h) could be ascribed to the differences in the experimental characteristics such as the type of wastewater or the rates of feed flow. In addition, the rejection efficiency in the current work is lower than previously reported, which could be solved through further enhancement of the produced membranes using hydrophilic additives such as PVA or PVP [25,57]. Based on previous studies, after saturation, membrane corroborated PAC can be washed back and reused [13,63].

Table 8. Comparison of performance with other modified PVDF membranes in wastewater treatment process.

Modification Agent	Pure Water Flux (L/m$^2\cdot$h)	Feed Type & Feeding Rate (L/min)	Removal Rate Avg. (%)	Roughness (nm)	Reference
Titanium dioxide (TiO$_2$)	143.24	Wastewater FR = 0.850	86.1	-	[62]
Granular activated carbon (GAC)	13.90	Berlin tap water (Gravity driven)	88.0	-	[64]
Silica nanoparticles (SiO$_2$)	-	Cooking wastewater FR = 48.96	66.7	R_a = 174	[65]
Reduced graphene oxide (rGO)	-	NaCl solution FR = 0.385	58.0	R_a = 84	[66]
Powdered activated carbon (PAC)	61.00	SLF leachate FR = 0.20	35.35	R_a = 36.39	Present study

4. Conclusions

The adsorbent material PAC was used to fabricate a novel PVDF membrane for the treatment of stabilized landfill leachate. The fabricated PVDF flat sheet membranes integrated with PAC showed better performance when compared with PVDF membrane (without PAC). The addition of PAC effectively enhanced the removal rate and the fouling control parameters of produced membranes. Increasing PAC content to a certain value has a positive influence on the removal efficiency of COD, colour, and NH$_3$-N, as well as on membrane characteristics. Operational optimization was performed using RSM to select the optimum membrane design in terms of the removal efficiency. The best membrane composition was found at (14.9 wt.%) PVDF and (1.0 wt.%) PAC, which removed 36.63% of COD, 49.50% of colour, and 23.84% of NH$_3$-N. This was in agreement with the predicted removals. The corresponding experimental values of water flux and max. TMP also agreed with the prediction, with the values of 61.10 L/m$^2\cdot$h and 0.64 bar, respectively. The performance and structure of fabricated membranes were investigated by filtration tests, FTIR, FESEM, and AFM spectroscopy. In general, this work shows the potential of treatment and hydrophilic improvement of hydrophobic PVDF polymer membranes using

PAC. For further removal efficiency, membrane properties or practice could be improved by either adding a hydrophilic material, or applying pre-treatment process such as adsorption via PAC.

Author Contributions: S.M.A.A.: experimental work, writing original draft, preparation, and revisions. Z.H.J.: visualization, investigation, and language reviewing. C.-A.N.: supervision. Y.-C.H.: funding and technical support. M.J.K.B.: supervision, conceptualization, methodology, software, and revision. All authors have read and agreed to the published version of the manuscript.

Funding: This research was funded by Higher Education Ministry for their fund (FRGS/1/2019/TK10/ UTAR/02/3 and PETRONAS through YUTP grant (015LC0-169).

Institutional Review Board Statement: Not applicable.

Informed Consent Statement: Not applicable.

Data Availability Statement: Not applicable.

Acknowledgments: The authors are thankful to the Higher Education Ministry for their fund (FRGS/1/2019/TK10/UTAR/02/3). This research was funded by PETRONAS through YUTP grant (015LC0-169).

Conflicts of Interest: The authors declare no conflict of interest.

References

1. Xu, Y.; Chen, C.; Li, X.; Lin, J.; Liao, Y.; Jin, Z. Recovery of humic substances from leachate nanofiltration concentrate by a two-stage process of tight ultrafiltration membrane. *J. Clean. Prod.* **2017**, *161*, 84–94. [CrossRef]
2. Bashir, M.J.K.; Jun, Y.Z.; Yi, L.J.; Abushammala, M.F.M.; Amr, S.S.A.; Pratt, L.M. Appraisal of student's awareness and practices on waste management and recycling in the Malaysian University's student hostel area. *J. Mater. Cycles Waste Manag.* **2020**, *22*, 916–927. [CrossRef]
3. Abuabdou, S.M.A.; Ahmad, W.; Aun, N.C.; Bashir, M.J.K. A review of anaerobic membrane bioreactors (AnMBR) for the treatment of highly contaminated landfill leachate and biogas production: Effectiveness, limitations and future perspectives. *J. Clean. Prod.* **2020**, *255*, 120215. [CrossRef]
4. Azmi, N.B.; Bashir, M.J.K.; Sethupathi, S.; Ng, C.A. Anaerobic stabilized landfill leachate treatment using chemically activated sugarcane bagasse activated carbon: Kinetic and equilibrium study. *Desalin. Water Treat.* **2016**, *57*, 3916–3927. [CrossRef]
5. Abuabdou, S.M.A.; Teng, O.W.; Bashir, M.J.K.; Aun, N.C.; Sethupathi, S.; Pratt, L.M. Treatment of tropical stabilised landfill leachate using palm oil fuel ashisothermal and kinetic studies. *Desalin. Water Treat.* **2019**, *144*, 201–210. [CrossRef]
6. Banch, T.J.H.; Hanafiah, M.M.; Alkarkhi, A.F.M.; Abu Amr, S.S. Factorial Design and Optimization of Landfill Leachate Treatment Using Tannin-Based Natural Coagulant. *Polymers* **2019**, *11*, 1349. [CrossRef]
7. Xu, Q.; Siracusa, G.; Di Gregorio, S.; Yuan, Q. COD removal from biologically stabilized landfill leachate using Advanced Oxidation Processes (AOPs). *Process. Saf. Environ. Prot.* **2018**, *120*, 278–285. [CrossRef]
8. Guvenc, S.Y.; Dincer, K.; Varank, G. Performance of electrocoagulation and electro-Fenton processes for treatment of nano fi ltration concentrate of biologically stabilized landfill leachate. *J. Water Process Eng.* **2019**, *31*, 100863–100875. [CrossRef]
9. Abu Amr, S.S.; Zakaria, S.N.F.; Aziz, H.A. Performance of combined ozone and zirconium tetrachloride in stabilized landfill leachate treatment. *J. Mater. Cycles Waste Manag.* **2016**, *19*, 1384–1390. [CrossRef]
10. Zielinska, M.; Kulikowska, D.; Stanczak, M. Adsorption—Membrane process for treatment of stabilized municipal landfill leachate. *Waste Manag.* **2020**, *114*, 174–182. [CrossRef]
11. Abuabdou, S.M.A.; Bashir, M.J.K.; Aun, N.C.; Sethupathi, S. Applicability of anaerobic membrane bioreactors for landfill leachate treatment: Review and opportunity. *IOP Conf. Ser. Earth Environ. Sci.* **2018**, *140*, 012033–012042. [CrossRef]
12. Shen, J.; Zhang, Q.; Yin, Q.; Cui, Z.; Li, W.; Xing, W. Fabrication and characterization of amphiphilic PVDF copolymer ultra fi ltration membrane with high anti-fouling property. *J. Memb. Sci.* **2017**, *521*, 95–103. [CrossRef]
13. Ng, C.A.; Wong, L.Y.; Bashir, M.J.K.; Ng, S.L. Development of Hybrid Polymeric Polyerthersulfone (PES) Membrane Incorporated with Powdered Activated Carbon (PAC) for Palm Oil Mill Effluent (POME) Treatment. *Int. J. Integr. Eng. Spec. Issue Civ. Environ. Eng.* **2018**, *10*, 137–141. [CrossRef]
14. Krzeminski, P.; Leverette, L.; Malamis, S.; Katsou, E. Membrane bioreactors—A review on recent developments in energy reduction, fouling control, novel configurations, LCA and market prospects. *J. Memb. Sci.* **2017**, *527*, 207–227. [CrossRef]
15. Shehzad, A.; Bashir, M.J.K.; Sethupathi, S.; Lim, J. Simultaneous Removal of Organic and Inorganic Pollutants From Landfill Leachate Using Sea Mango Derived Activated Carbon via Microwave Induced Activation. *Int. J. Chem. React. Eng.* **2016**, *14*, 991–1001. [CrossRef]
16. Miller, D.J.; Paul, D.R.; Freeman, B.D. An improved method for surface modification of porous water purification membranes. *Polymer* **2014**, *55*, 1375–1383. [CrossRef]
17. Jankhah, S.; Bérubé, P.R. Pulse bubble sparging for fouling control. *Sep. Purif. Technol.* **2014**, *134*, 58–65. [CrossRef]

18. Venkataganesh, B.; Maiti, A.; Bhattacharjee, S.; De, S. Electric field assisted cross flow micellar enhanced ultrafiltration for removal of naphthenic acid. *Sep. Purif. Technol.* **2012**, *98*, 36–45. [CrossRef]
19. Crittenden, J.C.; Trussell, R.R.; Hand, D.W.; Howe, K.J.; Tchobanoglous, G. *MWH's Water Treatment: Principles and Design*; John Wiley & Sons: Hoboken, NJ, USA, 2012.
20. Ismail, N.H.; Salleh, W.N.W.; Ismail, A.F.; Hasbullah, H.; Yusof, N.; Aziz, F.; Jaafar, J. Hydrophilic polymer-based membrane for oily wastewater treatment: A review. *Sep. Purif. Technol.* **2020**, *233*, 116007. [CrossRef]
21. Nawi, N.I.M.; Bilad, M.R.; Nordin, N.A.H.M.; Mavukkandy, M.O.; Putra, Z.A.; Wirzal, M.D.H.; Jaafar, J.; Khan, A.L. Exploiting the interplay between liquid-liquid demixing and crystallization of the PVDF membrane for membrane distillation. *Int. J. Polym. Sci.* **2018**, *2018*, 1–10. [CrossRef]
22. Ji, J.; Liu, F.; Hashim, N.A.; Abed, M.R.M.; Li, K. Reactive & Functional Polymers Poly (vinylidene fluoride) (PVDF) membranes for fluid separation. *React. Funct. Polym.* **2015**, *86*, 134–153. [CrossRef]
23. Bashir, M.J.K.; Xian, T.M.; Shehzad, A.; Sethupahi, S.; Aun, N.C.; Amr, S.A. Sequential treatment for landfill leachate by applying coagulation-adsorption process. *Geosystem Eng.* **2017**, *20*, 9–20. [CrossRef]
24. Lv, J.; Zhang, G.; Zhang, H.; Zhao, C.; Yang, F. Improvement of antifouling performances for modified PVDF ultrafiltration membrane with hydrophilic cellulose nanocrystal. *Appl. Surf. Sci.* **2018**, *440*, 1091–1100. [CrossRef]
25. Mokhtar, N.M.; Lau, W.J.; Ng, B.C.; Ismail, A.F.; Veerasamy, D. Preparation and characterization of PVDF membranes incorporated with different additives for dyeing solution treatment using membrane distillation. *Desalin. Water Treat.* **2015**, *56*, 1999–2012. [CrossRef]
26. Zhou, A.; Jia, R.; Wang, Y.; Sun, S.; Xin, X.; Wang, M.; Zhao, Q.; Zhu, H. Abatement of sulfadiazine in water under a modified ultrafiltration membrane (PVDF-PVP-TiO$_2$-dopamine) filtration-photocatalysis system. *Sep. Purif. Technol.* **2020**, *234*. [CrossRef]
27. Jhaveri, J.H.; Murthy, Z.V.P. A comprehensive review on anti-fouling nanocomposite membranes for pressure driven membrane separation processes. *Desalination* **2016**, *379*, 137–154. [CrossRef]
28. Zhang, J.; Wang, Q.; Wang, Z.; Zhu, C.; Wu, Z. Modification of poly (vinylidene fluoride)/polyethersulfone blend membrane with polyvinyl alcohol for improving antifouling ability. *J. Memb. Sci.* **2014**, *466*, 293–301. [CrossRef]
29. Wong, L.Y.; Ng, C.A.; Bashir, M.J.K.; Cheah, C.K.; Leong, K. Desalination and Water Treatment Membrane bioreactor performance improvement by adding adsorbent and coagulant: A comparative study. *Desalin. Water Treat.* **2015**, 37–41. [CrossRef]
30. Wong, S.; Ngadi, N.; Inuwa, I.M.; Hassan, O. Recent advances in applications of activated carbon from biowaste for wastewater treatment: A short review. *J. Clean. Prod.* **2018**, *175*, 361–375. [CrossRef]
31. Bashir, M.J.K.; Wong, J.W.; Sethupathi, S.; Aun, N.C.; Wei, L.J. Preparation of Palm Oil Mill Effluent Sludge Biochar for the Treatment of Landfill Leachate. *MATEC Web Conf.* **2017**, *103*, 1–8. [CrossRef]
32. APHA Water Environment Federation. *Standard Methods for the Examination of Water and Wastewater*, 22nd ed.; APHA: Washington, DC, USA, 2012.
33. Abuabdou, S.M.A.; Bashir, M.J.K.; Aun, N.C.; Sethupathi, S.; Yong, W.L. Development of a novel polyvinylidene fluoride membrane integrated with palm oil fuel ash for stabilized landfill leachate treatment. *J. Clean. Prod.* **2021**, *311*, 127677. [CrossRef]
34. Mertens, M.; Van Dyck, T.; Van Goethem, C.; Gebreyohannes, A.Y.; Vankelecom, I.F.J. Development of a polyvinylidene di fluoride membrane for nano filtration. *J. Memb. Sci.* **2018**, *557*, 24–29. [CrossRef]
35. Tan, S.T.; Ho, W.S.; Hashim, H.; Lee, C.T.; Taib, M.R.; Ho, C.S. Energy, economic and environmental (3E) analysis of waste-to-energy (WTE) strategies for municipal solid waste (MSW) management in Malaysia. *Energy Convers. Manag.* **2015**, *102*, 111–120. [CrossRef]
36. Abu Amr, S.S.; Aziz, H.A.; Bashir, M.J.K. Application of response surface methodology (RSM) for optimization of semi-aerobic landfill leachate treatment using ozone. *Appl. Water Sci.* **2014**, *4*, 231–239. [CrossRef]
37. Stat-Ease Inc. Multifactor RSM Tutorial. Available online: https://www.statease.com/ (accessed on 30 July 2021).
38. Naragintia, S.; Yua, Y.-Y.; Fanga, Z.; Yong, Y.-C. Novel tetrahedral Ag$_3$PO$_4$@N-rGO for photocatalytic detoxification of sulfamethoxazole: Process optimization, transformation pathways and biotoxicity assessment. *Chem. Eng. J.* **2019**, *375*, 122035. [CrossRef]
39. Ghani, Z.A.; Suffian, M.; Qamaruz, N.; Faiz, M.; Ahmad, M. Optimization of preparation conditions for activated carbon from banana pseudo-stem using response surface methodology on removal of color and COD from landfill leachate. *Waste Manag.* **2017**, *62*, 177–187. [CrossRef]
40. Ike, I.A.; Dumée, L.F.; Groth, A.; Orbell, J.D.; Duke, M. Effects of dope sonication and hydrophilic polymer addition on the properties of low pressure PVDF mixed matrix membranes. *J. Memb. Sci.* **2017**, *540*, 200–211. [CrossRef]
41. Nejati, S.; Boo, C.; Osuji, C.O.; Elimelech, M. Engineering flat sheet microporous PVDF films for membrane distillation. *J. Memb. Sci.* **2015**, *492*, 355–363. [CrossRef]
42. Fan, H.; Peng, Y. Application of PVDF membranes in desalination and comparison of the VMD and DCMD processes. *Chem. Eng. Sci.* **2012**, *79*, 94–102. [CrossRef]
43. Aziz, H.A.; Mojiri, A. *Wastewater Engineering: Advanced Wastewater Treatment Systems*, 1st ed.; IJSR Publications: Raipur, India, 2014; ISBN 2322-4657.
44. Chidambaram, T.; Oren, Y.; Noel, M. Fouling of nanofiltration membranes by dyes during brine recovery from textile dye bath wastewater. *Chem. Eng. J.* **2015**, *262*, 156–168. [CrossRef]

45. Zheng, Y.; Zhang, W.; Tang, B.; Ding, J.; Zheng, Y.; Zhang, Z. Membrane fouling mechanism of biofilm-membrane bioreactor (BF-MBR): Pore blocking model and membrane cleaning. *Bioresour. Technol.* **2018**, *250*, 398–405. [CrossRef]

46. Chen, X.; Zhao, B.; Zhao, L.; Bi, S.; Han, P.; Chen, L. Temperature- and pH-responsive properties of poly(vinylidene fluoride) membranes functionalized by blending microgels. *RSC Adv.* **2014**, *4*, 29933–29945. [CrossRef]

47. *Environmental Requirements: A Guide for Investors, Department of Environment*; Ministry of Natural Resources and Environment: Kuala Lumpur, Malaysia, 2010.

48. Kemal, A.; Siraj, K.; Michael, W.H. Adsorption of Cu (II) and Cd (II) onto Activated Carbon Prepared from Pumpkin Seed Shell. *J. Environ. Sci. Pollut. Res.* **2019**, *5*, 328–333. [CrossRef]

49. Aslam, M.; Ahmad, R.; Kim, J. Recent developments in biofouling control in membrane bioreactors for domestic wastewater treatment. *Sep. Purif. Technol.* **2018**, *206*, 297–315. [CrossRef]

50. Ali, I.; Bamaga, O.A.; Gzara, L.; Bassyouni, M. Assessment of Blend PVDF Membranes, and the Effect of Polymer Concentration and Blend Composition. *J. Membr.* **2018**, *8*, 13. [CrossRef] [PubMed]

51. Gunst, R.F.; Stein, R.; Hatcher, L.; Heckler, C.E.; Adams, B.M.; Kit, T.S.; Fink, A.; Lawson, J. *Response Surface Methodology: Process and Product Optimization Using Designed Experiments Book Reviews*; John Wiley & Sons: Hoboken, NJ, USA, 2012.

52. Abuabdou, S.M.A.; Teng, O.W.; Bashir, M.J.K.; Aun, N.C.; Sethupathi, S. Adsorptive treatment of stabilized landfill leachate using activated palm oil fuel ash (POFA). *AIP Conf. Proc.* **2019**, *2157*, 20002–20008. [CrossRef]

53. Khadijah, S.; Ha, M.; Othman, D.; Harun, Z.; Ismail, A.F.; Iwamoto, Y.; Honda, S.; Rahman, M.A.; Jaafar, J.; Gani, P.; et al. Effect of fabrication parameters on physical properties of metakaolin-based ceramic hollow fi bre membrane (CHFM). *Ceram. Int.* **2016**, *42*, 15547–15558. [CrossRef]

54. Fumoto, E.; Sato, S.; Takanohashi, T. Determination of Carbonyl Functional Groups in Heavy Oil Using Infrared Spectroscopy. *Energy Fuels* **2020**, *34*, 5231–5235. [CrossRef]

55. Gurkan, B.E.; de la Fuente, J.C.; Mindrup, E.M.; Ficke, L.E.; Goodrich, B.F.; Price, E.A.; Schneider, W.F.; Brennecke, A.J.F. Equimolar CO_2 Absorption by Anion-Functionalized Ionic Liquids. *J. Am. Chem. Soc.* **2010**, *132*, 2116–2117. [CrossRef] [PubMed]

56. Guo, D.; Xiao, Y.; Li, T.; Zhou, Q.; Shen, L.; Li, R.; Xu, Y.; Lin, H. Fabrication of high-performance composite nanofiltration membranes for dye wastewater treatment: Mussel-inspired layer-by-layer self-assembly. *J. Colloid Interface Sci.* **2020**, *560*, 273–283. [CrossRef]

57. Zhang, J.; Wang, Z.; Wang, Q.; Ma, J.; Cao, J.; Hu, W.; Wu, Z. Relationship between polymers compatibility and casting solution stability in fabricating PVDF/PVA membranes. *J. Memb. Sci.* **2017**, *537*, 263–271. [CrossRef]

58. Kunst, B.; Sourirajan, S. An approach to the development of cellulose acetate ultrafiltration membranes. *J. Appl. Polym. Sci.* **1974**, *18*, 3423–3434. [CrossRef]

59. Tae, J.; Kim, J.F.; Hyun, H.; Drioli, E.; Moo, Y. Understanding the non-solvent induced phase separation (NIPS) effect during the fabrication of microporous PVDF membranes via thermally induced phase separation (TIPS). *J. Memb. Sci.* **2016**, *514*, 250–263. [CrossRef]

60. Bezerra, M.A.; Santelli, R.E.; Oliveira, E.P.; Villar, L.S.; Escaleira, L.A. Response surface methodology (RSM) as a tool for optimization in analytical chemistry. *Talanta* **2008**, *76*, 965–977. [CrossRef]

61. Ayyaru, S.; Ahn, Y.H. Application of sulfonic acid group functionalized graphene oxide to improve hydrophilicity, permeability, and antifouling of PVDF nanocomposite ultrafiltration membranes. *J. Memb. Sci.* **2017**, *525*, 210–219. [CrossRef]

62. Penboon, L.; Khrueakham, A.; Sairiam, S. TiO₂ coated on PVDF membrane for dye wastewater treatment by a photocatalytic membrane. *Water Sci. Technol.* **2019**, *79*, 958–966. [CrossRef]

63. Tai, C.Y.; Abuabdou, S.M.A.; Ng, C.A.; Gan, C.H.; Bashir, M.J.K. Performance of hybrid anaerobic membrane bioreactors (AnMBRs) augmented with activated carbon in treating palm oil mill effluent (POME). *Desalin. Water Treat.* **2020**, *26156*, 1–8. [CrossRef]

64. Schumann, P.; Andrade, J.A.O.; Jekel, M.; Ruhl, A.S. Packing granular activated carbon into a submerged gravity-driven fl at sheet membrane module for decentralized water treatment. *J. Water Process Eng.* **2020**, *38*, 101517. [CrossRef]

65. Li, J.; Guo, S.; Xu, Z.; Li, J.; Pan, Z.; Du, Z.; Cheng, F. Preparation of omniphobic PVDF membranes with silica nanoparticles for treating coking wastewater using direct contact membrane distillation: Electrostatic adsorption vs. chemical bonding. *J. Memb. Sci.* **2019**, *574*, 349–357. [CrossRef]

66. Abdel-karim, A.; Luque-alled, J.M.; Leaper, S.; Alberto, M.; Fan, X.; Vijayaraghavan, A.; Gad-allah, T.A.; El-kalliny, A.S.; Szekely, G.; Ahmed, S.I.A.; et al. PVDF membranes containing reduced graphene oxide: Effect of degree of reduction on membrane distillation performance. *Desalination* **2019**, *452*, 196–207. [CrossRef]

Review

Treatment of Poultry Slaughterhouse Wastewater with Membrane Technologies: A Review

Faryal Fatima [1], **Hongbo Du** [1] and **Raghava R. Kommalapati** [2,*]

[1] Center for Energy and Environmental Sustainability, Prairie View A&M University, Prairie View, TX 77446, USA; ffatima@pvamu.edu (F.F.); hodu@pvamu.edu (H.D.)

[2] Center for Energy and Environmental Sustainability & Department of Civil and Environmental Engineering, Prairie View A&M University, Prairie View, TX 77446, USA

* Correspondence: rrkommalapati@pvamu.edu; Tel.: +1-936-261-1660

Abstract: Poultry slaughterhouses produce a large amount of wastewater, which is usually treated by conventional methods. The traditional techniques face some challenges, especially the incapability of recovering valuable nutrients and reusing the treated water. Therefore, membrane technology has been widely adopted by researchers due to its enormous advantages over conventional methods. Pressure-driven membranes, such as microfiltration (MF), ultrafiltration (UF), nanofiltration (NF), and reverse osmosis (RO), have been studied to purify poultry slaughterhouse wastewater (PSWW) as a standalone process or an integrated process with other procedures. Membrane technology showed excellent performance by providing high efficiency for pollutant removal and the recovery of water and valuable products. It may remove approximately all the pollutants from PSWW and purify the water to the required level for discharge to the environment and even reuse for industrial poultry processing purposes while being economically efficient. This article comprehensively reviews the treatment and reuse of PSWW with MF, UF, NF, and RO. Most valuable nutrients can be recovered by UF, and high-quality water for reuse in poultry processing can be produced by RO from PSWW. The incredible performance of membrane technology indicates that membrane technology is an alternative approach for treating PSWW.

Keywords: poultry slaughterhouse wastewater; microfiltration; ultrafiltration; nanofiltration; reverse osmosis

Citation: Fatima, F.; Du, H.; Kommalapati, R.R. Treatment of Poultry Slaughterhouse Wastewater with Membrane Technologies: A Review. *Water* **2021**, *13*, 1905. https://doi.org/10.3390/w13141905

Academic Editors: Amin Mojiri and Mohammed J. K. Bashir

Received: 1 June 2021
Accepted: 7 July 2021
Published: 9 July 2021

Publisher's Note: MDPI stays neutral with regard to jurisdictional claims in published maps and institutional affiliations.

1. Introduction

Water is essential for all lives and a natural resource at the core of sustainable development. It is critical for socio-economic prosperity, healthy ecosystems, and human survival. Unfortunately, water is a finite and irreplaceable resource in time and space. The increase in water consumption has made water management a priority. On the other hand, improper wastewater treatment in some regions has intensified the inadequate discharge of wastewater into the environment and augmented natural water resource pollution. As a result, progressively stricter standards for effluent discharge worldwide have changed the target from wastewater disposal to water reuse and recycling, leading to advanced wastewater treatment technologies, which can recycle and reuse wastewater [1].

Food industries such as dairy, beverage, vegetable, fruit, oilseed, seafood, poultry, and other types of meat consume a high volume of freshwater. Among them, the poultry industry is at the top [2]. From 2018 to 2019, the world poultry market increased by 6% due to an increase in the per capita poultry consumption, which corresponds to 58 kg per person in the U.S., 57 kg per person in Brazil, and 48 kg per person in Peru. The high demand for poultry meat correspondingly increases freshwater consumption by poultry processing plants [3].

Poultry slaughterhouses discharge massive amounts of wastewater into the environment because of their high freshwater usage for the continuous operations of cutting up, rinsing, and packaging meat. Other operations in poultry slaughterhouses such as scalding, de-feathering, evisceration, and bird wash are also water-intensive and generate a significant amount of wastewater. The eviscerating step and bird wash generate enormous wastewater, at 7.57 L/bird and 4.35 L/bird, respectively, as shown in Figure 1. On average, a 2.3 kg bird consumes 26.5 L of water [4,5]. The wastewater is highly contaminated with organic matter quantified as biochemical oxygen demand (BOD) and chemical oxygen demand (COD). It also contains high nitrogen and phosphorous constituents, including blood, fats, oil, grease, and proteins [6]. Thus, discharging improperly treated poultry slaughterhouse wastewater (PSWW) has a high risk of polluting freshwater sources. It can also cause serious environmental and health concerns such as deoxygenation of rivers, groundwater contamination, eutrophication, and the spread of water-borne diseases [7,8].

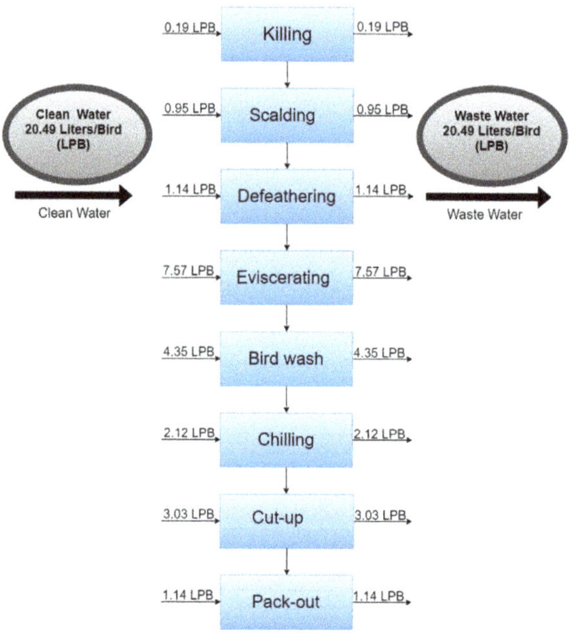

Figure 1. Water consumption during poultry processing in the poultry slaughterhouse.

Generally, PSWW is treated by physical, chemical, and biological methods. These conventional techniques are only responsible for discharging the treated water into the environment without recycling it. Besides, they face some challenges, such as lack of nutrient recovery, frequent use of chemical cleaning agents, and the degradation of valuable compounds in wastewater. Therefore, unconventional methods, e.g., pressure-driven membrane technologies, are being explored for PSWW treatment. These membrane filtration technologies can include microfiltration (MF), ultrafiltration (UF), nanofiltration (NF), and reverse osmosis (RO) [9,10]. They can overcome some limits of the conventional methods by removing colloids and suspended and macromolecular matter, and eliminating mineral substances and low-molecular organic compounds. Membrane filtration is a physical process that provides great separation efficiency and improves final product quality. Most importantly, membrane technology can produce water clean enough for the reuse of the treated water in industrial poultry processing. Furthermore, it can recover a fair amount of valuable nutrients, e.g., proteins, which could be utilized as animal feed, thus supplementing the global protein demand for animals [9,11,12].

This review identifies membrane technologies for advanced PSWW treatment. It evaluates the quality of treated water based on pollutant removal efficiency and the degree of permeate produced to meet the environmental regulation of discharging and recycling standards. In addition, we highlight the potency of membrane technologies to recover valuable nutrients from PSWW, which is not achieved by traditional methods.

2. Characteristics of Poultry Slaughterhouse Wastewater

Before any wastewater treatment, it is critical to characterize the wastewater to show pollutant levels in using some instruments and test reagents [7]. The parameters commonly used to describe PSWW are pH, COD, BOD, total organic carbon (TOC), total suspended solids (TSS), total nitrogen (TN), total phosphorus (TP), and pathogens [13,14]. COD indicates the amount of organic compounds in wastewater; a high concentration of COD suggests a large amount of oxidizable organic substances in the wastewater. Similarly, BOD indicates the biological oxidation of organic compounds, and a high BOD level also signifies large quantities of organic pollutants in wastewater. Nutrients in wastewater are TN and TP; nitrogen in wastewater is available in the organic form, primarily present in proteins, and the inorganic form, which includes nitrite (NO_2^-) and nitrate (NO_3^-). The most stable type of nitrogen in water is nitrate that originates from a natural decaying process of biological matter. Excessive nitrates in wastewater can lead to harmful algae bloom, oxygen depletion, fish poison, and putrid odors. Moreover, in wastewater, orthophosphate (PO_4^{3-}) is the most common type of phosphorus, originating from disinfectants and cleaning agents; high phosphorus constituents in wastewater may prompt eutrophication. To reduce phosphorus in wastewater, a practical and straightforward technique is chemical precipitation [15]. These parameters vary from one slaughterhouse to another due to many factors, such as system type, operation method, and processing capacity [9]. The characteristics of PSWW, effluent discharge standards, and water reuse applications are tabulated in Table 1. It is required to treat PSWW to or below the standard limits because these measured parameters of raw PSWW are much higher than the acceptable standards by the World Health Organization (WHO) and other regulations [16].

Table 1. PSWW characteristics, discharging limits, and reclaiming wastewater treatment goals [6,15,17–20].

Parameter	Significance in Wastewater Reclamation	Characteristic Type	Pollutant Level	EU Discharge Standard	US Discharge Standard	Treatment Goal in Reclaimed Wastewater	Water Reuse Application
BOD	Organic substrate for microbial or algal growth	Biological	925–5000 mg/L	25 mg/L	26 mg/L	≤10 mg/L	Urban use, agricultural irrigation, recreational use, environmental enhancement, and industrial reuse
COD	Measure of oxidizable substrate	Chemical	2133–12,490 mg/L	125 mg/L	-	-	-
TOC	Measure of organic carbon	Chemical	194.9–651.5 mg/L	-	-	<10 mg/L	-
TSS	Measure of particles in wastewater; can also be related to microbial contamination	Physical	313–8200 mg/L	35 mg/L	30 mg/L	≤30 mg/L	Urban use, agricultural irrigation, recreational use, environmental enhancement, and industrial reuse
Turbidity	Measure of particles in wastewater	Physical	237–997 mg/L	-	-	≤2 NTU	Urban use, agricultural irrigation, recreational use, environmental enhancement, and industrial reuse
Nitrogen	Nutrient source for irrigation; can also contribute to algal growth	Chemical	162.6–563.8 mg/L	10 mg/L	8 mg/L	<30 mg/L	-
Phosphorous	Nutrient source for irrigation; can also contribute to algal growth	Chemical	8–27 mg/L	1 mg/L	-	<20 mg/L	-
pH	Measure of acidity and basicity	Chemical	6.13–8.0 mg/L	-	6–9 mg/L	6–9	Urban use, agricultural irrigation, recreational use, environmental enhancement, and industrial reuse
Pathogen	Measure of the risk of microbial infection due to enteric viruses, pathogenic bacteria, and protozoa	Biological	30–4020 CFU/ 100 mL	-	-	≤200 CFU/100 mL	Urban use, agricultural irrigation, recreational use, environmental enhancement, and industrial reuse

-: Not reported; CFU = colony-forming unit.

3. Conventional Treatment of Poultry Slaughterhouse Wastewater as Pretreatment Prior to Membrane Separation

The conventional treatment for PSWW is similar to municipal wastewater treatment, consisting of preliminary, primary, and secondary treatments. There are various combined treatment methods after preliminary treatment, and the most common combination is physicochemical treatment as primary and biological treatment as secondary, as described below [18]. Prior to PSWW membrane filtration, some conventional treatment is necessary as pretreatment to alleviate membrane fouling and improve overall membrane performance. Without proper pretreatment, membranes will suffer severe membrane fouling, hindering the membrane performance, and some heavy fouling could even cause membrane failure. For example, Meiramkulova et al. [21] investigated the performance of an integrated membrane process with electrochemical pretreatment on PSWW purification, and their findings showed that the electrochemical pretreatment was highly efficient at reducing turbidity, color, TSS, COD, and BOD by 71–85%. In addition, the pretreatment resulted in a low rate of cake formation on the membrane [21].

3.1. Preliminary Treatment

The preliminary treatment removes suspended solids from PSWW; the most common preliminary treatment uses screeners, sieves, and strainers. A typical wire mesh screen retains solid fraction with a size of 10–30 mm. Rotary screeners extract solids greater than 0.5 mm diameters; they also protect the equipment from fouling, clogging, and jamming. In preliminary treatment, 60% of suspended solids and 30% of BOD are removed from PSWW. Other preliminary methods are catch basins, flotation, equalization, and settlers [1].

3.2. Primary Treatment

After the preliminary treatment, PSWW goes through primary treatment in which the physiochemical process eliminates BOD, COD, oil, grease, fats, and residual TSS. The typical primary methods are dissolved air flotation (DAF), electrocoagulation, coagulation-flocculation, and sedimentation [18].

3.2.1. Dissolved Air Flotation

DAF is a separation of solid from liquid in which air is introduced into PSWW from the bottom of the tank, resulting in moving the light solids, grease, and fats on the surface, creating a sludge blanket. The efficiency of a DAF system can be improved by adding polymers and other flocculants. Generally, DAF's efficiency at removing BOD and COD is 30–90% and 70–80%, respectively [1]. Additionally, DAF removes suspended solids from PSWW in the range of 38–70%, and it eliminates fats in the range of 63–95%. The drawbacks of DAF are regular malfunctioning and poor TSS separation [2].

3.2.2. Coagulation-Flocculation and Sedimentation

In the coagulation process, the colloidal particles present in the PSWW are grouped with large particles to form flocs. Those colloidal particles are nearly negatively charged, so they can be destabilized by adding positively charged coagulants to rescind the formation of flocs and ease the sedimentation process [18]. Previous research showed that this process could remove oil, grease, and TSS by up to 85%, and the removal efficiency was reported as 62–78.8% of BOD and 74.6–79.5% of COD [22]. However, this process results in toxicity and health hazards, incfficient removal of heavy metals and emerging contaminants, and an increase in effluent color [23].

3.2.3. Electrocoagulation

Electrocoagulation (EC) is an advanced method for removing large amounts of pollutants from wastewater, such as organics, heavy metals, and pathogens, using electric current. The EC process generates M^{3+}, Fe^{3+}, or Al^{3+} ions using different electrode materials; the most commonly used electrodes are Al and Fe [18]. EC is a three-step process.

147

In the first step, electrolytic oxidation forms metal hydroxides and oxyhydroxides at the sacrificial electrode. Then, the produced coagulants destabilize and adsorb the pollutants. Finally, flocs formed by aggregation of the destabilized phase are removed by a downstream sedimentation and filtration process [24]. The EC sets up sacrificial anodes that need to be changed regularly, and chlorinated toxic compounds can form if chlorine is present. In some regions where electricity is expensive, the cost of operating EC is high [25].

3.3. Secondary Treatment

The pollutants present in wastewater that are not removed by primary treatment are further treated by secondary treatment. The main goal of secondary treatment is the removal of organic compounds to reduce the BOD level. In the secondary treatment, the biological process, aerobic and anaerobic digesters are used for treating PSWW [1]. In both treatments, organic matter is degraded into simple compounds with the help of decomposers, where the efficiency of the decomposers depends on the quality of wastewater [2].

3.3.1. Anaerobic Digestion

In biological treatment, anaerobic digestion of organic waste, sludge, and high-strength wastewater is a widespread technique [26]. The anaerobic system's primary goal is to reduce high-level BOD [27]. Anaerobic digestion consists of hydrolysis, acidogenesis, acetogenesis, and methanogenesis steps. With the help of a diverse group of microorganisms (bacteria and archaea), complex organic compounds are degraded in the absence of oxygen. The degradation rate relies primarily on various bacterial activity rates [26]. In the anaerobic treatment, organic compounds are broken down into methane, water, and carbon dioxide by anaerobic bacteria in an anaerobic environment [28]. However, PSWW usually has high organic strength, which can negatively affect the anaerobic process's performance. Therefore, an anaerobic system for PSWW is often followed by additional treatment to remove TP, TN, and pathogenic microorganisms [2].

3.3.2. Aerobic Digestion

The main goal of aerobic digestion is nitrification [27]. Aerobic digestion uses oxygen to break down organic matter and other pollutants; it degrades ammonia or other organic matter into less harmful compounds like carbon dioxide, water, and nitrate. The oxygen and time needed for this treatment depend on the organic strength of PSWW. Aerobic digestion is usually applied as the last nutrient removal when using anaerobic techniques for the decontamination of sludge water. Some drawbacks of aerobic digestion are daily maintenance, excess biomass production, and increased demand for oxygen and electricity [26].

4. Membrane Technology for Poultry Slaughterhouse Wastewater

Pressure-driven membrane processes such as MF, UF, RO, and NF are widely studied for wastewater treatment throughout the world. Pressure-driven membranes rely on hydraulic pressure to achieve separation [29]. Membrane filtration is one of the most emerging technologies to produce high-quality water because it utilizes zero chemical constituents and offers enormous advantages over conventional methods. Several research groups have reported the use of membrane technology for PSWW treatment [12]. Jason et al. [30] first used membrane technology for PSWW treatment with recovering nutritional by-products in the 1980s; they experimented on a laboratory scale with a commercial tubular UF membrane with a molecular weight cut-off (MWCO) of 50 kDa. The study reported that the membrane technology produced permeate with a significant reduction of 85% of TS and 95% of COD; the permeate was believed to be safe for discharge and potential reuse. In addition, it recovered 24–45% of fat and 30–35% of protein as by-products. Since then, some scientists have experimented with membrane technology for PSWW treatment used as a standalone or an integrated process. The available literature on membrane applications in PSWW treatment is mainly based on UF membranes; in a few other research activities, such as MF, NF, RO, and membrane distillation, experiments have also been carried out. All the studies conducted on PSWW using membrane technology showed excellent performance in separation efficiency and compliance with environmental regulations.

4.1. Characteristics of Pressure-Driven Membrane

A pressure-driven membrane separates the feed into concentrate and permeate using the pressure difference as a driving force to transport the liquid or gas. Pressure-driven membranes such as MF, UF, NF, and RO membranes differentiate based on their characteristics [31]. The most important characteristics of pressure-driven membranes are pore size, structure, and operating pressure, as presented in Table 2. In reference to structure, membranes can be divided as symmetrical or asymmetrical; symmetrical membranes show uniform pore sizes in their cross-section, whereas asymmetrical membranes' pore size gets larger farther from the filter surface [11]. All the pressure-driven membranes are asymmetrical except for some MF membranes [32]. As shown in Figure 2, an MF membrane has the largest pore, highest permeability, and it can reject large suspended particles. A UF membrane has a smaller pore size and lower permeability than MF membranes and can separate small suspended particles and macromolecules. An NF membrane has the properties of the second smallest pore size, the high rejection of multivalent ions, but the low rejection of monovalent ions. In contrast, an RO membrane has a very high rejection of monovalent ions. It can be seen that as the pore size becomes smaller, the operating pressure increases [31]. For PSWW treatment, the membranes are chosen according to the pollutant levels in PSWW. A UF membrane is used to remove a substantial amount of suspended solids present in PSWW, and an RO membrane can eliminate all the pollutants, such as BOD, COD, TSS, salts, etc.

Table 2. Characteristics of pressure-driven membranes [29,32–34].

	Reverse Osmosis	Nanofiltration	Ultrafiltration	Microfiltration
Structure	Asymmetrical	Asymmetrical	Asymmetrical	Symmetrical, Asymmetrical
Pore size	<0.001 μm	0.001–0.002 μm	0.002–0.05 μm	0.05–10 μm
Rejection	All contaminants, including monovalent ions	Pigments, sulfates, divalent anions, divalent cations, lactose, sucrose, sodium chloride	Proteins, pigments, oils, sugar, micro-plastics	Bacteria, fat, oil, grease, colloids, microparticles
Membrane material(s)	CA, PS	CA, PA	PVDS, PS, poly (acrylonitrile), poly (ether sulfone)	PVDS, PS, poly (acrylonitrile), poly (ether sulfone), nylons, poly (tetrafluoroethylene), CA, cellulose nitrate
Membrane module	Tubular, spiral wound, plate-and-frame	Tubular, spiral wound, plate-and-frame	Tubular, hollow fiber spiral wound, plate-and-frame	Tubular, hollow fiber, plate-and-frame
Operating pressure	10–100 bar	5–20 bar	1–10 bar	0.1–2 bar

Most membranes for MF, UF, RO, and NF are made from synthetic organic polymers. MF and UF membranes are usually produced from the same materials under different membrane formation conditions to achieve different pore sizes. The commonly used polymers for MF and UF membranes are PVDF, PS, poly (acrylonitrile), poly (ether sulfone), and copolymers of poly (acrylonitrile) and PVDF. The materials used for MF membranes also include nylons, poly (tetrafluoroethylene), polypropylene, polyethylene, and blends of CA and cellulose nitrate. NF membranes are made from CA blends or PA composites, whereas RO membranes are produced from CA or PS coated with aromatic PA. Moreover, ceramic and metals are used to create inorganic membranes. Ceramic membranes are microporous, thermally stable, chemically resistant, and mostly used for MF, UF, and NF [31,33]. Metallic membranes are usually fabricated from stainless steel and can be very finely porous. The most common configurations of membrane modules are plate-and-frame, spiral wound, tubular, and hollow fiber.

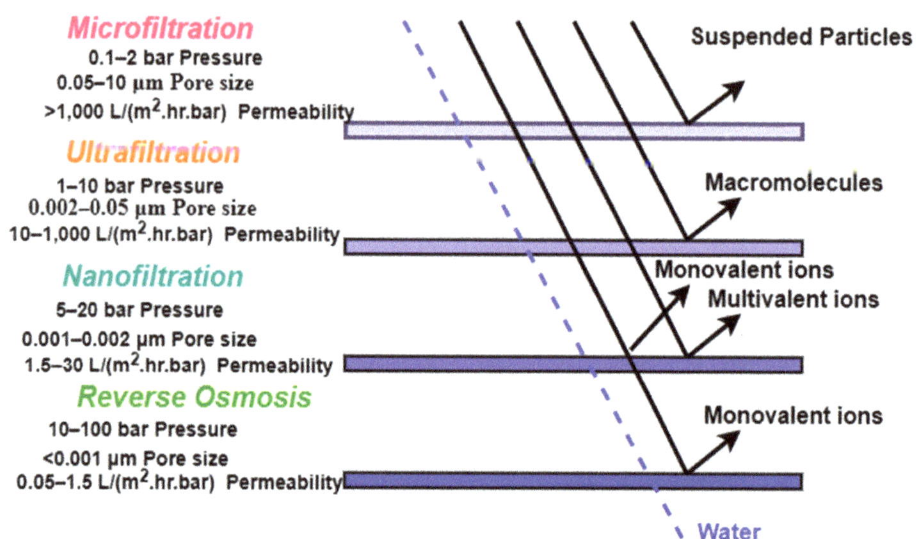

Figure 2. Schematic representation of pressure-driven membranes.

4.2. Microfiltration

The pore size of MF membranes is in the range of 0.05–10 µm, and the operating pressure of MF is between 0.1 and 2 bar for separating colloids and particles, reducing the effluent's turbidity and COD [34,35]. MF membranes have been used widely in industries such as liquid clarification and wastewater treatment, especially as an initial filtration stage for wastewater treatment [36]. Goswami and Pugazhenthi [12] evaluated the performance of fly ash tubular MF membranes with a pore size of 0.133 µm and porosity of 40.17% for PSWW treatment. The study revealed that the MF membrane produced a filtrate with zero turbidity and almost 100% removal of COD and TSS, thus satisfying the COD and TSS norms for discharging and reusing, which can help attenuate the water shortage crisis. A study was led by Marchesi et al. [37] to recycle the pre-chiller wastewater generated from the poultry carcass chilling process by using MF membranes. Both hollow fiber PA membrane with a pore size of 0.20 µm on a bench scale and spiral-wound membrane with a 0.1 µm pore size on a pilot scale provided complete retention of turbidity, apparent color, fat, and microorganisms. The rejection efficiency was up to 92.5% of COD, up to 89.1% of TOC, and 100% of microorganisms; it was stated that the use of membranes was a promising approach for the recycling and reuse of poultry pre-chiller wastewater. Abboah-afari and Kiepper [38] investigated the effects of pore size on the performance of MF membranes for the treatment of pre-DAF poultry processing wastewater. The results indicated that the 0.3 µm PVDF membrane was the most effective among the three tested membranes (0.3 µm PVDF, 0.1 µm PS, and 100,000 MWCO Ultrafilic). In detail, the 0.3 µm PVDF achieved a maximum mean permeate flux of 115 L/m^2/h, and the removal efficiency was 88% of COD and 34% of TS. Moreover, in their other research work, they identified the 0.3 µm PVDF membrane as an alternative to DAF in poultry processing wastewater [39].

4.3. Ultrafiltration

The pore size of UF membranes is in the range of 0.002–0.05 µm, and the operating pressure of UF is between 1 and 10 bar for separating macromolecules and suspended solids [34]. The UF process has been extensively explored for PSWW treatment due to its significant advantages, such as low pressure, high permeate flux, cost-effectiveness, and the capability to eliminate pathogens that are very harmful to PSWW recycling [40]. The UF's transport properties are influenced by concentration polarization, fouling, and interactions between the feed stream and the membrane [9]. The UF process is considered an economical and environmentally friendly substituent for conventional wastewater treatments by some researchers [9]. Coskun et al. [41] studied the PSWW treatment using laboratory-scale membrane processes. Their study reported that UF as pretreatment improved the removal efficiencies for NF and RO processes; NF reduced almost 90% of COD, RO removed 97.4% of COD, and the UF pretreatment resulted in higher final fluxes 8.1 and 5.7 times more for NF and RO, respectively, than for those without UF. Yordanov et al. [9] examined the efficiency of a UF 25-PAN membrane for PSWW treatment. The results indicated that UF exhibited excellent performance by removing 97% of BOD and 94% of COD and that it also reduced 99% of TSS and 98% of fats. Rinquest et al. [13] treated PSWW using a UF membrane system for the removal of organic matter and suspended solids coupled with aerobic single-stage nitrification-denitrification (SSND) and an anaerobic static granular bed reactor, as shown in Figure 3. The experimental efforts showed that the UF system further reduced COD and TSS by an average of 65% and 54%, respectively, after SSND. All the measured parameters of final effluent excluding PO_4^{3-} and NH_4^+-N satisfied industrial wastewater discharge requirements. Moreover, water flux higher than 200 L/m^2/h was obtained by identifying the optimal condition. Mannapperuma and Santos [42] assessed UF for reconditioning poultry chiller overflow. The UF operation reduced microbial counts by more than 5.4 log cycles and achieved rejection efficiency of over 73% for COD and above 99.2% for turbidity. Their results verified that UF produced water acceptable for reuse in the chiller to replace freshwater makeup. Meiramkulova et al. [20] evaluated the performance of an integrated PSWW treatment with electrolysis, UF, and ultraviolet radiation in terms of microbial inactivation from PSWW. The results showed that the integrated system achieved an overall microbial removal efficiency of 99.86–100%.

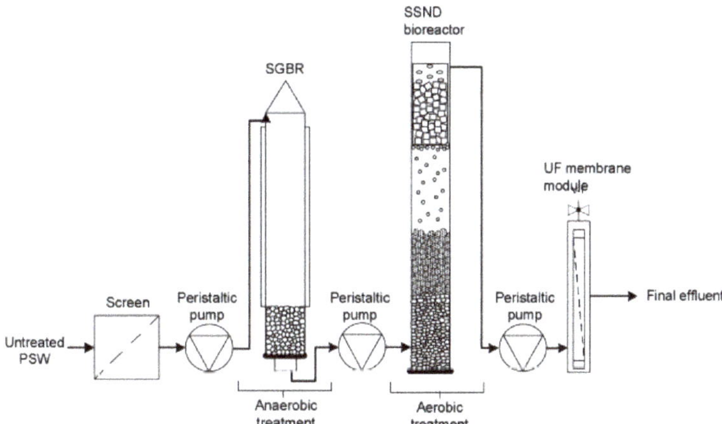

Figure 3. Static granular bed reactor coupled with single-stage nitrification-denitrification and ultrafiltration for PSWW treatment (reproduced with permission from [13]: Rinquest, Z.; Basitere, M.; Ntwampe, S. K. O.; Njoya, M. Poultry Slaughterhouse Wastewater Treatment Using a Static Granular Bed Reactor Coupled with Single-Stage Nitrification-Denitrification and Ultrafiltration Systems. *J. Water Process Eng.* **2019**, *29*, 100778. https://doi.org/10.1016/j.jwpe.2019.02.018. Accessed on 14 March 2021. Copyright (2019), Elsevier).

4.4. Nanofiltration

The pore size of NF membranes is in the range of 0.001–0.002 μm, and the operating pressure of NF is between 5 and 20 bar for separating low molecular weight particles [34]. The NF process is a great separation tool due to its versatile properties, which fall between UF and RO. It removes a large amount of multivalent inorganic salts and small organic molecules while operating at a moderate pressure; the moderate operating pressure makes the separation process consume little energy and is cost-effective [43]. Therefore, it has been applied to various industrial sectors for wastewater treatment, e.g., water recycling during fishmeal, lupin bean, and textile processing [44–46]. A few researchers explored NF for PSWW treatment by using standalone NF or combining NF with UF. Zhang et al. [10] evaluated some membrane filtration processes for poultry abattoir wastewater treatment to recycle the wastewater stream to meet the Canadian poultry wastewater reuse criteria. Their results showed that both NF membranes (DS: desal thin composite membrane and NF 45: thin-film composite membrane) produced permeate with less than 100 mg/L of TOC and gave a reasonable flux of 46 to 66 L/m^2/h. The TOC level of permeate produced by NF satisfied the Agriculture and Agri-Food Canada criteria for recycled water. However, the tested UF membranes did not meet the TOC criterion, although UF removed all bacteria and significantly reduced other organic species.

4.5. Reverse Osmosis

The operating pressure of RO is between 10 and 100 bar, and it can remove microparticles smaller than 0.001 μm, carbohydrates, amino acids, or even monovalent ions, including NH_4^+, from water [34]. The RO process has widely been used for seawater desalination since it is the most effective on a large scale [47]. It is shown that combining biological treatment with RO produces a permeate with a quality superior to the WHO standards for drinking water [48,49]. Bohdziewicz et al. [50] investigated the application of membrane processes such as UF and RO to treat the wastewater of the meat industry. The study revealed that the hybrid system consisting of RO was permissible in the removal potency of 100% phosphorus and 98.8% nitrogen compounds. The removal efficiency of COD and BOD both exceed 99%, and only the permeate produced by RO was satisfied for the reuse in the production cycle. Meiramkulova et al. [51] evaluated the performance of an integrated process for PSWW treatment on both laboratory and industrial scales; the RO step was designed to reduce the total salinity of the water. Their results showed that the removal efficiency was up to 100% of turbidity, color, and TSS, and 99.6% of BOD and COD for laboratory and industrial testing. Almost all the physical and chemical parameters of the produced water were within the recommended standards set by legislation. The water purified on an industrial scale was certified as excellent quality in terms of Kazakhstan's drinking water quality standards.

4.6. Membrane Bioreactor

A membrane bioreactor (MBR) is an integrated system with membrane filtration for the biological degradation of waste present in wastewater. Generally, it is composed of a biological unit and a membrane module, which separates water from the aerobically digested water and returns activated sludge to the biological unit, as shown in Figure 4. The MBRs can remove organic and inorganic contaminants and biological entities from wastewater [52]. They are widely used for recycling water in buildings, wastewater treatment for small communities, and industrial wastewater treatment by producing an effluent free of bacteria and pathogens. Williams [53] treated PSWW using MBR coupled with a single-stage nitrification-denitrification reactor and an expanded granular sludge bed reactor (EGSB). The study reported that the overall removal efficiency of the EGSB-SSND-MBR system was 99% of turbidity, 92% of TSS, and 99% of COD. Fuchs et al. [54] used a cross-flow MBR for PSWW treatment. The MBR produced effluent with a removal efficiency of over 90% of COD. Gürel and Büyükgüngör [55] investigated MBR to extract organic substances and nutrients from wastewater in the slaughterhouse plant. Hollow-fiber UF membranes with a pore size of 0.03 μm were used in the bioreactor. The removal efficiency of MBR was reported as 97% of COD, 96% of TOC, 65% of TP, and 44% of TN. The COD and TOC levels of permeate were 16 and 9 mg/L, respectively, which complied with the discharge limits of the slaughterhouse plant. Meyo et al. [56] treated PSWW using a pretreatment stage, an EGSB, and an MBR, and their results showed that MBR as a final stage treatment further reduced over 95% of TSS and COD.

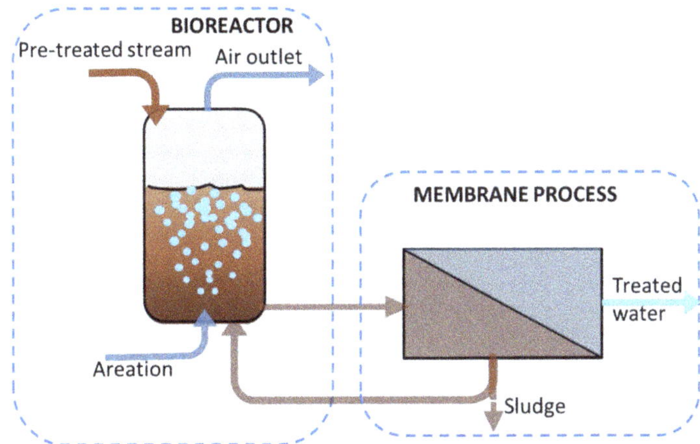

Figure 4. Schematic representation of an MBR process (reproduced with permission from [57]: Poerio, T.; Piacentini, E.; Mazzei, R. Membrane Processes for Microplastic Removal. *Molecules* 2019, 24, 4148. https://doi.org/10.3390/molecules24224148. Accessed on 15 March 2021. Copyright (2019), MDPI).

4.7. Vacuum Membrane Distillation

Membrane distillation (MD) is a thermal-based membrane separation process that was introduced in 1963. The driving force in MD is the vapor pressure difference across the hydrophobic membrane instead of the applied absolute pressure difference [58]. The MD process can be used for water desalination, removal of organic matter in drinking water production, water and wastewater treatment, recovery of valuable components, and treatment of radioactive wastes [59]. Vacuum membrane distillation (VMD) is one configuration of MD in which the permeate side is vapor or water under reduced pressure [60]. Bialas et al. [61] conducted protein and water recovery from poultry processing wastewater using an integrating process of MF, UF, and VMD. A hydrophilic PVDF MF membrane was used in the pretreatment to extract suspended solids from the processing water, and a cellulose UF membrane was used to isolate soluble protein. During UF, the COD concentration of the permeate consistently surpassed the maximum permissible level. As a result, at the final stage, VMD was used. The membrane was a flat-sheet hydrophobic membrane made of polypropylene with a pore size of 0.2 μm. The study showed that the increase in temperature and the decrease in downstream pressure led to a considerable increase in the permeate flux, as shown in Figure 5. The removal efficiency of COD, TSS, TN, and total organic matter extractable by petroleum ether (TOEM) were very high, exceeding 99%. Moreover, VMD retained 93.3% of protein; the 6.7% loss of protein could have been due to adsorption of proteins to the membrane surface and denaturation of the proteins due to the high temperature. In terms of consistency, the permeate obtained through VMD was comparable to the RO permeate. The integrated process comprising MF, UF, and VMD made it possible to recover 70% of the water. The performance of VMD along with other pressure-driven membrane technologies used in the PSWW treatment is summarized in Table 3.

Table 3. Summary of treatment of poultry slaughterhouse wastewater by membrane technology.

Membrane	Material	Module	Treatment	BOD	COD	TOC	TSS	TS	Turbidity	TP	TN	Reference
MF	Ceramic	Tubular	Standalone	-	100%	-	100%	-	≈100%	-	-	[12]
MF	PA	Hollow fiber	Standalone	-	63.3%	57.8%	-	-	98.8%	-	55.9%	[37]
MF	Polyether sulfone	Spiral	Standalone	-	92.5%	88.4%	-	-	100%	-	82.5%	[37]
MF	PVDF	Flat sheet	Standalone	-	88%	-	-	34%	-	-	-	[38]
MF	PVDF	-	Standalone	-	89%	-	-	35%	-	-	-	[39]
MF	Ceramic	Tubular	Standalone	≈70%	-	-	-	-	-	-	-	[62]
UF	Polyethersulfone	Hollow fiber	Standalone	-	76.7%	61.7%	-	-	99.6%	-	41.9%	[37]
UF	Polyethersulfone	Spiral	Standalone	-	94.2%	92.5%	-	-	100%	-	87.1%	[37]
UF	PS	-	Integrated	-	97.4%	-	-	-	-	-	-	[41]
UF	Polyacrylonitrile	Plate-and-frame	Standalone	>97%	>94%	-	99%	-	-	-	-	[9]
UF	Polyacrylonitrile	Flat sheet	Standalone	98%	96.6%	-	-	63%	-	-	-	[63]
UF	PS	Hollow fiber	Standalone	-	74%	-	-	-	99.9%	-	-	[42]
UF	PS	Flat sheet	Standalone	-	58.86%	-	-	-	-	-	-	[64]
UF	Ceramic	-	Integrated	-	98%	-	99.8%	-	-	-	-	[65]
UF	Ceramic	Hollow fiber	Integrated	-	91%	-	97%	-	-	-	-	[13]
UF	-	Flat sheet	Standalone	-	89%	-	-	22%	-	-	-	[38]
UF	-	-	Standalone	-	91%	-	-	22%	-	-	-	[39]
UF	Non-cellulosic	Tubular	Standalone	-	95%	-	-	85%	-	-	86%	[30]
UF	PS	-	-	-	-	55%	-	-	-	-	-	[10]
UF	Polyethersulfone	-	Standalone	93%	94%	-	100%	-	-	-	-	[66]
UF	-	-	-	-	-	-	100%	-	100%	-	-	[67]
NF	Thin film composite	-	Standalone	-	82%	-	-	-	-	-	-	[10]
NF	Thin film	-	Standalone	-	>85%	-	-	-	-	-	-	[41]
RO	PA	-	Standalone	-	≈90%	-	-	-	-	-	-	[41]
RO	-	-	Integrated	-	-	-	100%	-	100%	-	-	[51]
RO	-	-	Integrated	99.70%	99.76%	-	99.71%	-	99.88%	-	-	[21]
MBR	Ceramic	-	Integrated	-	-	62%	57%	-	97%	-	-	[53]
MBR	Ceramic	-	Integrated	-	>90%	-	-	-	-	-	-	[54]
MBR	polyether-sulfone	-	Integrated	-	>95%	-	>95%	-	-	-	-	[56]
VMD	Polypropylene	Flat sheet	Integrated	-	100%	-	100%	-	-	-	100%	[61]

-: Not reported; TP is presented by phosphates.

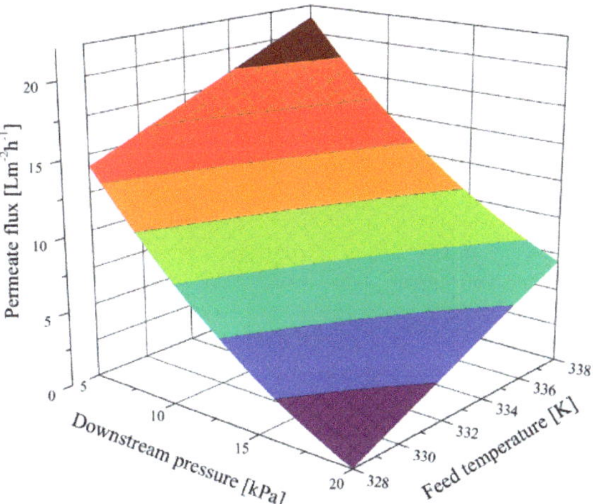

Figure 5. Response surface for permeate flux, J, as function of downstream pressure and feed temperature (reproduced with permission from [61]: Białas, W.; Stangierski, J.; Konieczny, P. Protein and Water Recovery from Poultry Processing Wastewater Integrating Microfiltration, Ultrafiltration and Vacuum Membrane Distillation. *Int. J. Environ. Sci. Technol.* **2015**, *12*, 1875–1888. https://doi.org/10.1007/s13762-014-0557-4. Accessed on 19 March 2021. Copyright (2015), Springer).

Summarized in Table 3, MF, UF, NF, and RO membranes have widely been used to treat PSWW, from lab tests to case studies on a large scale. Both polymeric and ceramic UF membranes can effectively remove organic matter at a low energy cost since MF membranes possess the largest pore size among these membranes. More impressive, a fly ash-based ceramic membrane recently developed by Goswami et al. [12] removed 100% of COD, TSS, and turbidity. Compared to other membranes, UF membranes have been used more frequently due to their capacity for removing suspended solids, proteins, and pathogens with high water flux and great energy efficiency. Although a study demonstrated that the TOC level of poultry abattoir wastewater treated by NF met the Canadian poultry wastewater reuse criteria, the quality of water produced by NF is not compatible with RO. RO membranes, which can even block monovalent ions, are ideal for producing qualified water for reuse in poultry processing. As an alternative to RO, it was confirmed that thermal-driven VMD equipped with a hydrophobic MF membrane could produce a similar quality of water as RO during PSWW treatment [41].

4.8. Nutrient Recovery from PSWW by Membrane Separation

In the stage of conventional treatment, valuable nutrients such as proteins could be recovered using coagulation or flocculation from PSWW. Unfortunately, the protein concentrate obtained by the traditional methods cannot be used as animal food because coagulants and flocculants introduce some harmful compounds and change protein properties. Pressure-driven membrane processes are good at protein recovery while keeping protein unchanged because membrane separation is a physical process. For example, Hart et al. [68], in their preliminary studies of MF for the reuse of food-processing water, concluded that MF is a suitable method for reconditioning processing water for reuse, leading to substantial energy savings and reducing disposal costs, and recovering by-products such as protein and fats. Lo et al. [64] investigated protein recovery from poultry processing wastewater using a PS UF membrane with MWCO at 30,000 Da. Their findings revealed that almost all crude proteins with a concentration of 390 ppm were retained, and it reduced 58.86% of COD in poultry processing wastewater. Bialas et al. [61] demonstrated a recovery of 84% of total protein using the integrated process of MF, UF, and VMD.

4.9. Membrane Fouling and Cleaning Methods

One of the biggest challenges of membrane technology applied to wastewater treatment is membrane fouling, which is caused by the deposition of molecules or particulates on the membrane surface or into membrane pores [69]. Fouling in the membrane is caused by various contaminants in water such as colloidal or particulate matter, dissolved organics, chemical reactants, and microorganisms and microbial products [70]. There is no unified statement regarding the mechanisms of membrane fouling. However, from the analysis of the causes of membrane fouling, four main reasons have been confirmed, including the blocking of membrane pores, adsorption of solute by the membrane, deposition of the activated sludge on the membrane surface, and compaction of the filter cake layer on the membrane surface [71]. Membrane fouling is mainly categorized into two types: reversible and irreversible. Reversible fouling occurs when there is no permanent permeate flux loss, whereas irreversible fouling is caused by permanent permeate flux loss. Other types of membrane fouling include organic fouling, scaling fouling, colloidal fouling, and biofouling. The main drawbacks of membrane fouling are that it could drastically reduce membrane lifetime, productivity, and permeate quality [72,73]. To control membrane fouling, several fouling control strategies have been explored. The commonly used technologies are chemical methods, including coagulation, chemical cleaning, and membrane surface modification; hydrodynamic methods such as a vibrating membrane, high shear, and rotating disk; and physical processes such as ultrasound and physical cleaning techniques [74]. Lo et al. [64] stated that for the treatment of PSWW, membrane fouling was inevitable after processing, and that flushing the UF membrane with a cleaning reagent containing 200 ppm sodium hypochlorite was found to be capable of effectively restoring 90% of membrane performance. Hart et al. [62] reported that the flux rate of MF membranes for PSWW treatment was restored by 15 min in-line cleaning with Micro brand detergent. Moreover, Marchesi et al. [37] recovered 95% of water flux by cleaning the MF and UF membranes sequentially with sodium hypochlorite, citric acid, sodium hydroxide, and ultrapure water. Some pretreatment approaches can be adopted to alleviate membrane fouling before the membrane filtration process. For example, Sardari et al. [40] reported that for the PSWW treatment, the EC as a pretreatment for UF significantly mitigated membrane fouling. Alternatively, Racar et al. [75], who worked on the treatment of rendering plant wastewater, confirmed that sand filtration was an effective pretreatment for later UF, decreasing fouling. Sand filtration primarily eliminated soluble microbial product from secondary effluents, thus improving the UF performance. Although current fouling control approaches are practical, further research shall be carried out to develop cost-effective pretreatments, advance membrane configuration, identify optimal membrane operating conditions, and design an effective hybrid physical/chemical cleaning process [74].

5. Economic Assessment

Achieving excellent performance with membrane technologies in PSWW treatment and making it cost-competitive with conventional methods are critical for the poultry processing industry. Białas et al. [61] reported that recycling 70% of water by integrating MF, UF, and VMD would yield a savings of EUR 10,850 per month. In addition, the use of clean water would decrease by 7000 m^3/month with a savings of EUR 6166 per month, and an 84% protein recovery would generate a product value of EUR 33,000 per month. Jason et al. [30] reported that protein recovery could lead to USD 424 income per day where 100,000 chickens are processed each day. This income could be used as a partial operating cost, thus making the membrane process economic and competitive. Houston et al. [76] analyzed the economic feasibility of incorporating a UF chiller water recycling unit in the pilot poultry processing plant and indicated positive impacts by attaining a profit of more than USD 60,000 a year. Their work reported that recycling the water would reduce the cost by USD 219,465 annually, which came from USD 84,600.75 in water savings, USD 90,695.00 in sewage cost savings, and USD 44,169.84 in energy savings. After detecting the cost of recycling, depreciation, labor, filter cleaning, and miscellaneous, a net income of USD 68,756 per year with a return rate of 45.6% was achieved. Coskun et al.'s [40] economic analysis for poultry processing wastewater treatment stated that raw water consumption with the conventional method was large as 1220 m^3/day, which was decreased to 396 m^3/day when using the UF+RO process, as the recycled water was used for various poultry processing purposes. The unit costs of the conventional treatment were reported as USD 738,600/year, the sum raw consumption costs were USD 4,402,500/year, and the treatment costs were USD 336,000/year. Coskun et al. [40] declared that the most economical process was UF+RO with a total cost of USD 295,700/year. As the processing water was recycled and reused, the raw water consumption and treatment costs were decreased to USD 130,800/year and USD 164,900/year, respectively. The experiments by Mannapperuma and Santos [42] verified that UF could treat the 480 L/min chiller overflow to produce 380 L/min reconditioned water at about 80% recovery. It would replace 346 L/min freshwater with the chiller based on guidelines. The use of this system resulted in total savings of USD 165,800/year, which includes savings in freshwater, disposal costs, and energy. This assessment indicates a 2.4-year payback period.

6. Future Perspective and Recommendation

Membrane technology is promising for PSWW treatment mainly due to its advantages for producing high-quality water for reuse, nutrient recovery, and the operational perspective of compactness and modularity. This review identifies that the research on pressure-driven membrane filtration and membrane distillation for PSWW treatment has mostly been done on a lab or pilot scale, making it unclear for application on a larger scale. Therefore, it is necessary to assess its properties and efficiency with an analysis of energy and operating costs on the industrial level to implement the membrane technologies in poultry slaughterhouse plants. Correspondingly, most research has been conducted only on UF for the PSWW treatment; thus, for a broad perspective, NF and RO should be examined often. In most research articles, the quality of the product water is certified by quantifying parameters such as BOD, COD, TSS, TOC, TN, and TP; therefore, to provide more authenticity to membrane technology, a wide range of physical, chemical, and biological parameters should also be considered.

As an alternative to the traditional membrane approaches, dynamic membrane technology (DMT) is an attractive method for municipal and industrial wastewater treatment and surface water treatment [77–79]. The concept is that when a cake layer forms on a support, such as a mesh or woven filter cloth instead of a conventional membrane, the cake layer acts as a dynamic membrane by properly controlling its thickness. There are some examples of DMT integration with anaerobic BMR for treating high-strength wastewater [80,81]. Recently, Mahat et al. [82] evaluated the 90-day performance of dynamic anaerobic MBR by utilizing low-cost non-woven filter cloth as the support material and producing biomethane as in situ renewable energy while treating high-strength food processing wastewater. Their success indicates that dynamic anaerobic MBR has great application potential for the treatment of high-strength wastewater, including PSWW, at a low operating cost. On the other hand, due to ceramic membranes having a super chemical/thermal stability, low fouling propensity, and long lifespan, the applications of ceramic membrane technology in water and wastewater are rapidly growing, even on a full-size/industrial scale [83,84]. The lower lifecycle cost of ceramic membranes than PVDF membranes in water treatment [85] also implies that the use of ceramic membranes is an excellent option for PSWW treatment in the near future. It is also possible to integrate PSWW treatment with electricity generation. Recently, Roshanravan et al. [86] conducted some tests of polymer-electrolyte membrane microbial fuel cells by feeding meat poultry wastewater, and they found that the cell equipped with SPSU20/MIL7 composite membrane could generate electric power at a power density of 27.50 mW/m^2 and Coulombic efficiency of 31.01% with a COD removal rate of 57.65%. In addition, we highly recommend conducting more research to purify PSWW for recycling water than discharging it into the environment by using RO membranes on a pilot or industrial scale. RO is a very suitable solution to alleviate water scarcity, especially for slaughterhouses located in arid or semiarid regions.

7. Conclusions

It is concluded that the poultry slaughterhouses produce a large amount of wastewater, which is generally treated by conventional methods. The traditional technologies' inability to recover water and nutrients has given great attention to membrane technology for PSWW treatment. Membrane technology is energy efficient, with a reduction in the number of processing steps, and it provides greater separation efficiency and improved final product quality. It produces water clean enough for reuse and recycling for industrial processing when appropriate two-stage or hybrid membrane separation, e.g., UF+RO, is used. The UF membranes are used as the pretreatment prior to RO by removing suspended solids and macromolecules and recover a good amount of valuable nutrients. Therefore, membrane separation is a promising approach for PSWW treatment. It exhibits excellent performance as a standalone or integrated process by providing high efficiency for pollutant removal and the recovery of valuable products. It removes almost all the pollutants and purifies the water as required to discharge into the environment and reuse for industrial poultry purposes. The summarized economic assessment shows that membrane technology is an economical alternative for the treatment of PSWW. In the near future, robust ceramic UF membranes and DMT using a low-cost mesh or woven filter cloth will have great potential for PSWW pretreatment. The integration of membrane separation with power generation for PSWW treatment is worth further exploration.

Author Contributions: Conceptualization, H.D. and R.R.K.; methodology, F.F.; resources, R.R.K.; data curation, F.F.; writing—original draft preparation, F.F.; writing—review and editing, H.D. and R.R.K.; supervision, R.R.K.; project administration, R.R.K.; funding acquisition, R.R.K. All authors have read and agreed to the published version of the manuscript.

Funding: This research was funded by the USDA-National Institute of Food and Agriculture, grant number 2020-38821-31091, and National Science Foundation CREST Center for Energy & Environmental Sustainability, NSF grant number 1914692.

Acknowledgments: This work was primarily funded through a grant from the USDA-National Institute of Food and Agriculture, Award #2020-38821-31091. Partial support was also received from National Science Foundation CREST Center for Energy & Environmental Sustainability, NSF grant #1914692.

Conflicts of Interest: The authors declare no conflict of interest. The funders had no role in the design of the study, in the collection, analyses, or interpretation of data, in the writing of the manuscript, or in the decision to publish the results.

References

1. Bustillo-Lecompte, C.F.; Mehrvar, M. Slaughterhouse Wastewater Characteristics, Treatment, and Management in the Meat Processing Industry: A Review on Trends and Advances. *J. Environ. Manag.* **2015**, *161*, 287–302. [CrossRef] [PubMed]
2. Baker, B.R.; Mohamed, R.; Al-Gheethi, A.; Aziz, H.A. Advanced Technologies for Poultry Slaughterhouse Wastewater Treatment: A Systematic Review. *J. Dispers. Sci. Technol.* **2020**, *42*, 880–899. [CrossRef]
3. Hilares, R.T.; Atoche-Garay, D.F.; Pagaza, D.A.P.; Ahmed, M.A.; Andrade, G.J.C.; Santos, J.C. Promising Physicochemical Technologies for Poultry Slaughterhouse Wastewater Treatment: A Critical Review. *J. Environ. Chem. Eng.* **2021**, *9*, 105174. [CrossRef]
4. Williams, Y.; Basitere, M.; Ntwampe, S.K.O.; Ngongang, M.; Njoya, M.; Kaskote, E. Application of Response Surface Methodology to Optimize the Cod Removal Efficiency of an Egsb Reactor Treating Poultry Slaughterhouse Wastewater. *Water Pract. Technol.* **2019**, *14*, 507–514. [CrossRef]
5. Avula, R.Y.; Nelson, H.M.; Singh, R.K. Recycling of Poultry Process Wastewater by Ultrafiltration. *Innov. Food Sci. Emerg. Technol.* **2009**, *10*, 1–8. [CrossRef]
6. Basitere, M.; Njoya, M.; Rinquest, Z.; Ntwampe, S.K.O.; Sheldon, M.S. Performance Evaluation and Kinetic Parameter Analysis for Static Granular Bed Reactor (SGBR) for Treating Poultry Slaughterhouse Wastewater at Mesophilic Condition. *Water Pract. Technol.* **2019**, *14*, 259–268. [CrossRef]

7. Njoya, M.; Basitere, M.; Ntwampe, S.K.O. Analysis of the Characteristics of Poultry Slaughterhouse Wastewater (PSW) and Its Treatability. *Water Pract. Technol.* **2019**, *14*, 959–970. [CrossRef]

8. Massé, D.I.; Masse, L. Characterization of Wastewater from Hog Slaughterhouses in Eastern Canada and Evaluation of Their In-Plant Wastewater Treatment Systems. *Can. Biosyst. Eng.* **2000**, *42*, 139–146.

9. Yordanov, D. Preliminary Study of the Efficiency of Ultrafiltration Treatment of Poultry Slaughterhouse Wastewater. *Bulg. J. Agric. Sci.* **2010**, *16*, 700–704.

10. Zhang, S.Q.; Kutowy, O.; Kumar, A.; Malcolm, I. A Laboratory Study of Poultry Abattoir Wastewater Treatment by Membrane Technology. *Can. Agric. Eng.* **1997**, *39*, 99–105.

11. de Morais Coutinho, C.; Chiu, M.C.; Basso, R.C.; Ribeiro, A.P.B.; Gonçalves, L.A.G.; Viotto, L.A. State of Art of the Application of Membrane Technology to Vegetable Oils: A Review. *Food Res. Int.* **2009**, *42*, 536–550. [CrossRef]

12. Goswami, K.P.; Pugazhenthi, G. Treatment of Poultry Slaughterhouse Wastewater Using Tubular Microfiltration Membrane with Fly Ash as Key Precursor. *J. Water Process Eng.* **2020**, *37*, 101361. [CrossRef]

13. Rinquest, Z.; Basitere, M.; Ntwampe, S.K.O.; Njoya, M. Poultry Slaughterhouse Wastewater Treatment Using a Static Granular Bed Reactor Coupled with Single Stage Nitrification-Denitrification and Ultrafiltration Systems. *J. Water Process Eng.* **2019**, *29*, 100778. [CrossRef]

14. Bustillo-Lecompte, C.; Mehrvar, M.; Quiñones-Bolaños, E. Slaughterhouse Wastewater Characterization and Treatment: An Economic and Public Health Necessity of the Meat Processing Industry in Ontario, Canada. *J. Geosci. Environ. Prot.* **2016**, *4*, 175–186. [CrossRef]

15. Yaakob, M.A.; Mohamed, R.M.S.R.; Al-Gheethi, A.A.S.; Kassim, A.H.M. Characteristics of Chicken Slaughterhouse Wastewater. *Chem. Eng. Trans.* **2018**, *63*, 637–642. [CrossRef]

16. Rahmanian, N.; Ali, S.H.B.; Homayoonfard, M.; Ali, N.J.; Rehan, M.; Sadef, Y.; Nizami, A.S. Analysis of Physiochemical Parameters to Evaluate the Drinking Water Quality in the State of Perak, Malaysia. *J. Chem.* **2015**, *2015*, 716125. [CrossRef]

17. Guidelines for Water Reuse. Available online: https://www.epa.gov/waterreuse/guidelines-water-reuse (accessed on 26 May 2021).

18. Bustillo-Lecompte, C.; Mehrvar, M. Slaughterhouse Wastewater: Treatment, Management and Resource Recovery. In *Physico-Chemical Treatment of Wastewater and Resource Recovery*; Farooq, R., Ahmad, Z., Eds.; InTech: Rijeka, Croatia, 2017; pp. 153–173.

19. Basitere, M.; Williams, Y.; Sheldon, M.S.; Ntwampe, S.K.O.; De Jager, D.; Dlangamandla, C. Performance of an Expanded Granular Sludge Bed (EGSB) Reactor Coupled with Anoxic and Aerobic Bioreactors for Treating Poultry Slaughterhouse Wastewater. *Water Pract. Technol.* **2016**, *11*, 86–92. [CrossRef]

20. Meiramkulova, K.; Temirbekova, A.; Saspugayeva, G.; Kydyrbekova, A.; Devrishov, D.; Tulegenova, Z.; Aubakirova, K.; Kovalchuk, N.; Meirbekov, A.; Mkilima, T. Performance of a Combined Treatment Approach on the Elimination of Microbes from Poultry Slaughterhouse Wastewater. *Sustainability* **2021**, *13*, 3467. [CrossRef]

21. Meiramkulova, K.; Devrishov, D.; Zhumagulov, M.; Arystanova, S.; Karagoishin, Z.; Marzanova, S.; Kydyrbekova, A.; Mkilima, T.; Li, J. Performance of an Integrated Membrane Process with Electrochemical Pre-Treatment on Poultry Slaughterhouse Wastewater Purification. *Membranes* **2020**, *10*, 256. [CrossRef]

22. de Sena, R.F.; Moreira, R.F.P.M.; José, H.J. Comparison of Coagulants and Coagulation Aids for Treatment of Meat Processing Wastewater by Column Flotation. *Bioresour. Technol.* **2008**, *99*, 8221–8225. [CrossRef]

23. Teh, C.Y.; Budiman, P.M.; Shak, K.P.Y.; Wu, T.Y. Recent Advancement of Coagulation-Flocculation and Its Application in Wastewater Treatment. *Ind. Eng. Chem. Res.* **2016**, *55*, 4363–4389. [CrossRef]

24. Fan, T.; Deng, W.; Feng, X.; Pan, F.; Li, Y. An Integrated Electrocoagulation—Electrocatalysis Water Treatment Process Using Stainless Steel Cathodes Coated with Ultrathin TiO2 Nanofilms. *Chemosphere* **2020**, *254*, 126776. [CrossRef] [PubMed]

25. Mollah, M.Y.A.; Morkovsky, P.; Gomes, J.A.G.; Kesmez, M.; Parga, J.; Cocke, D.L. Fundamentals, Present and Future Perspectives of Electrocoagulation. *J. Hazard. Mater.* **2004**, *114*, 199–210. [CrossRef] [PubMed]

26. Aziz, A.; Basheer, F.; Sengar, A.; Khan, S.U.; Farooqi, I.H. Biological Wastewater Treatment (Anaerobic-Aerobic) Technologies for Safe Discharge of Treated Slaughterhouse and Meat Processing Wastewater. *Sci. Total Environ.* **2019**, *686*, 681–708. [CrossRef]

27. Johns, M.R. Developments in Wastewater Treatment in the Meat Processing Industry: A Review. *Bioresour. Technol.* **1995**, *54*, 203–216. [CrossRef]

28. Chan, Y.J.; Chong, M.F.; Law, C.L.; Hassell, D.G. A Review on Anaerobic-Aerobic Treatment of Industrial and Municipal Wastewater. *Chem. Eng. J.* **2009**, *155*, 1–18. [CrossRef]

29. Ezugbe, E.O.; Rathilal, S. Membrane Technologies in Wastewater Treatment: A Review. *Membranes* **2020**, *10*, 89. [CrossRef]

30. Shih, J.C.H.; Kozink, M.B. Ultrafiltration Treatment of Poultry Processing Wastewater and Recovery of a Nutritional By-Product. *Poult. Sci.* **1980**, *59*, 247–252. [CrossRef]

31. Van Der Bruggen, B.; Vandecasteele, C.; Van Gestel, T.; Doyenb, W.; Leysenb, R. Review of Pressure-Driven Membrane Processes. *Environ. Prog.* **2003**, *22*, 46–56. [CrossRef]

32. Eyvaz, M.; Arslan, S.; İmer, D.; Yüksel, E.; İsmail, K. Forward Osmosis Membranes—A Review: Part I. In *Osmotically Driven Membrane Processes*; Du, H., Thompson, A., Wang, X., Eds.; InTech: Rijeka, Croatia, 2018; pp. 11–40.

33. El-Ghaffar, M.A.A.; Tieama, H.A. A Review of Membranes Classifications, Configurations, Surface Modifications, Characteristics and Its Applications in Water Purification. *J. Chem. Biomol. Eng.* **2017**, *2*, 57. [CrossRef]

34. Beier, S.P. *Pressure Driven Membrane Processes*, 2nd ed.; Bookboon: London, UK, 2007; pp. 1–22.

35. Cheryan, M. *Ultrafiltration and Microfiltration Handbook*, 2nd ed.; CRC Press: Washington, DC, USA, 1998; pp. 1–552.

36. Charcosset, C. *Membrane Processes in Biotechnology and Pharmaceutics*; Elsevier: Amsterdam, The Netherlands, 2012; pp. 1–330.

37. Marchesi, C.M.; Paliga, M.; Oro, C.E.D.; Dallago, R.M.; Zin, G.; Di Luccio, M.; Oliveira, J.V.; Tres, M.V. Use of Membranes for the Treatment and Reuse of Water from the Pre-Cooling System of Chicken Carcasses. *Environ. Technol.* **2019**, *42*, 126–133. [CrossRef] [PubMed]

38. Abboah-afari, E.; Kiepper, B.H. Membrane Filtration of Poultry Processing Wastewater: I. Pre-DAF (Dissolved Air Flotation). *Appl. Eng. Agric.* **2012**, *28*, 231–236. [CrossRef]

39. Abboah-afari, E.; Kiepper, B.H. The Use of Membrane Filtration as an Alternative Pretreatment Method for Poultry Processing Wastewater. In Proceedings of the 2011 Georgia Water Resources Conference, the University of Georgia, Athens, GA, USA, 11–13 April 2011.

40. Sardari, K.; Askegaard, J.; Chiao, Y.H.; Darvishmanesh, S.; Kamaz, M.; Wickramasinghe, S.R. Electrocoagulation Followed by Ultrafiltration for Treating Poultry Processing Wastewater. *J. Environ. Chem. Eng.* **2018**, *6*, 4937–4944. [CrossRef]

41. Coskun, T.; Debik, E.; Kabuk, H.A.; Manav Demir, N.; Basturk, I.; Yildirim, B.; Temizel, D.; Kucuk, S. Treatment of Poultry Slaughterhouse Wastewater Using a Membrane Process, Water Reuse, and Economic Analysis. *Desalin. Water Treat.* **2016**, *57*, 4944–4951. [CrossRef]

42. Mannapperuma, J.D.; Santos, M.R. Reconditioning of Poultry Chiller Overflow by Ultrafiltration. *J. Food Process Eng.* **2004**, *27*, 497–516. [CrossRef]

43. Oatley-Radcliffe, D.L.; Walters, M.; Ainscough, T.J.; Williams, P.M.; Mohammad, A.W.; Hilal, N. Nanofiltration Membranes and Processes: A Review of Research Trends over the Past Decade. *J. Water Process Eng.* **2017**, *19*, 164–171. [CrossRef]

44. Esteves, T.; Mota, A.T.; Barbeitos, C.; Andrade, K.; Carlos, A.M.; Ferreira, F.C. A Study on Lupin Beans Process Wastewater Nanofiltration Treatment and Lupanine Recovery. *J. Clean. Prod.* **2020**, *277*, 123349. [CrossRef]

45. Afonso, M.D.; Yaez, R.B. Nanofiltration of Wastewater from the Fishmeal Industry. *Desalination* **2001**, *139*, 429. [CrossRef]

46. Tang, C.; Chen, V. Nanofiltration of Textile Wastewater for Water Reuse. *Desalination* **2002**, *143*, 11–20. [CrossRef]

47. Xu, G.R.; Wang, J.N.; Li, C.J. Strategies for Improving the Performance of the Polyamide thin Film Composite (PA-TFC) Reverse Osmosis (RO) Membranes: Surface Modifications and Nanoparticles Incorporations. *Desalination* **2013**, *328*, 83–100. [CrossRef]

48. Zhang, X.; Liu, Y. Reverse Osmosis Concentrate: An Essential Link for Closing Loop of Municipal Wastewater Reclamation towards Urban Sustainability. *Chem. Eng. J.* **2020**, *421*, 127773. [CrossRef]

49. Suwaileh, W.; Johnson, D.; Hilal, N. Membrane Desalination and Water Re-Use for Agriculture: State of the Art and Future Outlook. *Desalination* **2020**, *491*, 114559. [CrossRef]

50. Bohdziewicz, J.; Sroka, E.; Korus, I. Application of Ultrafiltration and Reverse Osmosis to the Treatment of the Wastewater Produced by the Meat Industry. *Pol. J. Environ. Stud.* **2003**, *12*, 269–274.

51. Meiramkulova, K.; Zorpas, A.A.; Orynbekov, D.; Zhumagulov, M.; Saspugayeva, G.; Kydyrbekova, A.; Mkilima, T.; Inglezakis, V.J. The Effect of Scale on the Performance of an Integrated Poultry Slaughterhouse Wastewater Treatment Process. *Sustainability* **2020**, *12*, 4679. [CrossRef]

52. Cicek, N. A Review of Membrane Bioreactors and Their Potential Application in the Treatment of Agricultural Wastewater. *Can. Biosyst. Eng.* **2003**, *45*, 37–49.

53. Williams, Y. Treatment of Poultry Slaughterhouse Wastewater Using an Expanded Granular Sludge Bed Anaerobic Digester Coupled with Anoxic/Aerobic Hybrid Side Stream Ultrafiltration Membrane Bioreactor. Master's Thesis, Cape Peninsula University of Technology, Cape Town, South Africa, December 2017.

54. Fuchs, W.; Binder, H.; Mavrias, G.; Braun, R. Anaerobic Treatment of Wastewater with High Organic Content Using a Stirred Tank Reactor Coupled with a Membrane Filtration Unit. *Water Res.* **2003**, *37*, 902–908. [CrossRef]

55. Gürel, L.; Büyükgüngör, H. Treatment of Slaughterhouse Plant Wastewater by Using a Membrane Bioreactor. *Water Sci. Technol.* **2011**, *64*, 214–219. [CrossRef] [PubMed]

56. Meyo, H.B.; Njoya, M.; Basitere, M.; Ntwampe, S.K.O.; Kaskote, E. Treatment of Poultry Slaughterhouse Wastewater (PSW) Using a Pretreatment Stage, an Expanded Granular Sludge Bed Reactor (EGSB), and a Membrane Bioreactor (MBR). *Membranes* **2021**, *11*, 345. [CrossRef]

57. Poerio, T.; Piacentini, E.; Mazzei, R. Membrane Processes for Microplastic Removal. *Molecules* **2019**, *24*, 4148. [CrossRef] [PubMed]

58. Zou, T.; Kang, G.; Zhou, M.; Li, M.; Cao, Y. Investigation of Flux Attenuation and Crystallization Behavior in Submerged Vacuum Membrane Distillation (SVMD) for SWRO Brine Concentration. *Chem. Eng. Process. Process Intensif.* **2019**, *143*, 107567. [CrossRef]

59. Tijing, L.D.; Woo, Y.C.; Choi, J.S.; Lee, S.; Kim, S.H.; Shon, H.K. Fouling and Its Control in Membrane Distillation-A Review. *J. Memb. Sci.* **2015**, *475*, 215–244. [CrossRef]

60. Camacho, L.M.; Dumée, L.; Zhang, J.; de Li, J.; Duke, M.; Gomez, J.; Gray, S. Advances in Membrane Distillation for Water Desalination and Purification Applications. *Water* **2013**, *5*, 94–196. [CrossRef]

61. Białas, W.; Stangierski, J.; Konieczny, P. Protein and Water Recovery from Poultry Processing Wastewater Integrating Microfiltration, Ultrafiltration and Vacuum Membrane Distillation. *Int. J. Environ. Sci. Technol.* **2015**, *12*, 1875–1888. [CrossRef]

62. Hart, M.R.; Huxsoll, C.C.; Tsai, L.-S.; NG, K.C. Preliminary Studies of Microfiltration for Food Processing Water Reuse. *J. Food Prot.* **1988**, *51*, 269–276. [CrossRef]

63. Nelson, H.M. Performance of Polymeric Membrane Systems in the Treatment of Poultry Processing Plant Waste Effluent. Master's Thesis, The University of Georgia, Athens, GA, USA, December 2006.

64. Lo, Y.M.; Cao, D.; Argin-Soysal, S.; Wang, J.; Hahm, T.S. Recovery of Protein from Poultry Processing Wastewater Using Membrane Ultrafiltration. *Bioresour. Technol.* **2005**, *96*, 687–698. [CrossRef]
65. Basitere, M.; Rinquest, Z.; Njoya, M.; Sheldon, M.S.; Ntwampe, S.K. Treatment of Poultry Slaughterhouse Wastewater Using a Static Granular Bed Reactor (SGBR) Coupled with Ultrafiltration (UF) Membrane System. *Water Sci. Technol.* **2017**, *76*, 106–114. [CrossRef]
66. Malmali, M.; Askegaard, J.; Sardari, K.; Eswaranandam, S.; Sengupta, A.; Wickramasinghe, S.R. Evaluation of Ultrafiltration Membranes for Treating Poultry Processing Wastewater. *J. Water Process Eng.* **2018**, *22*, 218–226. [CrossRef]
67. Meiramkulova, K.; Orynbekov, D.; Saspugayeva, G.; Aubakirova, K.; Arystanova, S.; Kydyrbekova, A.; Tashenov, E.; Nurlan, K.; Mkilima, T. The Effect of Mixing Ratios on the Performance of an Integrated Poultry Slaughterhouse Wastewater Treatment Plant for a Recyclable High-Quality Effluent. *Sustainability* **2020**, *12*, 6097. [CrossRef]
68. Hart, M.R.; Huxsoll, C.C.; Tsai, L.S.; NG, K.C.; King, A.D.; Jones, C.C.; Halbrook, W.U. Microfiltration of Chicken Process Waters for Reuse: Plant Studies and Projected Operating Costs. *J. Food Process Eng.* **1990**, *12*, 191–210. [CrossRef]
69. Cui, Z.F.; Muralidhara, H.S. *Membrane Technology: A Practical Guide to Membrane Technology and Applications in Food and Bioprocessing*; Elsevier: New York, NY, USA, 2010; pp. 1–312.
70. Chun, Y.; Mulcahy, D.; Zou, L.; Kim, I.S. A Short Review of Membrane Fouling in Forward Osmosis Processes. *Membranes* **2017**, *7*, 30. [CrossRef] [PubMed]
71. Du, X.; Shi, Y.; Jegatheesan, V.; Haq, I.U. A Review on the Mechanism, Impacts and Control Methods of Membrane Fouling in Mbr System. *Membranes* **2020**, *10*, 24. [CrossRef]
72. Yousuf, A. *Microalgae Cultivation for Biofuels Production*; Academic Press: San Diego, CA, USA, 2020; pp. 1–382.
73. Jiang, S.; Li, Y.; Ladewig, B.P. A Review of Reverse Osmosis Membrane Fouling and Control Strategies. *Sci. Total Environ.* **2017**, *595*, 567–583. [CrossRef] [PubMed]
74. Bokhary, A.; Tikka, A.; Leitch, M.; Liao, B. Membrane Fouling Prevention and Control Strategies in Pulp and Paper Industry Applications: A Review. *J. Membr. Sci. Res.* **2018**, *4*, 181–197. [CrossRef]
75. Racar, M.; Dolar, D.; Špehar, A.; Košutić, K. Application of UF/NF/RO Membranes for Treatment and Reuse of Rendering Plant Wastewater. *Process Saf. Environ. Prot.* **2017**, *105*, 386–392. [CrossRef]
76. Saravia, H.; Houston, J.E.; Toledo, R.; Nelson, H.M. Economic Feasibility of Recycling Chiller Water in Poultry Processing Plants By Ultrafiltration. In Proceedings of the 2005 Georgia Water Resources Conference, the University of Georgia, Athens, GA, USA, 25–27 April 2005.
77. Ma, J.; Wang, Z.; Zou, X.; Feng, J.; Wu, Z. Microbial Communities in an Anaerobic Dynamic Membrane Bioreactor (AnDMBR) for Municipal Wastewater Treatment: Comparison of Bulk Sludge and Cake Layer. *Process Biochem.* **2013**, *48*, 510–516. [CrossRef]
78. Yi, Z.; Shibin, X.; Feng, H.; Dong, X.; Lingwei, K.; Zhenbin, W. Phosphate Removal of Acid Wastewater from High-Phosphate Hematite Pickling Process by in-Situ Self-Formed Dynamic Membrane Technology. *Desalin. Water Treat.* **2012**, *37*, 77–83. [CrossRef]
79. Yang, T.; Ma, Z.F.; Yang, Q.Y. Formation and Performance of Kaolin/MnO2 Bi-Layer Composite Dynamic Membrane for Oily Wastewater Treatment: Effect of Solution Conditions. *Desalination* **2011**, *270*, 50–56. [CrossRef]
80. Ersahin, M.E.; Ozgun, H.; Tao, Y.; van Lier, J.B. Applicability of Dynamic Membrane Technology in Anaerobic Membrane Bioreactors. *Water Res.* **2014**, *48*, 420–429. [CrossRef]
81. Hu, Y.; Wang, X.C.; Ngo, H.H.; Sun, Q.; Yang, Y. Anaerobic Dynamic Membrane Bioreactor (AnDMBR) for Wastewater Treatment: A Review. *Bioresour. Technol.* **2018**, *247*, 1107–1118. [CrossRef]
82. Mahat, S.B.; Omar, R.; Che Man, H.; Mohamad Idris, A.I.; Mustapa Kamal, S.M.; Idris, A.; Shreeshivadasan, C.; Jamali, N.S.; Abdullah, L.C. Performance of Dynamic Anaerobic Membrane Bioreactor (DAnMBR) with Phase Separation in Treating High Strength Food Processing Wastewater. *J. Environ. Chem. Eng.* **2021**, *9*, 105245. [CrossRef]
83. Noguchi, H.; Oo, M.H.; Niwa, T.; Fong, E.; Yin, R.; Supaat, N. Applications of Flat Sheet Ceramic Membrane for Surface Water and Seawater Treatments—Introduction of Performance in Large-Scale Drinking Water Plant and Seawater Pretreatment Pilot System in Singapore. *Water Pract. Technol.* **2019**, *14*, 289–296. [CrossRef]
84. Asif, M.B.; Zhang, Z. Ceramic Membrane Technology for Water and Wastewater Treatment: A Critical Review of Performance, Full-Scale Applications, Membrane Fouling and Prospects. *Chem. Eng. J.* **2021**, *418*, 129481. [CrossRef]
85. Kurth, C.J.; Wise, B.L.; Smith, S. Design Considerations for Implementing Ceramics in New and Existing Polymeric UF Systems. *Water Pract. Technol.* **2018**, *13*, 725–737. [CrossRef]
86. Roshanravan, B.; Younesi, H.; Abdollahi, M.; Rahimnejad, M.; Pyo, S.H. Application of Proton-Conducting Sulfonated Polysulfone Incorporated MIL-100(Fe) Composite Materials for Polymer-Electrolyte Membrane Microbial Fuel Cells. *J. Clean. Prod.* **2021**, *300*, 126963. [CrossRef]

Review

Multi-Integrated Systems for Treatment of Abattoir Wastewater: A Review

Larryngeai Gutu [1], Moses Basitere [2,*], Theo Harding [3], David Ikumi [3], Mahomet Njoya [1] and Chris Gaszynski [3]

[1] Department of Chemical Engineering, Bioresource Engineering Research Group (BioERG), Cape Peninsula University of Technology, P.O. Box 1906, Bellville 7535, South Africa; 216000157@mycput.ac.za (L.G.); mahomet.njoya@gmail.com (M.N.)

[2] Academic Support Programme for Engineering in Cape Town (ASPECT) & Water Research Group, University of Cape Town, Rondebosch, Cape Town 7700, South Africa

[3] Department of Civil Engineering, Water Research Group, University of Cape Town, Rondebosch, Cape Town 7700, South Africa; theo.harding@uct.ac.za (T.H.); david.ikumi@uct.ac.za (D.I.); chris.gaszynski@uct.ac.za (C.G.)

* Correspondence: moses.basitere@uct.ac.za

Citation: Gutu, L.; Basitere, M.; Harding, T.; Ikumi, D.; Njoya, M.; Gaszynski, C. Multi-Integrated Systems for Treatment of Abattoir Wastewater: A Review. *Water* **2021**, *13*, 2462. https://doi.org/10.3390/w13182462

Academic Editors: Amin Mojiri and Mohammed J.K. Bashir

Received: 21 July 2021
Accepted: 30 August 2021
Published: 7 September 2021

Publisher's Note: MDPI stays neutral with regard to jurisdictional claims in published maps and institutional affiliations.

Abstract: Biological wastewater treatment processes such as activated sludge and anaerobic digestion remain the most favorable when compared to processes such as chemical precipitation and ion exchange due to their cost-effectiveness, eco-friendliness, ease of operation, and low maintenance. Since Abattoir Wastewater (AWW) is characterized as having high organic content, anaerobic digestion is slow and inadequate for complete removal of all nutrients and organic matter when required to produce a high-quality effluent that satisfies discharge standards. Multi-integrated systems can be designed in which additional stages are added before the anaerobic digester (pre-treatment), as well as after the digester (post-treatment) for nutrient recovery and pathogen removal. This can aid the water treatment plant effluent to meet the discharge regulations imposed by the legislator and allow the possibility for reuse on-site. This review aims to provide information on the principles of anaerobic digestion, aeration pre-treatment technology using enzymes and a hybrid membrane bioreactor, describing their various roles in AWW treatment. Simultaneous nitrification and denitrification are essential to add after anaerobic digestion for nutrient recovery utilizing a single step process. Nutrient recovery has become more favorable than nutrient removal in wastewater treatment because it consumes less energy, making the process cost-effective. In addition, recovered nutrients can be used to make nutrient-based fertilizers, reducing the effects of eutrophication and land degradation. The downflow expanded granular bed reactor is also compared to other high-rate anaerobic reactors, such as the up-flow anaerobic sludge blanket (UASB) and the expanded granular sludge bed reactor (EGSB).

Keywords: bio-membrane; multi-integrated system; expanded granular bed reactor; anaerobic digestion; activated sludge; membrane bioreactor

1. Introduction

The continuous influx and increase in urbanization and industrialization have led to an increase in the consumption of goods and services. Relative to other commodities such as winery and car manufacturing, the abattoir industries have also increased and doubled in production in the past decade, increasing water consumption. This increase in water consumption inevitably poses a threat to the environment due to added pollution and increasing water scarcity such that by 2050 global water demand is projected to be 20–30% higher than current levels given both population growth and socio-economic development [1]. This is caused by the presence of organic matter such as chemical oxygen demand (COD), which poses a threat to the environment by accelerating the deoxygenation of rivers and contamination of ground water [2]. Abattoir industries consume about

26 L of potable water per bird to clean the blood off of slaughtered animals, clean off the slaughtering surfaces, cleaning of by-products, steam generation, and for chilling [3]. The slaughtering process and the periodic washing of residue particles in the slaughterhouse result in large quantities of water containing high amounts of biodegradable organic matter [4,5]. The contribution of organic load to these effluents usually comes from different materials such as undigested food, blood, fats, oil, and grease (FOG) and lard, loose meat, paunch, colloidal particles, soluble proteins, manure, grit, and suspended materials [4]. Farzadkia, Vanani [6] stated that the characterization of abattoir wastewater contaminants is influenced by the type of treated water, the kind of animals that have been slaughtered for the particular time frame leading up to water collection, the sampling techniques of the individuals involved, as well as the cleaning and sanitizing procedures of a specific abattoir. These wastewater contaminants can be further characterized into three categories, as shown in Figure 1. Biological oxygen demand (BOD), chemical oxygen demand (COD), and total organic carbon (TOC) are the most widely used parameters for testing effluent quality before discharge according to discharge standards, as shown in Table 1.

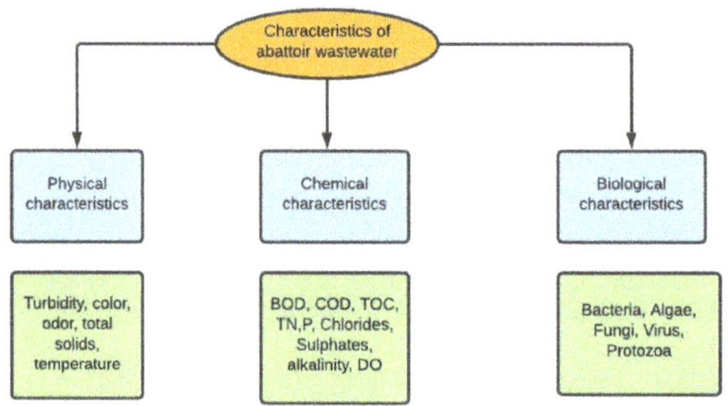

Figure 1. Characteristics of abattoir wastewater [Abbreviations: BOD biological oxygen demand; COD chemical oxygen demand; TOC total organic carbon; TN total nitrogen, P phosphorus; DO dissolved oxygen] [7].

Table 1. Maximum limits permitted by the City of Cape Town: Wastewater and Industrial Effluent By-law 2013 and the characteristics of different abattoir wastewater.

		CCT Industrial Effluent by-Law (2013)	SANS 214:2011 (Portable Water Quality)	AWW [8]	AWW [9]	AWW [10]
Parameter	Unit	Maximum limits		Range	Range	Range
General limits						
Temperature	°C	40	-	-	-	-
Conductivity at 25 °C	mS/m	500	170	-	-	-
pH at 25 °C	n/a	12	9.7	6.5–8.0	5–7.8	6.5
Chemical oxygen demand (COD)	Mg/L	5000	-	2133–10,655	1100–15,000	8575
Turbidity	NTU	-	1/5	-	-	-
Chemical substance limits						
Total dissolved solids (TDS)	mg/L	4000	1200	-		-
Total suspended solids (TSS)	mg/L	1000	-	315–1273	220–6400	1550
Fats, oils and greases (FOGs)	mg/L	400	-	131–684	40–1385	121.5
Ammonium as (N)	mg/L	-	1.5	29–51	20–300	-
Nitrates as (N)	mg/L	-	11	-	50–840	455
Nitrites as (N)	mg/L	-	0.9	-	40–700	455
Total phosphates as (P)	mg/L	25	-	8–30	15–200	112.5

The discharge of untreated water not only poses a severe threat to public health but also causes the death of aquatic species and eutrophication, leading to the depletion of dissolved oxygen (DO) and possible emanation of harmful gases [9,11]. Blood and fat are a major problem in contaminated AWW. Blood has a COD of 375 000 mg/L which is considered very high and on the other hand, fats cause physical problems in treatment plants such as blockages, clogging, scum formation and possible shut downs [2]. Governments have imposed strict regulations on the discharge of water to mitigate the expenses of pollution, for which non-compliance results in heavy penalties. Each municipality in South Africa has regulation standards for water discharge, whether it is into the sewers, land applications or for onsite reuse.

Due to the high costs associated with the efforts to reduce and handle waste, abattoirs are aiming to treat the wastewater onsite with the possibility of reusing and recycling to reduce plant running costs, have a smaller footprint, as well as upgrading to newer cost-effective technologies. The increase in onsite treatment and waste eradication requires advanced refuse-handling equipment and methods to produce organic-rich and less bio-toxic waste [12]. The wastewater can be treated using biological and chemical treatment. Recently, chemical treatment has become less popular as the use of chemicals increases the cost of treatment, leaves the difficult task of disposing of the chemical sludge and is environmentally unfriendly, making this option uneconomical and unfavorable [9]. As a result, aerobic and anaerobic treatment systems have become dominant and favorable options [9,13]. AWW contains high concentrations of organic contaminants and is rich in proteins and lipids, making it ideal for biogas production [4], as well as being a good candidate for the highly attractive anaerobic digestion [5]. According to Ozdemir and Yetilmezsoy [14], analysis confirmed that the bio-diesel produced from the waste fats, oils and grease (FOG) obtained from slaughterhouse waste showed excellent fuel properties when compared to biodiesel produced from other common crop-based feedstocks. This is because AWW is protein and lipid rich and has great potential to produce high methane yields at different concentrations of volatile solids.

Anaerobic treatment is advantageous as it has excellent eco-friendly organic matter removal, less sludge production, lower energy consumption, execution of higher organic loading rates (OLR), fewer nutrients and chemical requirements, high COD and BOD removal efficiency, and requires a smaller footprint as well as the considerable production of renewable energy in the form of biogas [15,16]. However, anaerobic digestion poses some limitations, such as having longer start-up and running periods, sensitivity to higher temperature conditions and the inability to effectively remove nutrients such as nitrogen and phosphates, which results in low to moderate effluent quality [17]. Additionally, the process often faces operational challenges due to the difficulties related to the treatment of suspended solids, fats, oils and grease (FOGs) accumulating in the reactors, leading to reduced methanogenic activity, as well as sludge and biomass washout [4,18,19]. These challenges result in process failure, hence the need to incorporate pre-treatment for FOG removal, initiate hydrolysis, and remove solid particles and feathers.

Mondal, Jana [20] stated that aerobic treatment is superior to anaerobic treatment for treating water with a high organic content because it is quicker and more effective for degrading contaminants. However, aerobic digestion also has its flaws, such as high energy requirements for aeration compared to anaerobic, which adds to running costs. Hence a combination of both anaerobic and aerobic processes must be employed to tackle this predicament and effectively remove the nutrients and organic matter [4,9]. The fraction of lipids presents in AWW poses a threat due to their slow hydrolysis rate [21]. Typically, induced and dissolved air flotation is used to remove the oils and grease before aerobic–anaerobic digestion. However, the costs of the air and reagents used, if chemically assisted, tend to make this process uneconomical and expensive. Additionally, the removal efficiency is low and sometimes produces difficult sludges to treat [15]. Other methods tested include alkaline, thermal [22,23] and ultrasonic [24] pre-treatment; however, these all fall short in one way or another. Enzymatic pre-treatment is a good option to satisfy the concerns of improving methane production, reducing the number of suspended solids before anaerobic digestion and is environmentally friendly [19]. Enzymes hydrolyse the triglycerides to fatty acids and glycerol, which improves the efficiency of biodegradation by microorganisms and eases operation during biological treatment [15]. A study done by Zhang, Zou [19] compared the stability of anaerobic digestion by feeding enzyme pre-treated water vs non-pre-treated water. The reactor containing the enzyme pre-treated feed showed higher stability during operation, even at higher organic loading rates (OLR).

Although it may be a great option, it is not economically feasible to use commercial enzymes practically in engineering practice, as most enzymes have to be significantly monitored as they are sensitive to temperature and pH, and some cannot digest all the organic matter present [19]. An economic and feasibility study done by [15], without considering the ability of methane production to offset costs, revealed that using enzymes to pre-treat wastewater with high fat content has lower installation and operational costs than the traditional technologies. Therefore, it is still a better and cheaper alternative with great potential, despite its complex operation if methane generation is considered as an income generating byproduct. Alternatively, the application of biosurfactants produced by micro-organisms has recently been reported in studies as a cheaper alternative to commercial enzymes [25]. The biosurfactants enhance biodegradation by dissolving FOGs and can be incorporated simultaneously into the biological aeration process, reducing the number of stages for pre-treatment. Other advantages include lower capital and operation costs, reduction in operational problems, as well as an increase in methane production through anaerobic digestion [15,26].

This review highlights the importance of using biological processes in wastewater treatment. The use of a bioremediation agent known as the eco-flush, a product developed by Mavu Biotechnologies (Pty) limited during aerobic treatment, is a novel method that has not been extensively researched. Still, it can pose as an economical and more preferable approach when compared to pure commercial enzymes. Since biological processes are generally slow and not adequate, a multi-integrated system approach can be used, where each stage focuses solely on removing a particular nutrient or pathogen.

2. Analytical Methods for Testing Water Quality

Measurements need to be performed to check if the treated water complies with municipal discharge regulations. The analytical methods are all outlined in the Environmental protection agency (EPA) handbook, and each analysis is specifically coded. Analysis can be tested on: pH (EPA 9040C), temperature, total dissolved solids (TDS) (EPA 160.1), salinity (EPA 320), turbidity (EPA 180.1), total suspended solids (TSS) (EPA 160.2), volatile suspended solids (VSS) (EPA 1684), COD (EPA 410.4), ammonium (EPA 350.1), nitrates concentration (EPA 353.4), biological oxygen demand (BOD) (EPA 405.1), volatile fatty acids (VFAs) (EPA 8260D) and fats, oils and grease (FOGs) (EPA 1664A). Monitoring the efficiency of a wastewater treatment plant is essential. One of the widely used methods for presenting water quality data is the water quality index (WQI) approach. A summary of different water quality parameters is calculated to a single number, which helps define the general quality status of water and its suitability for various purposes like drinking, irrigation, fishing etc., [27].

3. Aerobic Treatment

Aerobic treatments involve the treatment of sludge with air in the presence of aerobic or facultative anaerobic microbes before anaerobic digestion [9]. Oxygen is injected into the treatment system, which accelerates the hydrolysis rate of the organics by enhancing the activity of the micro-organisms [28]. Aerobic treatment prior to anaerobic digestion improves the hydrolysis stage, the sludge solubilization, accelerates hydrolytic activities, increases the methane yield by 20–50%, and decreases VS by 21–64% [28]. This suggests that aerobic pre-treatment does not decrease the methanogenic activity of methanogens within the anaerobic digester and can be a great addition to a multi-stage system [28]. Besides being used in the pre-treatment stages, aerobic processes can be employed after anaerobic digestion to enhance nutrient removal. The required oxygen and treatment time correlate with the strength of the AWW being treated [29]. Due to the expenses incurred during the pumping of artificial oxygen and maintenance, using aerobic treatment for extended periods becomes uneconomical and produces large volumes of biomass. Furthermore, due to the benefits of aerobic treatment, it can be incorporated for shorter processes such as before anaerobic digestion and for nutrient removal after the digester. This will ensure maximum organic matter removal and lower costs, as the processes are relatively short. Despite the higher running costs compared to anaerobic digestion, aerobic treatment has some advantages, such as low odor production and a fast-biological growth rate [30].

4. Aeration Pretreatment Using Enzymes

Enzymes are used to accelerate the hydrolysis of macromolecules to enhance anaerobic digestion [21]. Pre-treatment is included to ensure complete degradation during anaerobic digestion at shorter hydraulic retention times (HRT). Enzymes breakdown the bonds between the triglycerides and hydrolyze them to basic components of fatty acids and glycerol, thereby giving the aerobic micro-organisms a higher chance to biodegrade the FOGs [15]. An eco-flush is an Ergofito's commercially manufactured bioremediation agent supplied in South Africa by Mavu Biotechnologies. An eco-flush is a mixture of natural ingredients and bacteria with the ability to remain dormant until a rich organic source, which acts as a substrate (such as AWW), is applied to activate them, primarily producing enzymes for hydrolysing FOGs [31,32]. Its natural ingredients are derived from glaucids and essential amino acids, which form powerful decomposing agents that stimulate the natural predisposition of certain bacteria to produce enzymes. These enzymes are capable of breaking down the hydrocarbon chains in FOGs and also compete with the bacteria that are responsible for producing Ammonia (NH_3) and Hydrogen Sulphide (H_2S), which results in no to less odor during pre-treatment [31]. The eco-flush can be added to raw AWW at the desired ratio as shown in Figure 2, a systematic diagram representing the pretreatment stage before anaerobic digestion. Artificial aeration is required to facilitate the bacteria to produce enzymes to degrade the FOGs by providing oxygen as an electron acceptor. For successful enzymatic pre-treatment, several parameters such as temperature, pH, substrate quantity and enzymes stability have to be assessed and optimized [28].

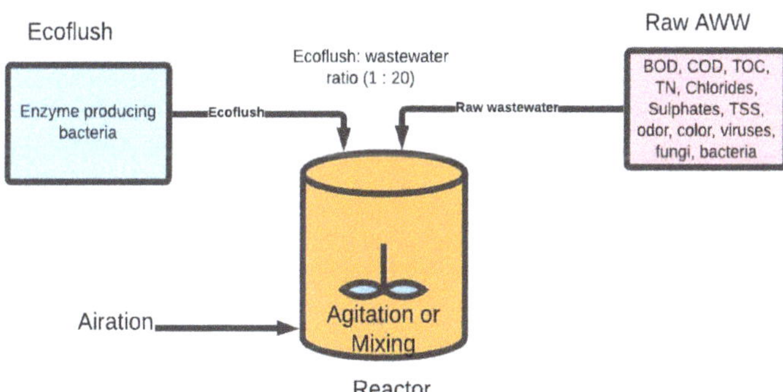

Figure 2. AWW pre-treatment stage using an eco-flush.

Generally, the oxidation of 1 kg of COD requires 1 kWh of aeration energy when the aerobic treatment is selected for wastewater treatment [33]. Oxygen is slightly soluble in water and has to be transferred from the gas phase to the liquid phase, which is called absorption, driven by the concentration gradient between the atmosphere and the bulk liquid [33,34]. The aeration requirement results in the need for a large surface for efficient oxidation of the organic matter, which increases the running costs [33].

Although an aerated pre-treatment stage improves the anaerobic digestion as mentioned previously, the presence of dissolved oxygen in the treated wastewater can also inhibit the methanogenic activity in the anaerobic digestion stage. One critical parameter for a good performance of anaerobic treatment is the lack of oxygen. This is usually determined through the redox potential that should remain < -50 mV for anaerobic digestion and < -300 mV for a good methanogenic activity [35]. For a hermetically closed digester, there is usually no need to attempt to remove the oxygen present, as the BOD in the wastewater consumes the oxygen present rapidly since aerobes and facultative aerobes normally use 100 mg/L of dissolved oxygen to degrade 100 mg/L of BOD [33]. Furthermore, for lab studies and industrial scales, oxygen removal must be implemented through nitrogen purging, which includes three main methods [35], namely: Displacement purging, Pressurizing purging, and Dilution purging. Purging consists of the replacement of one gas by another in an enclosed chamber or space, e.g., removal of oxygen and replacing it with nitrogen gas in anaerobic digestion [34]. Therefore, before the pre-treated water is fed into the anaerobic digester, the Dissolved Oxygen must be monitored.

A study done by [36] a pre-treatment using an Ecoflush bioremediation agent was implemented and resulted in FOG removal of 80% and the TSS and COD removal which reached 38% and 56%, respectively, before feeding the slaughter wastewater into the anaerobic digester. Meyo, Njoya [32] also did a similar study on the pre-treatment of Poultry Slaughter Wastewater (PSW), and the removal percentages varied between 20 and 50% for total suspended solids (TSS), 20 and 70% for chemical oxygen demand (COD), and 50 and 83% for fats, oil, and grease (FOG) before anaerobic treatment using an EGSB reactor. These studies are among the few that reported the use of an Ecoflush reagent. The removal efficiencies do suggest there is potential in bioremediation technology as a pre-treatment stage for high fat content wastewater.

5. Anaerobic Digestion

Anaerobic digestion is a degradation process that occurs in the absence of oxygen to produce methane and carbon dioxide. It consists of four stages: hydrolysis, acidogenesis, acetogenesis, and methanogenesis, as shown in Figure 3. The hydrolysis stage reduces insoluble organic matter and high molecular weight compounds such as polysaccharides, proteins, and lipids into monosaccharides and amino and fatty acids [37]. During acidogenesis, acidogenic bacteria produce volatile fatty acids, carbon dioxide, hydrogen sulphide, ammonia and other by-products, using the components formed during hydrolysis [29]. Acetogenesis is the third stage in which acetic acid, carbon dioxide and hydrogen are produced from the digestion of higher alcohols and organic acids. Methanogenesis is the last and final step in which methane gas is produced by methanogenic bacteria [37]. The production and accumulation of volatile fatty acids (VFAs) can cause a drop in pH, which can affect methane production. Consequently, the VFA: alkalinity ratio is a critical factor in determining reactor performance and should in no case exceed 0.3 [9,38]. Besides a pH range of 6.8–7.2, the organic matter loading/substrate ratio largely affects biogas production, where either too little or too much can cause a slow digestion process and should in no case be >0.3 [9].

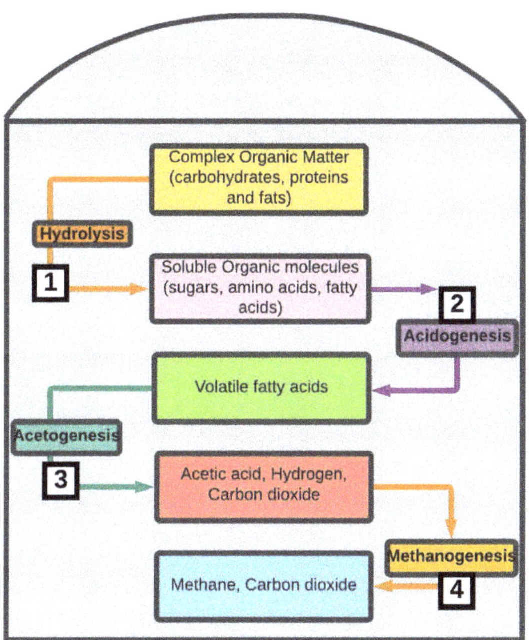

Figure 3. Anaerobic digestion stages in an anaerobic reactor.

6. High-Rate Anaerobic Reactors (HRABS)

High-rate anaerobic digesters have been a subject of increasing interest, due to their high loading capacity and low sludge production. The commonly used high-rate anaerobic digesters include: anaerobic filters, up flow anaerobic sludge blanket (UASB) reactors, anaerobic baffled, fluidized beds, expanded granular sludge beds (EGSB), sequencing batch reactors, anaerobic hybrid/hybrid up-flow anaerobic sludge blanket reactors and the downflow expanded granular bed reactors (DEGBR) which is a hybrid of the EGSB and static granular bed reactor (SGBR) which is shown in Figure 4 [39].

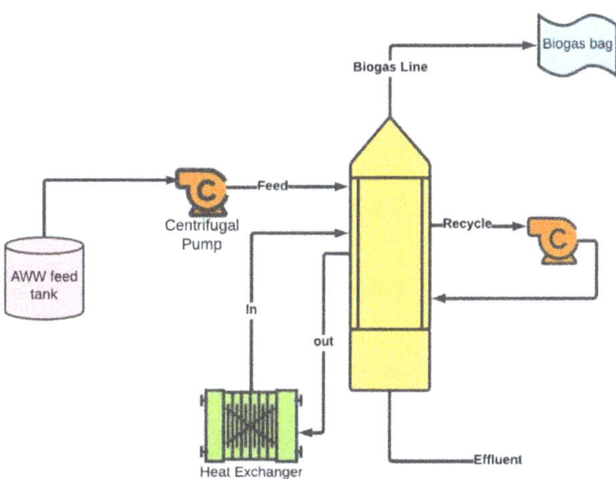

Figure 4. High-rate anaerobic reactor (Downflow Expanded Granular Bed Reactor).

Biological processes heavily rely on the growth and bio-preservation of the required microorganisms through controlling essential operational parameters such as temperature, pH, organic loading rate, carbon to nitrogen ratio, inoculation and start-up of the biodigester, mixing, and inhibition factors [34]. The stability of the HRABS is usually reliant on the maintenance of the mentioned operational parameters within a specific prescribed range for growth of microorganisms [33,35]. Table 2 below describes some of the inhibition parameters for anaerobic digestion and how they affect methanogenic activity.

Table 2. Inhibition factors in anaerobic digestion.

Inhibition Parameter	Operational Range	References
Oxygen concentration	Oxygen concentration is measured as ORP which serves as a relative quantity of oxidised materials i.e., NO_3^-, NH_4^+, SO_4^{2-} ORP between −200 mV and −400 mV is ideal for anaerobic conditions. An ORP of +50 mV suggests a high presence of molecular oxygen and affects the anaerobic microorganisms.	[34,35]
Temperature	Psychrophilic (0 °C–15 °C), mesophilic (20 °C–40 °C), Thermophilic (45 °C–60 °C) and hyper thermophilic > 65 °C. Mesophilic and thermophilic temperatures offer better organic biodegradation and biogas production. Mesophilic is most stable, requires less energy and there is less dominant ammonium inhibition as compared to thermophilic	[33,34,40]
pH	Prescribed range for anaerobic digestion is 6.5–8 Hydrolysis and acetogenesis favours pH range of 5.5–6.5 A pH range of 6.5–8.2 favours methanogenic activity and promotes methane producing bacteria i.e., methanogenium, methanolobus	[34,35]
Nutrients concentration	Nutrients are required to promote growth of microorganisms and results in efficiency of treatment process Some nutrients required are N (65 g/kg TSS), P (16 g/kg TSS), and Mg (3 g/kg TSS) and these quantities correlate to the chemical composition of the methanogenic microorganisms	[34]

Good methanogenic activity in HRABS results in the production of biogas and biogas production can be used as a direct measure of biodegradability efficiency. However, there were instances where a good removal of the substrate from the influent, which usually translates to a good COD or BOD5 removal percentage, didn't align with consequent production of biogas [8,41]. This may have been due to biogas entrapment within the anaerobic granular bed as a result of loss in kinetic energy due to friction losses, a weak connected porosity of the anaerobic granular bed or high surface tensions weakening the emergence of biogas bubbles [8,42].

Numerous studies have been carried out to develop high-rate bioreactors; however, most studies show various drawbacks, ranging from large space requirements, a massive volume of sludge generation, intensive use of energy, and the high overall cost of maintenance [11]. For instance, in the expanded granular sludge bed reactor (EGSB) and the up-flow anaerobic sludge blanket (UASB), the liquid up-flow velocity causes low and inadequate removal of nutrients, pathogens and suspended solids, which results in the requirement of post-treatment for compliance with environmental regulations [43]. Unlike the EGSB and UASB, the downflow expanded granular bed reactor (DEGBR) as shown in Figure 4, takes advantage of gravity as a supplementary force through the granular bed, hence using less energy, as there are no gravitational forces or upward frictional forces to compensate for [42]. The DEGBR consists of a recycle stream, which aids in wastewater distribution of the influent to the anaerobic biomass, and also develops a counter-current flow inside the bioreactor for enhanced mixing of its contents [42,43]. Furthermore, the downflow configuration results in the effluent being collected at the bottom and the gas naturally rising to the top, which eliminates the need for a three-phase separator to separate the gas and biomass compared to the UASB and EGSB [42]. Moreover, the DEGBR also has several advantages like design simplicity, low anaerobic granular sludge (AGS) production, high treatment efficiency, and low operating costs, all of which have turned this bioreactor into a sustainable alternative to mitigate the crisis of water pollution [16].

7. Multi-Integrated Systems

Anaerobic treatment does not produce discharge compliant effluent on its own. The complete degradation of the organic matter is difficult due to the high organic content levels in AWW, the long hydraulic retention times (HRTs) required to remove all the organics as well as the anaerobic process being slow as compared to aerobic processes. An additional treatment stage(s) is/are recommended to remove the organic matter, nutrients, and pathogens that remain after anaerobic treatment [30]. The integration of multi-stage systems can be used to remove pollutants such as heavy metals, grease and oils, color, BOD, TSS, COD and can be handled within one system with multiple stages [18]. Several studies have been done to incorporate additional stages after anaerobic digestion, as shown in Table 3. Comparing single systems and multi-stage systems shows that the latter provides higher removal efficiencies. The data from Table 3 was used to plot a graph, as shown in Figure 5.

Table 3. Comparison between effluent qualities of single systems vs. multi-integrated systems.

Process	Influent Characteristics					Effluent Characteristics (Removal Efficiency)				
	HRT (h)	TOC (mg/L)	BOD (mg/L)	TN (mg/L)	COD (mg/L)	TOC (%)	BOD (%)	TN (%)	COD (%)	REFERENCES
Anaerobic + aerobic + uv	96.0	1.0	640.0	200.0	-	99.98	99.69	82.84	-	[13]
Anaerobic + aerobic + advanced oxidation	75.0–168.0	941.0–1009.0	630.0–650.0	254.0–428.0	-	89.5–99.9	99.70	76.40–81.60	-	[13]
Aerobic	3.0–96.0	-	-	1950–3400	6185–6840	-	-	8.81–93.22	9.42–80.11	[44]
Reverse osmosis	-	-	10.0	13.0	76.0	-	50.0	90.0	85.8	[45]
Anaerobic	24	-	30.0–76.0	6.1–27.0	49.0–137.0	-	11.30	42.30–77.20	13.90	[46]
Anaerobic + aerobic + chemical coagulation	16.0–72.0	-	5143–8360	46.6–138.0	6363–11,000	-	97.76–98.92	73.48–92.72	50.10–97.42	[47]

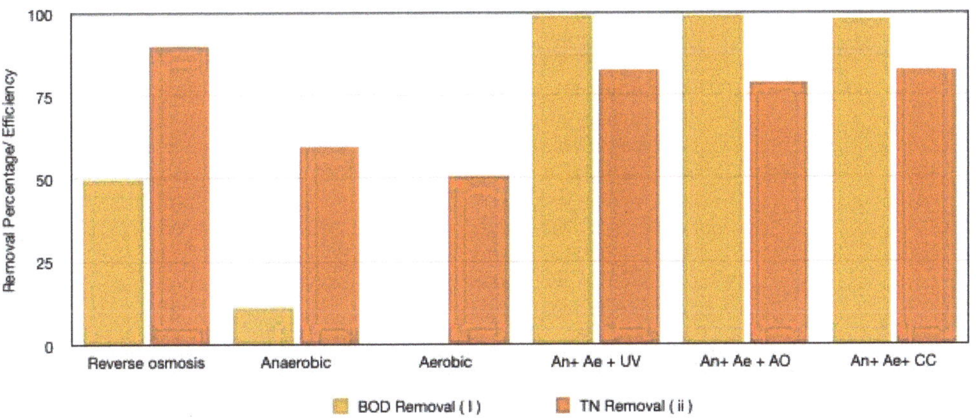

Figure 5. Comparison between single systems and MIS in removing BOD & TN [Abbreviations: An—anaerobic process; Ae—aerobic process; AO—advanced oxidation process, CC—chemical coagulation].

Treatment technologies such as (i) membrane separation using reverse osmosis, (ii) anaerobic, (iii) aerobic, (iv) anaerobic–aerobic–UV, (v) anaerobic–aerobic advanced oxidation, and (vi) anaerobic–aerobic chemical coagulation were compared graphically to show the effect of introducing multiple stages. Figure 5 shows that all single-stage processes have a BOD removal efficiency below 50%, whilst in multi-integrated systems, the values are above 90%. The TN removal follows the same trend, with reverse osmosis having the highest efficiency despite being a single-stage process. This further supports why membranes are necessary for nutrient recovery after anaerobic digestion as a separation process. Although multi-integrated systems offer many benefits, the type of water, cost, and effluent quality will determine the number of stages and processes to be used.

The use of multi integrated systems provide a significant impact on the effluent quality. Dyosile, Mdladla [36] had a higher overall removal efficiency when an integrated system of using enzymatic pre-treatment–DEGBR–MBR was analyzed as compared to anaerobically digesting the poultry slaughterhouse wastewater (PSW) with the DEGBR with no prior or post treatment. The pre-treatment had FOG removal of 80% and the TSS and COD removal reached 38% and 56%, respectively. The removal results on the DEGBR, at an OLR of 18–45 g COD/L.d, was 87%, 93%, and 90% for COD, TSS, and FOG, respectively. The total removal efficiency across the pre-treatment–DEGBR–MBR units was 99% for COD, TSS, and FOG which is much higher than the single stages. Their effluent quality also met requirements for effluent discharge after post treatment using a membrane bioreactor (MBR).

A similar setup of incorporating pre-treatment–EGSB digester–MBR system was used by [32] to reduce the concentration of organic matter in PSW. The pre-treatment stage resulted in a 50% for TSS removal, 80% for COD removal, and 82% for FOG removal. The EGSB effluent had removal percentages of 90% for TSS, >70% for COD, and >90% for FOG. Further removal was also observed using the MBR with the removal performance being >95% for both TSS and COD and 80% for FOG. Their effluent after the MBR process met the discharge standards. These studies add to the fact that single stages alone do not possess the ability to treat AWW to the required discharge standards. Pre and post treatment is required with any anaerobic processes.

Figure 6 shows a proposed process flow diagram of a multi-integrated system to treat AWW. The raw wastewater is first aerobically pre-treated to remove suspended solids and FOGs and enhance anaerobic digestion. Oxygen is artificially added using an adjustable pump. A stainless-steel sieve is used to filter out any suspended solids remaining from pre-treatment. The pre-treated wastewater is added to a holding tank, which feeds into the DEGBR at the desired organic loading rate. The DEGBR operates anaerobically to biodegrade the nutrients, and biogas is produced as a by-product. The effluent from the DEGBR does not meet the required discharge standards as mentioned previously. The effluent becomes the feed to the membrane bioreactor (MBR) where nitrification and denitrification takes place. The micropollutants that pass through membranes can be disinfected using the ultraviolet system (UV). Bustillo-Lecompte, Mehrvar [13] and Bustillo-Lecompte and Mehrvar [30] demonstrate an evaluation on treating AWW using combined advanced oxidation processes. The evaluation factored in treatment capability and overall costs for treatment technologies, including ABR, AS and UV. It was proven that the combined process of the ABR–AS–UV system was the most cost-effective solution compared to single processes for TOC removal under optimal conditions. However, as this may be a guide, different wastewaters have different characteristics, and analysis must be done to find the best method possible.

Ultraviolet (UV) light is frequently used for pathogen inactivation in wastewater treatment [48–51]. UV light effectively inactivates viruses, bacteria, and cysts by penetrating cell walls and damaging DNA or RNA without chemical addition. Traditional UV lamps are low-cost and accessible in developing economies, but also contain toxic mercury vapor. On the other hand, UV LEDs are more expensive but mercury-free [1,52].

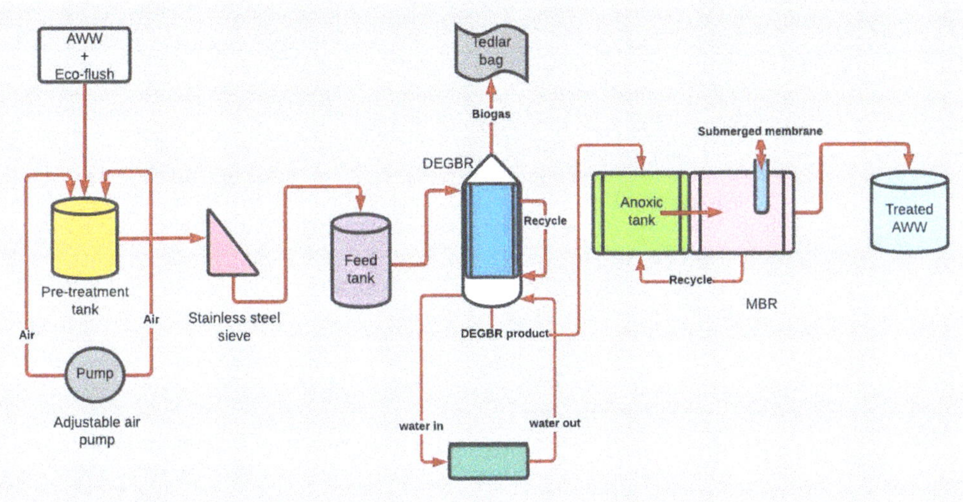

Figure 6. Process flow diagram of a proposed multi-stage integrated system to treat AWW.

The study by Beck, Suwan [1] evaluated a cost-effective, user-friendly, and relatively fast treatment process involving a woven-fiber microfiltration (WFMF) membrane to filter domestic wastewater followed by UV disinfection to disinfect the permeate. With an effective pore size of 1–3 μm [53] the membrane was capable of removing *Ascaris lumbricoides* eggs (50 mm) and *Giardia* cysts (10 μm), whereas bacteria (1–2 μm), viruses, and *Cryptosporidum* oocysts (3 μm), which are small enough to pass through the filter pores, were inactivated by exposure to UV light. The bacteria (total coliform and *Escherichia coli*) and viruses (MS2 bacteriophage) passing through the membrane were disinfected by flow-through UV reactors containing either a low-pressure mercury lamp or light-emitting diodes (LEDs) emitting an average peak wavelength of 276 nm. For domestic wastewater from a university campus that they used in their study, the membrane reduced TSS (by 79.8%), turbidity by 76.5%, COD by 38.5%, BOD by 47.8%, and NO_3 by 41.4%. UVT at 254 nm improved by 19.4%, and UVT at 280 nm by 12.4% [1]. Following UV disinfection, wastewater quality met the WHO standards for unrestricted irrigation. UV lambs can succumb to fouling and scaling after extensive use and it is reversible through citric acid circulation [54].

8. Hybrid Membrane Bioreactor

A membrane bioreactor (MBR) is an integrated system with membrane filtration for the biological degradation of waste present in wastewater. Generally, it is composed of a biological unit and a membrane module, which separates water from the aerobically digested water and returns activated sludge to the biological unit [55].

Anaerobic membrane bioreactors (AnMBRs) retain solids selectively through micro-filtration membranes which offer an alternative to lagoons and granule based high-rate anaerobic treatments [56]. They produce an excellent effluent quality, have a high tolerance to OLR variations, as well as the ability to produce a solid free effluent for the purposes of reuse [57]. The hybrid membrane bioreactor consists of (i) an anoxic stage and (ii) an aerobic membrane filtration stage. Since the DEGBR operates anaerobically, it has two significant drawbacks, (i) it is ineffective in removing nitrates and phosphorous, and (ii) it reduces the organic nitrogen and sulphur to ammonia and hydrogen sulphide, which are toxic, hence the need for incorporating a membrane bioreactor stage as post-treatment. The advantages of MBR compared with a conventional activated sludge process include high effluent quality, decreased reactor volume, elevated solid retention time (SRT) and high mixed liquor suspended solids (MLSS), low sludge yield, and easier operation [58,59]. However there are still some drawbacks associated with MBRs such as; membrane fouling, high energy consumption and low removal efficiency of poorly biodegradable micropollutants like diclofenac, atrazine, and carbamazepine [58].

MBR technology has been widely used recently for nutrient recovery. Coagulation or flocculation can be used to recover valuable nutrients in the conventional process. Unfortunately, the protein concentrate obtained by the traditional methods cannot be used as animal food because coagulants and flocculants can introduce some harmful compounds and change protein properties. Pressure-driven membrane processes are good at protein recovery while keeping protein unchanged since membrane separation is a physical process [55].

Recovering nutrients from wastewater reduces the environmental effects of wastewater treatment, and subsequently, the recovered nutrients can be used to produce fertilizers. Phosphorus and nitrogen are essential for organism growth and result in eutrophication in surface water sources, leading to the death of aquatic life [56]. If ammonium and phosphate ions were to be removed from wastewater using processes such as chemical precipitation, it would consume a large amount of electricity and cost about 4% extra compared to nutrient recovery [60]. Besides the extra costs involved, nutrient recovery is better than complete removal because i) nutrient-based fertilizers can be produced for agricultural purposes, (ii) the environmental impact from wastewater discharged is reduced, hence less eutrophication occurs, and iii) N recovery can reduce the consumption of natural resources and save costs on nitrogen fixation [56].

The hybrid membrane bioreactor is required to provide an anoxic–aerobic stage where oxygen is utilized by bacteria to oxidize the ammonia and hydrogen sulphide to less harmful substances. Nitrification occurs due to two specific autotrophic bacteria, the ammonia oxidising organisms (ANOs) and the nitrite oxidising organisms (NNOs), and occurs in two steps. The ANOs convert free and saline ammonia to nitrite. In the second step, the NNOs convert nitrite to nitrate. Ammonia and nitrite are used for catabolism [33]. Nitrification is a prerequisite for denitrification, and without it, biological N removal is not possible. Denitrification becomes possible once nitrification takes place by incorporating anoxic zones in the reactor. The denitrification occurs anoxically via facultative heterotrophic biomass [33]. During nitrification, the N remains in the liquid phase because it is transformed from ammonia to nitrate. In the denitrification step, the N is transferred from the liquid to the gas phase and escapes to the atmosphere.

The proposed study referred to in Figure 7 employs the modified Ludzack–Ettinger system (MLE), which separates the anoxic and aerobic reactors by putting them in series, as shown in Figure 7 below. It also consists of a recycle for the underflow feeding back to the first anoxic reactor as well as a mixed liquor recycle from the aerobic to the primary anoxic reactor. The influent is discharged to the first or primary anoxic reactor, which is maintained in an anoxic state by mixing without aeration and provides conducive conditions for denitrification. The second reactor is aerated and is where nitrification takes place. However, the MLE system has one major drawback: complete nitrate removal cannot be achieved because a part of the total flow from the aerobic reactor is not recycled to the anoxic reactor but instead exits the system with the effluent [33].

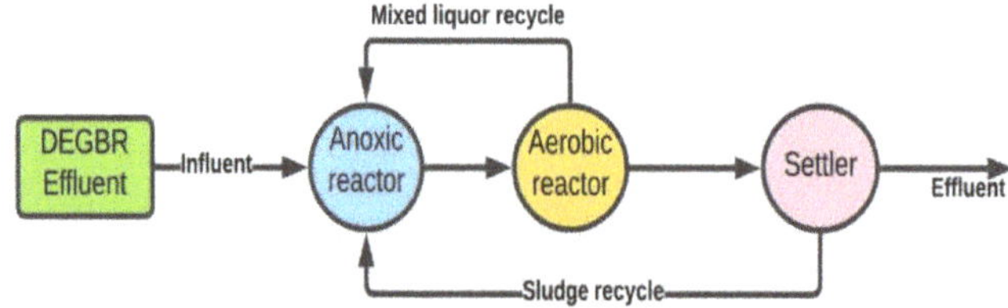

Figure 7. MLE system for nitrification and denitrification.

Phosphorus can be removed biologically through enhanced biological phosphorus removal, exploiting the ability of polyphosphate-accumulating organisms (PAOs) to take up P in excess of metabolic requirements and accumulate it intracellularly as polyphosphate [61]. This metabolic phenotype is facilitated by a continuing cycle of provision of organic carbon, mainly in the form of volatile fatty acids (VFAs) to the microorganisms, and then exposure of the organisms to first anaerobic and then aerobic conditions. Organic carbon is often the limiting substrate for both denitrification and P removal, and many wastewater treatment plants add extra carbon for denitrification to balance the processes. A combination of denitrification and enhanced biological phosphorus removal in one process could offer substantial savings on carbon for the overall nutrient removal process, which makes this approach highly attractive [62].

The performance of the membrane is mainly characterized by the permeate flux and retention properties [5]. Membrane separation has one particular advantage over other separation processes such as distillation, crystallization and adsorption because it relies on physical separation without phase change and usually no addition of chemicals. Therefore, energy consumption is usually much lower compared to distillation and crystallization [63]. Two main MBR configurations exist: side stream and submerged, as shown in Figure 8. A recirculation pump provides cross-flow velocity in the side stream configuration to reduce blockage by suspended solids on the membrane surface. The side stream MBR is widely used in industrial wastewater treatment but has a higher energy demand. On the other hand, the submerged MBR operates at lower flux and offers higher permeability. They are often used in municipal wastewater treatment. Coarse aeration is provided to the system to reduce fouling as well as provide oxygen to the biomass [64].

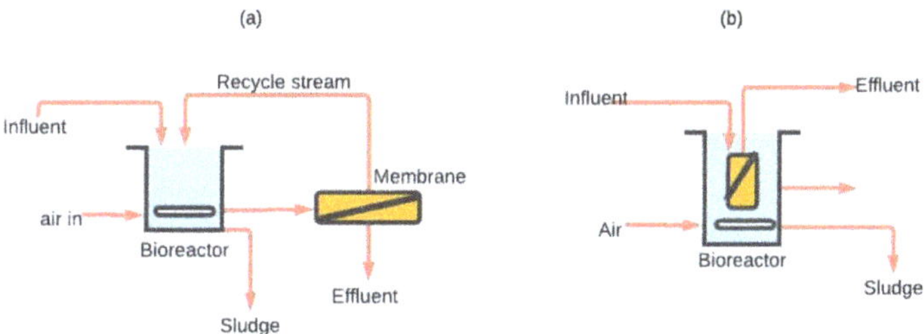

Figure 8. AnMBR configurations: (**a**) side stream configuration, (**b**) submerged configuration.

9. Applications of Membranes in Wastewater Treatment

Pressure-driven membrane processes such as microfiltration (MF), ultrafiltration (UF), reverse osmosis (RO), and nanofiltration (NF) are widely studied for wastewater treatment and they rely on hydraulic pressure to achieve separation [65]. Membrane filtration is one of the most emerging technologies to produce high-quality water because it utilizes zero chemical constituents and offers enormous advantages over conventional methods [55]. Mostly used membrane filtration in wastewater treatments are RO, UF and MF. Although reverse osmosis (RO) is a well-established technology for water reuse and desalination [66,67] it is still limited by its high energy consumption and operating costs as the flow is against the pressure gradient.

An alternative is the use of low-pressure RO (LPRO) membranes which have been developed to reduce the RO operation pressure when maintaining high rejections to small soluble organic molecules and ionic species [68–70]. The operation pressure is an important operation parameter of LPRO, which affects the filtration productivity (flux), membrane fouling, and energy consumption. The performance of RO in the treatment of secondary effluent of SWW was reported by [45] to remove organic matter and the removal efficiencies of BOD, COD, TN, and TP were found to be 50.0, 85.8, 90.0, and 97.5%, respectively. It was concluded that LPRO was a suitable technique for the post-treatment of AWW effluent.

A study done by [71] on the performance of the UF membrane treating AWW showed BOD and COD removal efficiencies of around 97.8–97.89% to 94.52–94.74%, whereas TSS and FOG removal were 98% and 99%, respectively [72]. Pressure driven membrane processes have proven to be successful in the separation of valuable organic and inorganic compounds in black liquor as well as being energy-efficient in several studies [73–75]. In recent studies, separation processes are being coupled to improve effluent quality. For example, UF/NF combinations have been reported to be a promising solution in wastewater with high amounts of organic material such as black liquor. In these cases, UF is used as a pre-treatment for NF [73,76].

The ultrafiltration (UF) pre-treatment and the control of the operation pressure were found to be essential for mitigating LPRO membrane fouling. Water quality analyses showed that an integrated process of the UASB + UF + LPRO could achieve an effluent quality characterized by concentrations of 10.4–12.5 mg/L of chemical oxygen demand (COD), 1.8–2.1 mg/L of total nitrogen (TN), 1.3–1.8 mg/L of ammonia nitrogen (NH3-N), and 0.8–1.2 mg/L of total phosphorus (TP) [70].

Coskun, Debik [77] studied the PSW treatment using laboratory-scale membrane processes. Their study reported that UF as pretreatment improved the removal efficiencies for NF and RO processes; NF reduced almost 90% of COD, RO removed 97.4% of COD, and the UF pretreatment resulted in higher final fluxes 8.1 and 5.7 times more for NF and RO, respectively, than for those without UF.

Ionic species can be removed to meet the reuse requirements of brewery wastewater effluent discharge by the inclusion of reverse osmosis into the treatment chain [66,70]. Verhuelsdonk, Glas [78] did an economic analysis on brewery wastewater reuse and reported that UASB wastewater could be treated to drinking water quality with a yield of 63% by using an MBR + UF + RO system.

A comparison study was done by Skoczko, Puzowski [59] to compare the modernized vs conventional treatment methods on a newly modernized wastewater treatment plant. On the basis of the conducted research, it was noted that the operation of the plant after modernization was more cost-intensive. There were additional electricity costs due to ensuring adequate pressure on the membranes. Nevertheless, the obtained results of the removal of contaminants showed BOD removal of over 99.0%, COD removal of 99.0%, TSS removal of 99.5%, and removal degree of biogenic compounds also increased and exceeded 98%. Although the membranes have been well researched and are still being improved, it continues to show high operational costs due to aeration and membrane fouling, which constitutes a major drawback.

10. Membrane Preservation, Fouling and Cleaning Methods

The accumulation of particulates such as fats, grease, protein, and organic matter can cause build-up on the membrane material resulting in membrane fouling and wetting which is a huge economic influence on the use of membranes as they account for 72% of the capital investment [10]. The types of foulants which may interfere with membrane performance include chemical foulants such as scaling, physical foulants such as deposition of particles, biological foulants such as microbes, and organic foulants which interact with the membrane material [79]. Membrane wetting is the process in which membrane materials lose their hydrophobicity and allow for liquids to penetrate the membrane pores resulting in a direct liquid flow from feed through the wetted pores, substantially deteriorating permeate quality [79]. The fouling and wetting of membrane materials impairs the membrane performance and shortens membrane lifetime, thereby reducing NH3 recovery from AWW.

To reduce fouling and wetting, membranes can be cleaned. Several chemical and physical cleaning methods were developed to remove membrane fouling. The membrane cleaning process is affected by different factors. The type and mode of cleaning for example, physical cleaning, doesn't really retrieve membrane permeability effectively as it only removes loose particles. Temperature is considered as another factor that may take effect on the membrane cleaning strategy. Increasing temperature is substantial for cleaning the fouling membrane by increasing solubility due to reactivity of functional groups at high temperatures of the organic matters and increasing mass transfer dispersion with mechanical destabilization of biofilm layers on the membrane surface [58]. Increasing the pH also has a directly proportional relationship with membrane cleaning efficiency [80]. For instance, increasing pH from 4.9–11.0 will affect the cleaning percentage from 25%–44% and, at pH 11, it is very easy to break down the gel layer on the membrane surface when compared to the lower pH [58]. Table 4 below shows some of the membrane cleaning methods used to reduce fouling and improve membrane life in membrane technology.

Table 4. Membrane cleaning methods used in membrane technology.

Industry	Membrane Process Type	Chemicals Used	Result	Reference
Municipal (drinking water treatment systems)	Ultrafiltration (UF)	Membrane vibration + coagulation. coagulants, such as Al (III) and Fe (III) compounds, were added to the influent	Membrane rotation speed of 60 r/min, the permeate flux increased by 90% and the organic removal by 35%, with a 40 mg/L coagulant dosage, with an additional 70% increase of flux and a 5% increment of organic removal to 80% was obtained.	[81]
Food industry (fruit juice concentration)	Forward Osmosis	Pretreatment by microfiltration before FO process	There was an attractive interaction between the FO membrane and orange juice foulants. Eliminating those foulants using the microfiltration pre-treatment weakened such an attractive interaction and effectively prevented the fouling layer from growing, leading to a lower process resistance and, finally, resulting in a great improvement of concentration efficiency	[82]
Food industry	Electrodialysis with bipolar membranes (EDBM)	Pulsed electric field (PEF) mode, which consists in the application of constant current density pulses during a fixed time (Ton) alternated with pause lapses (Toff)	Both a long pause and high flow rate contribute to a more effective decrease in the concentration of protons and caseinate anions at the BPM surface: a very good membrane performance was achieved with 50 s of pause duration of PEF and a flow rate corresponding to Reynolds number = 374	[83]
Municipal wastewater	Membrane Bio-Reactors (MBRs)	Examines the effect of operating conditions on fouling of membrane Bio-Reactors (MBRs). Conditions such as: diminishing DO, recirculating rate and controlled growth of filamentous microorganisms were optimised	The diminishing of DO in the recirculated sludge improved denitrification, and resulted in lower concentrations of $N\text{-}NO_3^-$ and TN in the effluent of the Control-MBR. Furthermore, the recirculation rate of $Q_r = 2.6 \cdot Q_{in}$, resulted in improved performance regarding the removal of $N\text{-}NH_4^+$	[84]
Second effluent of sewage with Activated sludge	FO	Physical cleaning Air scouring	-	[85]
PSW	UF, MF	Sodium hypochlorite, citric acid, sodium hydroxide, and ultrapure water	Recovered 95% of water flux	[5]
PSW	UF	Electrocoagulation pre-treatment	Pre-treatment approaches can be adopted to alleviate fouling before the membrane filtration process.	[86]

Recommendations and future perspectives:

1. There are numerous NF membranes available in the market, but only some of them can resist harsh operating conditions (such as extreme pH) [73]. Further studies can be carried out to produce membranes that are stable and not susceptible to fouling in high pH conditions.

2. For high quality effluents, a novel MBR called the osmosis membrane bioreactor (OMBR) has been developed to promote wastewater treatment and reuse [58]. In OMBR, the FO membrane module is displaced in the wastewater. Combined with

biological treatment, water from the mixed liquor is forced to transfer through the semipermeable membrane to the draw side under the osmotic pressure gradient. The pollutants, activated sludge, and solids are all rejected by the membrane. The superior performance of OMBR over conventional MBR has been demonstrated in previous research [58]. This OMBR can be integrated into the proposed system of this review instead of UF. This will reduce overall running costs incurred through high energy consumption, the cost for chemical cleaning, and membrane life which are limitations in pressure-driven membrane processes.

3. Several studies reported that chemical cleaning could achieve highly efficient membrane cleaning from organic foulants, which may have a strong interaction with the membrane surface [87–90]. Although chemical cleaning is a viable option, it does not provide the eco-friendliness and biological treatment options the world is moving towards and this might cause an environmental problem as the effluent stream may be discharged containing chemicals. Hence more physio-biological pretreatment options and parameter optimization can be a way to ensure limited fouling and maintain a minimal pollution footprint.

11. Conclusions

Whilst biological processes such as anaerobic and aerobic digestion provide the much-needed benefits of being environmentally friendly and economical, they still fall short in nutrient removal, digesting FOGs, and removing suspended solids. The choice of reactor also affects the composition of effluent, the costs incurred during operation, and the space required. Anaerobic digestion is very sensitive, involving different bacterial groups (methanogenic, acetogenic, etc.), which all have different optimum conditions. These bacteria are inhibited by process parameters such as temperature, pH, VFA concentrations, etc. Therefore, it is paramount to maintain stable operating conditions in the digester. The DEGBR gives numerous advantages, such as ease of operation, and lowers energy requirements for pumping, as the water is aided by gravity and also provides turbulent mixing through the recycled stream. In contrast, the up-flow reactors such as the EGSB and the UASB experience poor reactor performance caused by a high up-flow velocity, biomass washout and higher energy requirements to oppose gravity and compensate for head losses to friction. The DEGBR has become more favorable for treating high strength wastewater. Adding a pre-treatment stage before anaerobic digestion, where enzymes are used to hydrolyze and break down FOGs increases biogas production, improves reactor performance, and results in ease of operation. Other post anaerobic digestion treatment stages such as nitrification, denitrification, membrane filtration, and ultraviolet radiation can be added to improve the removal efficiency of P, C, and N, as well as help meet the regulation standards.

Author Contributions: Writing—original draft preparation, L.G.; Supervision, M.B.; Funding acquisition, M.B.; Writing—review and editing, M.B. & L.G.; Reviewing—T.H., D.I., C.G., and M.N. All authors have read and agreed to the published version of the manuscript.

Funding: This research was funded by National Research Foundation, Thuthuka Funding, R017, the Cape Peninsula University of Technology University Research Fund (URF), and the Bioresource Engineering Research Group (CPUT, BioERG) subsidy cost centers RK45 and R971.

Acknowledgments: The authors wish to acknowledge the National Research Foundation Thuthuka Funding, R017, for their financial contribution to this work, the South African Breweries (SAB).

Conflicts of Interest: The authors declare no conflict of interest.

Abbreviations

COD	Chemical oxygen demand
BOD	Biological oxygen demand
TSS	Total suspended solids
VSS	Volatile suspended solids

DEGBR	Downflow expanded granular bed reactor
EGSB	Expanded granular sludge bed reactor
AnMBR	Anaerobic membrane bioreactor
VFA	Volatile fatty acids
FOG	Fats oils and grease
WQI	Water quality index
TDS	Total dissolved solids
UASB	Up-flow anaerobic sludge bed
PSW	Poultry slaughterhouse wastewater
HRABS	High rate anaerobic bioreactors
AGS	Anaerobic granular sludge
HRT	Hydraulic retention time
OLR	Organic loading rate
UV	Ultraviolet
SRT	Solids retention time
MLSS	Mixed liquor suspended solids
ANOs	Ammonia oxidising organisms
NNOs	Nitrite oxidising organisms
MLE	Modified Ludzack Ettinger
PAO	Phosphorus accumulating organisms
LPRO	Low pressure reverse osmosis
RO	Reverse osmosis
UF	Ultrafiltration
MF	Microfiltration
FO	Forward osmosis
OMBR	Osmosis membrane bioreactor
EDBM	Electrodialysis bipolar membrane
NF	Nanofiltration

References

1. Beck, S.E.; Suwan, P.; Rathnayeke, T.; Nguyen, T.M.H.; Huanambal-Sovero, V.A.; Boonyapalanant, B.; Hull, N.M.; Koottatep, T. Woven-Fiber Microfiltration (WFMF) and Ultraviolet Light Emitting Diodes (UV LEDs) for Treating Wastewater and Septic Tank Effluent. *Water* **2021**, *13*, 1564. [CrossRef]
2. Abdelhay, A.; Othman, A.A.; Albsoul, A. Treatment of slaughterhouse wastewater using high-frequency ultrasound: Optimization of operating conditions by RSM. *Environ. Technol.* **2020**, 1–9. [CrossRef]
3. Barbut, S. *The Science of Poultry and Meat Processing*; University of Guelph: Guelph, ON, Canada, 2015; pp. 1–704.
4. Aziz, H.A.; Puat, N.N.A.; Alazaiza, M.Y.D.; Hung, Y.-T. Poultry Slaughterhouse Wastewater Treatment Using Submerged Fibers in an Attached Growth Sequential Batch Reactor. *Int. J. Environ. Res. Public Health* **2018**, *15*, 1734. [CrossRef] [PubMed]
5. Marchesi, C.M.; Paliga, M.; Oro, C.E.D.; Dallago, R.M.; Zin, G.; Di Luccio, M.; Oliveira, J.V.; Tres, M.V. Use of membranes for the treatment and reuse of water from the pre-cooling system of chicken carcasses. *Environ. Technol.* **2021**, *42*, 126–133. [CrossRef] [PubMed]
6. Farzadkia, M.; Vanani, A.; Golbaz, S.; Sajadi, H.; Bazrafshan, E. Characterization and evaluation of treatability of wastewater generated in Khuzestan livestock slaughterhouses and assessing of their wastewater treatment systems. *Glob. Nest J.* **2016**, *18*, 108–118.
7. Rabah, F. Physical, Chemical and Biological Characteristics of Waste Water. In *Advanced Sanitary Engineering*; The Islamic University of Gaza: Gaza, Palestine, 2018; Available online: http://site.iugaza.edu.ps/halnajar/files/2010/02/unit-1.-Itroduction-to-wastewater-treatment.pdf.
8. Basitere, M. Performance Evaluation of an up-and down-flow Anaerobic Reactor for the Treatment of Poultry Slaughterhouse Wastewater in South Africa. Ph.D. Thesis, Cape Peninsula University of Technology, Cape Town, South Africa, 2017.
9. Aziz, A.; Basheer, F.; Sengar, A.; Irfanullah, A.; Khan, S.U.; Farooqi, I.H. Biological wastewater treatment (anaerobic-aerobic) technologies for safe discharge of treated slaughterhouse and meat processing wastewater. *Sci. Total Environ.* **2019**, *686*, 681–708. [CrossRef] [PubMed]
10. Brennan, B.; Gunes, B.; Jacobs, M.R.; Lawler, J.; Regan, F. Potential Viable Products Identified from Characterisation of Agricultural Slaughterhouse Rendering Wastewater. *Water* **2021**, *13*, 352. [CrossRef]
11. Musa, M.A.; Idrus, S.; Che Man, H.; Nik Daud, N.N. Performance Comparison of Conventional and Modified Upflow Anaerobic Sludge Blanket (UASB) Reactors Treating High-Strength Cattle Slaughterhouse Wastewater. *Water* **2019**, *11*, 806. [CrossRef]
12. Chen, Y.; Cheng, J.J.; Creamer, K.S. Inhibition of anaerobic digestion process: A review. *Bioresour. Technol.* **2008**, *99*, 4044–4064. [CrossRef]

13. Bustillo-Lecompte, C.F.; Mehrvar, M.; Quiñones-Bolaños, E. Combined anaerobic-aerobic and UV/H$_2$O$_2$ processes for the treatment of synthetic slaughterhouse wastewater. *J. Environ. Sci. Health Part A* **2013**, *48*, 1122–1135. [CrossRef]
14. Ozdemir, S.; Yetilmezsoy, K. A mini literature review on sustainable management of poultry abattoir wastes. *J. Mater. Cycles Waste Manag.* **2020**, *22*, 11–21. [CrossRef]
15. Damasceno, F.R.; Cavalcanti-Oliveira, E.D.; Kookos, I.K.; Koutinas, A.A.; Cammarota, M.C.; Freire, D.M. Treatment of wastewater with high fat content employing an enzyme pool and biosurfactant: Technical and economic feasibility. *Braz. J. Chem. Eng.* **2018**, *35*, 531–542. [CrossRef]
16. Basitere, M.; Njoya, M.; Ntwampe, S.K.O.; Sheldon, M.S. Up-flow vs downflow anaerobic digester reactor configurations for treatment of fats-oil-grease laden poultry slaughterhouse wastewater: A review. *Water Pract. Technol.* **2020**, *15*, 248–260. [CrossRef]
17. Liew, Y.X.; Chan, Y.J.; Manickam, S.; Chong, M.F.; Chong, S.; Tiong, T.J.; Lim, J.W.; Pan, G.-T. Enzymatic pretreatment to enhance anaerobic bioconversion of high strength wastewater to biogas: A review. *Sci. Total Environ.* **2020**, *713*, 136373. [CrossRef] [PubMed]
18. Meiramkulova, K.; Zorpas, A.A.; Orynbekov, D.; Zhumagulov, M.; Saspugayeva, G.; Kydyrbekova, A.; Mkilima, T.; Inglezakis, V.J. The Effect of Scale on the Performance of an Integrated Poultry Slaughterhouse Wastewater Treatment Process. *Sustainability* **2020**, *12*, 4679. [CrossRef]
19. Zhang, S.; Zou, L.; Wan, Y.; Ye, M.; Ye, J.; Li, Y.-Y.; Liu, J. Using an expended granular sludge bed reactor for advanced anaerobic digestion of food waste pretreated with enzyme: The feasibility and its performance. *Bioresour. Technol.* **2020**, *311*, 123504. [CrossRef]
20. Mondal, T.; Jana, A.; Kundu, D. Aerobic wastewater treatment technologies: A mini. *Int. J. Env. Tech. Sci.* **2017**, *4*, 135–140.
21. Affes, M.; Aloui, F.; Hadrich, F.; Loukil, S.; Sayadi, S. Effect of bacterial lipase on anaerobic co-digestion of slaughterhouse wastewater and grease in batch condition and continuous fixed-bed reactor. *Lipids Health Dis.* **2017**, *16*, 195. [CrossRef] [PubMed]
22. Carrère, H.; Dumas, C.; Battimelli, A.; Batstone, D.J.; Delgenès, J.P.; Steyer, J.P.; Ferrer, I. Pretreatment methods to improve sludge anaerobic degradability: A review. *J. Hazard. Mater.* **2010**, *183*, 1–15. [CrossRef]
23. Pilli, S.; Yan, S.; Tyagi, R.D.; Surampalli, R.Y. Thermal Pretreatment of Sewage Sludge to Enhance Anaerobic Digestion: A Review. *Crit. Rev. Environ. Sci. Technol.* **2015**, *45*, 669–702. [CrossRef]
24. Doosti, M.; Kargar, R.; Sayadi, M. Water treatment using ultrasonic assistance: A review. *Proc. Int. Acad. Ecol. Environ. Sci.* **2012**, *2*, 96–110.
25. Sanghamitra, P.; Mazumder, D.; Mukherjee, S. Treatment of wastewater containing oil and grease by biological method—A review. *J. Environ. Sci. Health Part A* **2021**, *56*, 394–412. [CrossRef] [PubMed]
26. Nakhla, G.; Farooq, S. Simultaneous nitrification–denitrification in slow sand filters. *J. Hazard. Mater.* **2003**, *96*, 291–303. [CrossRef]
27. Bora, M.; Goswami, D.C. Water quality assessment in terms of water quality index (WQI): Case study of the Kolong River, Assam, India. *Appl. Water Sci.* **2017**, *7*, 3125–3135. [CrossRef]
28. Nguyen, V.K.; Kumar Chaudhary, D.; Hari Dahal, R.; Hoang Trinh, N.; Kim, J.; Chang, S.W.; Hong, Y.; Duc La, D.; Nguyen, X.C.; Hao Ngo, H.; et al. Review on pretreatment techniques to improve anaerobic digestion of sewage sludge. *Fuel* **2021**, *285*, 119105. [CrossRef]
29. Yuan, Y.; Hu, X.; Chen, H.; Zhou, Y.; Zhou, Y.; Wang, D. Advances in enhanced volatile fatty acid production from anaerobic fermentation of waste activated sludge. *Sci. Total Environ.* **2019**, *694*, 133741. [CrossRef]
30. Bustillo-Lecompte, C.; Mehrvar, M. Slaughterhouse wastewater: Treatment, management and resource recovery. In *Physico-Chemical Wastewater Treatment Resource Recovery*; Farroq, R., Ahmad, Z., Eds.; Intech: Rijeka, Croatia, 2017; pp. 153–174.
31. Ergofito. Ecoflush Eliminated Ammonia. Ecoflush Eliminated Odours. Available online: https://www.ergofito.co.za/home (accessed on 25 July 2021).
32. Meyo, H.B.; Njoya, M.; Basitere, M.; Ntwampe, S.K.O.; Kaskote, E. Treatment of Poultry Slaughterhouse Wastewater (PSW) Using a Pretreatment Stage, an Expanded Granular Sludge Bed Reactor (EGSB), and a Membrane Bioreactor (MBR). *Membranes* **2021**, *11*, 345. [CrossRef]
33. Henze, M.; van Loosdrecht, M.C.; Ekama, G.A.; Brdjanovic, D. *Biological Wastewater Treatment*; IWA publishing: London, UK, 2008.
34. Njoya, M.; Basitere, M.; Ntwampe, S.K.O. *High Rate Anaerobic Treatment of Poultry Slaughterhouse Wastewater (PSW)*; Nova Science Publishers, Inc.: New York, NY, USA, 2019; Volume 38.
35. Gerardi, M.H. *The Microbiology of Anaerobic Digesters*; John Wiley & Sons: Hoboken, NJ, USA, 2003.
36. Dyosile, P.A.; Mdladla, C.; Njoya, M.; Basitere, M.; Ntwampe, S.K.O.; Kaskote, E. Assessment of an Integrated and Sustainable Multistage System for the Treatment of Poultry Slaughterhouse Wastewater. *Membranes* **2021**, *11*, 582. [CrossRef] [PubMed]
37. Appels, L.; Baevens, J.; Degrève, J.; Dewil, R. Principios y potencial de la digestión anaerobia de lodos activados por residuos. *Prog. Combust. De Energía. Sci.* **2008**, *34*, 755–781. [CrossRef]
38. Del Nery, V.; de Nardi, I.R.; Damianovic, M.H.R.Z.; Pozzi, E.; Amorim, A.K.B.; Zaiat, M. Long-term operating performance of a poultry slaughterhouse wastewater treatment plant. *Resour. Conserv. Recycl.* **2007**, *50*, 102–114. [CrossRef]
39. Rajagopal, R.; Saady, N.M.C.; Torrijos, M.; Thanikal, J.V.; Hung, Y.-T. Sustainable Agro-Food Industrial Wastewater Treatment Using High Rate Anaerobic Process. *Water* **2013**, *5*, 292–311. [CrossRef]
40. Burton, F.L.; Stensel, H.D.; Tchobanoglous, G.; Metcalf & Eddy, Inc. *Wastewater Engineering: Treatment and Reuse*; McGraw Hill: New York, NY, USA, 2003.

41. Basitere, M.; Williams, Y.; Sheldon, M.S.; Ntwampe, S.K.O.; De Jager, D.; Dlangamandla, C. Performance of an expanded granular sludge bed (EGSB) reactor coupled with anoxic and aerobic bioreactors for treating poultry slaughterhouse wastewater. *Water Pr. Technol.* **2016**, *11*, 86–92. [CrossRef]

42. Njoya, M.; Basitere, M.; Ntwampe, S.K.O. Treatment of poultry slaughterhouse wastewater using a down-flow expanded granular bed reactor. *Water Pr. Technol.* **2019**, *14*, 549–559. [CrossRef]

43. Cruz-Salomón, A.; Ríos-Valdovinos, E.; Pola-Albores, F.; Lagunas-Rivera, S.; Meza-Gordillo, R.; Ruíz-Valdiviezo, V.M.; Cruz-Salomón, K.C. Expanded granular sludge bed bioreactor in wastewater treatment. *Glob. J. Environ. Sci. Manag.* **2019**, *5*, 119–138. [CrossRef]

44. Kundu, P.; Debsarkar, A.; Mukherjee, S. Kinetic Modeling for Simultaneous Organic Carbon Oxidation, Nitrification, and Denitrification of Abattoir Wastewater in Sequencing Batch Reactor. *Bioremediat. J.* **2014**, *18*, 267–286. [CrossRef]

45. Bohdziewicz, J.; Sroka, E. Integrated system of activated sludge–reverse osmosis in the treatment of the wastewater from the meat industry. *Process Biochem.* **2005**, *40*, 1517–1523. [CrossRef]

46. Luu, H.M.; Nguyen, N.X.D.; Bui, T.L.M. Treatment of wastewater from slaughterhouse by biodigester and *Vetiveria zizanioides* L. *Livest. Res. Rural Dev.* **2014**, *26*, 68.

47. López-López, A.; Vallejo-Rodríguez, R.; Méndez-Romero, D.C. Evaluation of a combined anaerobic and aerobic system for the treatment of slaughterhouse wastewater. *Environ. Technol.* **2010**, *31*, 319–326. [CrossRef]

48. Song, K.; Mohseni, M.; Taghipour, F. Application of ultraviolet light-emitting diodes (UV-LEDs) for water disinfection: A review. *Water Res.* **2016**, *94*, 341–349. [CrossRef]

49. Hijnen, W.A.M.; Beerendonk, E.F.; Medema, G.J. Inactivation credit of UV radiation for viruses, bacteria and protozoan (oo)cysts in water: A review. *Water Res.* **2006**, *40*, 3–22. [CrossRef]

50. Gibson, J.; Drake, J.; Karney, B. UV Disinfection of Wastewater and Combined Sewer Overflows. *Adv. Exp. Med. Biol.* **2017**, *996*, 267–275. [CrossRef] [PubMed]

51. Chevremont, A.C.; Farnet, A.M.; Coulomb, B.; Boudenne, J.L. Effect of coupled UV-A and UV-C LEDs on both microbiological and chemical pollution of urban wastewaters. *Sci. Total Environ.* **2012**, *426*, 304–310. [CrossRef]

52. Azaizeh, H.; Linden, K.; Barstow, C.; Kalbouneh, S.; Tellawi, A.; Albalawneh, A.; Gerchman, Y. Constructed wetlands combined with UV disinfection systems for removal of enteric pathogens and wastewater contaminants. *Water Sci. Technol.* **2013**, *67*, 651–657. [CrossRef]

53. Vongsayalath, T. *Development of Woven Fiber Microfiltration Membrane System for Water and Wastewater Treatment*; Asian Institute of Technology: Pathum Thani, Thailand, 2015.

54. Nguyen, T.M.H.; Suwan, P.; Koottatep, T.; Beck, S.E. Application of a novel, continuous-feeding ultraviolet light emitting diode (UV-LED) system to disinfect domestic wastewater for discharge or agricultural reuse. *Water Res.* **2019**, *153*, 53–62. [CrossRef] [PubMed]

55. Fatima, F.; Du, H.; Kommalapati, R.R. Treatment of Poultry Slaughterhouse Wastewater with Membrane Technologies: A Review. *Water* **2021**, *13*, 1905. [CrossRef]

56. Yan, T.; Ye, Y.; Ma, H.; Zhang, Y.; Guo, W.; Du, B.; Wei, Q.; Wei, D.; Ngo, H.H. A critical review on membrane hybrid system for nutrient recovery from wastewater. *Chem. Eng. J.* **2018**, *348*, 143–156. [CrossRef]

57. Jensen, P.D.; Yap, S.D.; Boyle-Gotla, A.; Janoschka, J.; Carney, C.; Pidou, M.; Batstone, D.J. Anaerobic membrane bioreactors enable high rate treatment of slaughterhouse wastewater. *Biochem. Eng. J.* **2015**, *97*, 132–141. [CrossRef]

58. Yadav, S.; Ibrar, I.; Bakly, S.; Khanafer, D.; Altaee, A.; Padmanaban, V.C.; Samal, A.K.; Hawari, A.H. Organic Fouling in Forward Osmosis: A Comprehensive Review. *Water* **2020**, *12*, 1505. [CrossRef]

59. Skoczko, I.; Puzowski, P.; Szatyłowicz, E. Experience from the Implementation and Operation of the Biological Membrane Reactor (MBR) at the Modernized Wastewater Treatment Plant in Wydminy. *Water* **2020**, *12*, 3410. [CrossRef]

60. Svardal, K.; Kroiss, H. Energy requirements for waste water treatment. *Water Sci. Technol.* **2011**, *64*, 1355–1361. [CrossRef] [PubMed]

61. Mino, T.; Liu, W.-T.; Kurisu, F.; Matsuo, T. Modelling glycogen storage and denitrification capability of microorganisms in enhanced biological phosphate removal processes. *Water Sci. Technol.* **1995**, *31*, 25–34. [CrossRef]

62. Meyer, R.L.; Zeng, R.J.; Giugliano, V.; Blackall, L.L. Challenges for simultaneous nitrification, denitrification, and phosphorus removal in microbial aggregates: Mass transfer limitation and nitrous oxide production. *FEMS Microbiol. Ecol.* **2005**, *52*, 329–338. [CrossRef]

63. Mai, Z. Membrane Processes for Water and Wastewater Treatment: Study and Modeling of Interactions between Membrane and Organic Matter. Ph.D. Thesis, Ecole Centrale Paris, Chatenay-Malabry, France, 2013.

64. Le-Clech, P.; Jefferson, B.; Judd, S.J. A comparison of submerged and sidestream tubular membrane bioreactor configurations. *Desalination* **2005**, *173*, 113–122. [CrossRef]

65. Ezugbe Obotey, E.; Rathilal, S. Membrane Technologies in Wastewater Treatment: A Review. *Membranes* **2020**, *10*, 89. [CrossRef]

66. Bunani, S.; Yörükoğlu, E.; Yüksel, Ü.; Kabay, N.; Yüksel, M.; Sert, G. Application of reverse osmosis for reuse of secondary treated urban wastewater in agricultural irrigation. *Desalination* **2015**, *364*, 68–74. [CrossRef]

67. Tchobanoglus, G.; Burton, F.; Stensel, H.D.J.A.W.W.A.J. Wastewater engineering. *Treat. Reuse* **2003**, *95*, 201.

68. Ozaki, H.; Sharma, K.; Saktaywin, W.; Wang, D.; Yu, Y. Application of ultra low pressure reverse osmosis (ULPRO) membrane to water and wastewater. *Water Sci. Technol.* **2000**, *42*, 123–135. [CrossRef]

69. Venzke, C.D.; Rodrigues, M.A.S.; Giacobbo, A.; Bacher, L.E.; Lemmertz, I.S.; Viegas, C.; Striving, J.; Pozzebon, S. Application of reverse osmosis to petrochemical industry wastewater treatment aimed at water reuse. *Manag. Environ. Qual. Int. J.* **2017**, *28*, 70–77. [CrossRef]

70. Innes, P.; Chang, S.; Rahaman, M.S. Treatment of Effluent of Upflow Anaerobic Sludge Blanket Bioreactor for Water Reuse. *Water* **2021**, *13*, 2123. [CrossRef]

71. Yordanov, D. Preliminary study of the efficiency of ultrafiltration treatment of poultry slaughterhouse wastewater. *Bulg. J. Agric. Sci.* **2010**, *16*, 700–704.

72. Musa, M.A.; Idrus, S. Physical and Biological Treatment Technologies of Slaughterhouse Wastewater: A Review. *Sustainability* **2021**, *13*, 4656. [CrossRef]

73. Valderrama, O.J.; Zedda, K.L.; Velizarov, S. Membrane Filtration Opportunities for the Treatment of Black Liquor in the Paper and Pulp Industry. *Water* **2021**, *13*, 2270. [CrossRef]

74. Zhang, S.; Kutowy, O.; Kumar, A.; Malcolm, I. A laboratory study of poultry abattoir wastewater treatment by membrane technology. *Can. Agric. Eng.* **1997**, *39*, 99–106.

75. de Morais Coutinho, C.; Chiu, M.C.; Basso, R.C.; Ribeiro, A.P.B.; Gonçalves, L.A.G.; Viotto, L.A. State of art of the application of membrane technology to vegetable oils: A review. *Food Res. Int.* **2009**, *42*, 536–550. [CrossRef]

76. Beier, S.P. *Pressure Driven Membrane Processes: Downstream Processing*, 3rd ed.; Bookboon: London, UK, 2007.

77. Coskun, T.; Debik, E.; Kabuk, H.A.; Manav Demir, N.; Basturk, I.; Yildirim, B.; Temizel, D.; Kucuk, S. Treatment of poultry slaughterhouse wastewater using a membrane process, water reuse, and economic analysis. *Desalination Water Treat.* **2016**, *57*, 4944–4951. [CrossRef]

78. Verhuelsdonk, M.; Glas, K.; Parlar, H. Economic evaluation of the reuse of brewery wastewater. *J. Environ. Manag.* **2021**, *281*, 111804. [CrossRef] [PubMed]

79. Brennan, B.; Lawler, J.; Regan, F. Recovery of viable ammonia–nitrogen products from agricultural slaughterhouse wastewater by membrane contactors: A review. *Environ. Sci. Water Res. Technol.* **2021**, *7*, 259–273. [CrossRef]

80. Ang, W.S.; Lee, S.; Elimelech, M. Chemical and physical aspects of cleaning of organic-fouled reverse osmosis membranes. *J. Membr. Sci.* **2006**, *272*, 198–210. [CrossRef]

81. Yu, H.; Huang, W.; Liu, H.; Li, T.; Chi, N.; Chu, H.; Dong, B. Application of Coagulation–Membrane Rotation to Improve Ultrafiltration Performance in Drinking Water Treatment. *Membranes* **2021**, *11*, 643. [CrossRef] [PubMed]

82. Li, Z.; Wu, C.; Huang, J.; Zhou, R.; Jin, Y. Membrane Fouling Behavior of Forward Osmosis for Fruit Juice Concentration. *Membranes* **2021**, *11*, 611. [CrossRef]

83. Nichka, V.S.; Nikonenko, V.V.; Bazinet, L. Fouling Mitigation by Optimizing Flow Rate and Pulsed Electric Field during Bipolar Membrane Electroacidification of Caseinate Solution. *Membranes* **2021**, *11*, 534. [CrossRef] [PubMed]

84. Gkotsis, P.; Banti, D.; Pritsa, A.; Mitrakas, M.; Samaras, P.; Peleka, E.; Zouboulis, A. Effect of Operating Conditions on Membrane Fouling in Pilot-Scale MBRs: Filaments Growth, Diminishing Dissolved Oxygen and Recirculation Rate of the Activated Sludge. *Membranes* **2021**, *11*, 490. [CrossRef] [PubMed]

85. Yu, Y.; Lee, S.; Maeng, S.K. Forward osmosis membrane fouling and cleaning for wastewater reuse. *J. Water Reuse Desalination* **2016**, *7*, 111–120. [CrossRef]

86. Sardari, K.; Askegaard, J.; Chiao, Y.-H.; Darvishmanesh, S.; Kamaz, M.; Wickramasinghe, S.R. Electrocoagulation followed by ultrafiltration for treating poultry processing wastewater. *J. Environ. Chem. Eng.* **2018**, *6*, 4937–4944. [CrossRef]

87. Valladares Linares, R.; Li, Z.; Yangali-Quintanilla, V.; Li, Q.; Amy, G. Cleaning protocol for a FO membrane fouled in wastewater reuse. *Desalination Water Treat.* **2013**, *51*, 4821–4824. [CrossRef]

88. Yoon, H.; Baek, Y.; Yu, J.; Yoon, J. Biofouling occurrence process and its control in the forward osmosis. *Desalination* **2013**, *325*, 30–36. [CrossRef]

89. Valladares Linares, R.; Yangali-Quintanilla, V.; Li, Z.; Amy, G. NOM and TEP fouling of a forward osmosis (FO) membrane: Foulant identification and cleaning. *J. Membr. Sci.* **2012**, *421–422*, 217–224. [CrossRef]

90. Wang, Z.; Tang, J.; Zhu, C.; Dong, Y.; Wang, Q.; Wu, Z. Chemical cleaning protocols for thin film composite (TFC) polyamide forward osmosis membranes used for municipal wastewater treatment. *J. Membr. Sci.* **2015**, *475*, 184–192. [CrossRef]

Review

Recent Advances of Nanoremediation Technologies for Soil and Groundwater Remediation: A Review

Motasem Y. D. Alazaiza [1,*], Ahmed Albahnasawi [2], Gomaa A. M. Ali [3], Mohammed J. K. Bashir [4], Nadim K. Copty [5], Salem S. Abu Amr [6], Mohammed F. M. Abushammala [7] and Tahra Al Maskari [1]

[1] Department of Civil and Environmental Engineering, College of Engineering, A'Sharqiyah University, Ibra 400, Oman; tahra.almaskari@asu.edu.om
[2] Department of Environmental Engineering-Water Center (SUMER), Gebze Technical University, Kocaeli 41400, Turkey; ahmedalbahnasawi@gmail.com
[3] Chemistry Department, Faculty of Science, Al-Azhar University, Assiut 71524, Egypt; gomaasanad@azhar.edu.eg
[4] Department of Environmental Engineering, Faculty of Engineering and Green Technology (FEGT), Universiti Tunku Abdul Rahman, Kampar 31900, Malaysia; jkbashir@utar.edu.my
[5] Institute of Environmental Sciences, Bogazici University, Istanbul 34342, Turkey; ncopty@boun.edu.tr
[6] Faculty of Engineering, Demir Campus, Karabuk University, Karabuk 78050, Turkey; sabuamr@hotmail.com
[7] Department of Civil Engineering, Middle East College, Knowledge Oasis Muscat, Muscat 135, Oman; mabushammala@mec.edu.om
* Correspondence: my.azaiza@gmail.com

Citation: Alazaiza, M.Y.D.; Albahnasawi, A.; Ali, G.A.M.; Bashir, M.J.K.; Copty, N.K.; Amr, S.S.A.; Abushammala, M.F.M.; Al Maskari, T. Recent Advances of Nanoremediation Technologies for Soil and Groundwater Remediation: A Review. *Water* **2021**, *13*, 2186. https://doi.org/10.3390/w13162186

Academic Editor: Sergi Garcia-Segura and Chin-Pao Huang

Received: 29 May 2021
Accepted: 5 August 2021
Published: 10 August 2021

Publisher's Note: MDPI stays neutral with regard to jurisdictional claims in published maps and institutional affiliations.

Abstract: Nanotechnology has been widely used in many fields including in soil and groundwater remediation. Nanoremediation has emerged as an effective, rapid, and efficient technology for soil and groundwater contaminated with petroleum pollutants and heavy metals. This review provides an overview of the application of nanomaterials for environmental cleanup, such as soil and groundwater remediation. Four types of nanomaterials, namely nanoscale zero-valent iron (nZVI), carbon nanotubes (CNTs), and metallic and magnetic nanoparticles (MNPs), are presented and discussed. In addition, the potential environmental risks of the nanomaterial application in soil remediation are highlighted. Moreover, this review provides insight into the combination of nanoremediation with other remediation technologies. The study demonstrates that nZVI had been widely studied for high-efficiency environmental remediation due to its high reactivity and excellent contaminant immobilization capability. CNTs have received more attention for remediation of organic and inorganic contaminants because of their unique adsorption characteristics. Environmental remediations using metal and MNPs are also favorable due to their facile magnetic separation and unique metal-ion adsorption. The modified nZVI showed less toxicity towards soil bacteria than bare nZVI; thus, modifying or coating nZVI could reduce its ecotoxicity. The combination of nanoremediation with other remediation technology is shown to be a valuable soil remediation technique as the synergetic effects may increase the sustainability of the applied process towards green technology for soil remediation.

Keywords: environmental ecotoxicity; nanoremediation; nZVI; CNTs; remediation process; soil remediation

1. Introduction

Contaminated soil and groundwater, especially in industrialized and urban areas, is a widespread problem that presents extreme risks to the environment and humans [1,2]. Numerous studies have focused on the remediation of soil, groundwater, wastewater, and landfill leachate polluted by various contaminants [3,4]. Soil and groundwater remediation can be broadly classified according to the place of remediation, which can be ex situ or in situ. For ex situ remediation, the polluted soil or groundwater is recovered from the subsurface and treated on the same site or transferred to another site for treatment [5]. In contrast, in situ remediation is when the contaminated soil or groundwater is remediated directly in the subsurface. The in situ remediation process is often preferred because it is cheaper than the ex situ remediation process [2,6]. For example, according to Chany et al. [7], the remediation cost of removal and replacement of contaminated soil is very expensive (on the order of $3 million/ha), which is considered a big challenge for developing countries in terms of environmental sustainability practice [7].

Reducing pollution to a desirable and safe level is the main target of soil and groundwater remediation processes. Physical, chemical, and biological technologies have been used to achieve this goal for soil and groundwater remediation. In general, several factors play a significant role in the selection of the optimal soil and groundwater remediation, including soil properties and contaminants and the nature of selected and designed remediation technology [8]. Conventional methods such as pump-and-treat involve pumping groundwater by wells and removal of contaminants from the extracted groundwater by ex situ methods such as carbon adsorption, air stripping, chemical oxidation/precipitation, or biological reactors. However, these methods are associated with high operating costs and contaminated waste production [4]. For groundwater and soil contaminated with organic contaminants in the form of dense nonaqueous phase liquids (DNAPLs), emergent remediation technology such as surfactant enhanced remediation (SER) has been shown to be effective. Nevertheless, these technologies are associated with risks; with the decrease in interfacial force of DNAPLs, uncontrolled vertical movement may occur [9].

In recent years, nanotechnology has been increasingly considered in a broad range of fields. Nanoparticles (NPs) have many essential and promising properties due to their enabled functions in many sectors [10–13]. NPs are produced by combining multidisciplinary fields such as molecular level manufacturing principles and engineering. Generally, nanotechnology is a technique that constructs particles in a size range of 1–100 nanometers, studies the physical phenomena related to those particles and applying these in many sectors [4]. Nanotechnology is being used in many sectors such as the chemical, electrical, biomedical, and biotechnology industries. While many industries produce and use various forms of nanomaterials, there are many attempts to use nanotechnology for environmental protection activities such as water and wastewater treatment, pollution control, and treatment/remediation of contaminated soil and groundwater [14].

Technologies that apply nanoremediation for contaminated sites have been used in recent years (2009 till now). Studies conducted to evaluate nanoremediation technologies are mostly bench-scale with few field-scale applications [2]. The main advantages of using nanoremediation for soil and groundwater remediation, especially for extensive site cleaning, are reduced cost and cleanup time, complete degradation of some contaminants without the need for the disposal of polluted soil and without the need to transfer the soil or pump groundwater [14,15].

Nanoremediation technologies involve the use of reactive NPs for conversion and detoxification of contaminants. The main mechanisms for remediation by NPs are catalysis and chemical reduction [14,16]. In addition, adsorption is another removal mechanism facilitated by the NPs since NPs have high surface-area-to-mass ratios and different distribution of active sites, increasing the adsorption ability [17]. Many engineering NPs have highly feasible characteristics for in situ remediation applications due to their innovative surface coating and minute size. In addition, NPs can diffuse and penetrate the tiny spaces in the subsurface and be suspended in groundwater for a long time; compared to microparticles, NPs can potentially travel long distances and achieve larger spatial distribution [14].

The physical movement of NPs and/or transport in groundwater is dominated by random motion or Brownian movement rather than the wall effect as a result of their nanoscale characteristics [18]. Thus, compared to microscale particles, which are strongly influenced by gravity sedimentation due to their density and large size, the movement of NPs is not controlled by gravity sedimentation, remaining suspended in groundwater during the remediation process. Thus, NPs afford a functional treatment approach allowing direct injection into the subsurface where pollutants are present [14].

Several studies have revealed the potential use of nanoremediation for soil and groundwater [19–22]. However, the environmental effects of those NPs are still unclear and need more investigation to understand the environmental fate and toxicity of NPs, as these issues are crucial for environmental protection practice.

The use of nanomaterials for soil and groundwater remediation has been widely tested at the laboratory level against a large number of contaminants, offering promising results [23,24]. However, nanomaterials may pose positive or negative impacts on living organisms, the environment, society, and the economy, which should be evaluated in a case-specific context. Appropriate documentation of nanoremediation risks, field-scale validation of remediation results, science–policy interface consultations, and suitable market development initiatives are ways to increase the popularity and acceptability of nanoremediation technologies [25]. Savolainen et al. [26] stated that the fundamental elements of risk assessment are likely to remain and will continue to include the elements designed for other chemicals and particles, notably (1) hazard identification, (2) hazard characterization, (3) exposure assessment, and (4) risk characterization, which are the four steps of the risk assessment process [26]. However, the environmental effects of those NPs are still unclear and need more investigation to understand the environmental fate and toxicity of NPs as these issues are crucial for environmental protection practice [15,27].

Various nanomaterials have been investigated for soil and groundwater remediation, such as metal oxides, nanoscale zeolites, enzymes, carbon nanotubes and fibers, titanium dioxide, and noble metals [14]. Generally, zero-valent iron (nZVI) is most widely used for soil and groundwater remediation as nZVI is considered a suitable electron donor and highly reactive [28]. The use of these different nanomaterials will be discussed in detail in this review.

The main objective of this review is to present the recent studies and development regarding the application of nanotechnology for the remediation of soil and groundwater that are contaminated by a wide range of compounds such as hydrocarbons and heavy metals. The focus is primarily on the developments of the last decade, which has witnessed a substantial increase relating in the number of studies examine nanotechnology for the remediation of soil and groundwater. The potential impact of NP use on the environment is also presented and discussed. Finally, the feasibility of combining nanoremediation with other remediation technologies is also discussed.

2. Relationship between Soil and Groundwater: Contaminants and Remediation

Soil and groundwater are susceptible to pollution by a wide array of pollutants such as petroleum hydrocarbon, chlorinated solvents, and heavy metals. [29]. Selecting a proper remediation technology for a contaminated environment usually depends on contaminant characteristics and contaminated site characteristics such as physical, chemical, and biological properties. All these factors should be considered during the remediation process, design, and implication. Moreover, the time/cost constraints, the regulatory requirements, and the remediation mechanisms should be considered in the selection process.

Nevertheless, adopting risk-based management approaches is increasingly a focus of environmental researchers due to the high demand for sustainable responses to environmental pollutions [14,30]. The polluted environments are usually surface water, sediments, soil, and groundwater, which are mainly contaminated with low and high molecular weight petroleum hydrocarbon compounds, semi-volatile organic compounds, volatile organic compounds, polycyclic aromatic hydrocarbons (PAHs), polychlorinated biphenyls, persistent organic pollutants, organochlorinated pesticides, NAPL, hydrophobic organic compounds, heavy metals, and xenobiotics [31–33]. These pollutants may migrate or spread far from the source and seriously affect flora, fauna, and the ecosystem [30]. Managing the polluted environments requires the selection of the proper remediation technology for the pollutants, destruction, and separation methods according to many ex situ and in situ remediation methods for surface water, sediments, soil, and groundwater comprising physicochemical, biological, chemical, thermal, electromagnetic, electric, and ultrasonic remediation technologies [34–36]. Remediation in an aqueous environment includes remediation of groundwater and surface water polluted by contaminants, whereas soil remediation includes remediation of sediment subsoil and topsoil polluted by contaminates [37]. Soil and groundwater remediation could be conducted separately or together, depending on the concentration of contaminants and the extent of pollution. The efficiency of remediation technology depends on the design and implication based on the characteristics of polluted soil and the remediation technique. Combining one remediation technology with others sequentially or simultaneously may enhance the overall remediation process through combined or synergistic effects [38].

3. Nanomaterials

NPs are particles with a dimension between 1 and 100 nm, whereas nanomaterials are materials with a dimension of 100 nm or less in one dimension at least [39–46]. NPs have many reaction/adsorption sites on their surface due to the large proportion of atoms [10,47]. This unique property makes NPs highly reactive with surrounding contaminants cumbered to the composition materials in the macro scale [2]. Nanomaterials can be classified as natural or manufactured. Clay, iron oxide, and organic matter are examples of naturally occurring NPs in soil composition. Manufactured NPs are either developed or synthesized with a unique property to enhance their industrial or technological applications [32,48–50]. Generally, nanomaterials can be produced by two methods; the first method is from outside to inside or from top to bottom, whereby a significant part transfers into the minor parts. The second method is from bottom to top, whereby small parts are buildup into more extensive parts [51].

Several nanomaterials have been developed for contaminant remediation, such as nZVI, nanoscale zeolites, carbon nanotubes, metal oxides, bimetallic nanoparticles (BNPs), enzymes, and titanium dioxide (TiO_2) [52–57]. Soil remediation using these three nanoremediation materials (nZVI, TiO_2, and CNTs) can be found in [58].

The unique characteristics of the nanomaterials, such as high surface area, quantum size effect, ease of separation and recycling, etc., support their usage as adsorbents. For example, the ferromagnetism of iron-doped nanomaterials supports their recycling and reuse [11,59]. The potential for nanomaterials recyclability makes them economically attractive. The hydrophobicity of fullerene is the key factor for its adsorption properties and ease of recycling [60].

4. Nanoremediation

Many researchers have focused on the use and development of nanoremediation technologies for soil and groundwater remediation [4]. Nanoremediation is considered an eco-friendly technology. As a result, it is considered a feasible choice for conventional site remediation technology [4,17,61].

Nanoremediation may provide a cost-effective and faster solution for site remediation. Various NPs have been used for nanoremediation, such as metal oxides, nanoscale zeolites, nanometals, carbon nanotubes, and titanium dioxide. In this section, recently published studies relating to soil and groundwater remediation using four nanotechnologies are presented. The four technologies are nanoscale nZVI, carbon nanotube, metal nanoparticle, and magnetic nanoparticle.

4.1. Soil Nanoremediation

The first implementation of NPs on the field scale for soil and groundwater remediation was reported 20 years ago and revealed that NPs could remain active in injected soil for up to 56 days and move with groundwater for more than 20 m [62]. Zhang et al. [62] reported that more than 99% of trichloroethene (TCE) could be removed from contaminated sites within a few days.

4.1.1. Nanoscale Zero-Valent Iron

Injection of nZVI is well-suited to soil remediation because of its limited disruption to the environment, fast kinetics, cost-effectiveness, and non-toxic nature [63]. According to Karn et al. [63], the first synthesis and use of nZVI were reported in the 1990s. Iron nanoscale was synthesized from Fe^{2+} and Fe^{3+} to produce particles ranging from 10 to 100 nm [14]. nZVI was used to remove many contaminants from water, mainly halogenated organic compounds that usually contaminate soil and groundwater. They reported for the first time the effectiveness of using nZVI for detoxification and transformation of many environmental contaminants such as chlorinated organic solvents, polychlorinated biphenyls, and organochlorine pesticides.

Moreover, the authors also showed that modifying nZVI may increase process speed and efficiency. In a recent study, Tian et al. [64] characterized and investigated the application of polyvinylpyrrolidone (PVP)-enhanced nZVI to remediate TCE-contaminated soil [64]. The results showed that the size of prepared PVP-nZVI was around 70 nm when the isoelectric point was around 8.5. In terms of TCE removal efficiency of the investigated system, the removal of TCE was around 84.73%. They concluded that the PVP-nZVI technology was promising to remediate TCE-contaminated soil. Subsequently, Reginatto et al. [65] investigated the performance of nZVI for the removal of hexavalent chromium (Cr(VI)) from clayey residual soil [65]. Five different nZVI materials to contaminant ratios were used, and three different nZVI injections pressure were applied. The result showed that the ratio between nZVI and Cr(VI) significantly affected removal efficiency. The removal efficiency at (1000/11) mg mg^{-1} ratio was 98%, whereas at (1000/140) mg mg^{-1}, it was 18%. As the pressure increased, the contaminant leaching increased; thus, the pressure of 30 kPa was more efficient.

In another study, Shubair et al. [66] investigated nitrate removal in porous media using nZVI and modified nZVI using Cu in upflow packed sand column containing a multilayer system [66]. The results revealed the optimal condition for high nitrate removal when 10 cm of nZVI/sand was used, where the nitrate removal efficiency was around 97%. On the other hand, for Cu-modified nZVI/sand, the best condition was noted when a double 5 cm layer was used, where complete nitrate removal was observed. The result suggests that using nZVI in a single layer or Cu-modified nZVI in a multilayer could achieve high nitrate removal. In a subsequent study, Xue et al. [67] investigated the performance of rhamnolipid modified nZVI (R-nZVI) to reduce lead (Pb) and cadmium (Cd) in river sediments by immobilization [67]. They demonstrated that after 42 days, R-nZVI transformed unstable Pb and Cd to stable fractions as the residual percentage of Pb and Cd increased to reach 43.10 and 56.40%, respectively. In a recent study, Blundell and Owens [19] investigated the performance of nZVI for 1,1,1-Trichloro-2,2-bis(4-chlorophenyl) ethane (DDT) removal from contaminated soil [19]. They compared the efficacy of nZVI to microscale zero-valent (μZVI). They found that samples treated with nZVI showed around 85% reduction in DDT concentration, whereas about 11% reduction in DDT was observed when μZVI was used. The result clearly shows the superiority of using nZVI over μZVI on DDT removal from contaminated soil. Table 1 summarizes the main results of the recent studies conducted for soil remediation by nanoremediation technologies. The mechanism of metal ions removal using nZVI which involve reduction, oxidation, adsorption and/or precipitation, as shown in Figure 1 [19].

Table 1. Recent studies that employed nanoremediation methods for soil remediation.

Nanomaterial	Contaminant	Experimental Condition	Important Results	Ref.
PVP-nZVI	Trichloroethylene	TCE initial concentration 1 mg kg^{-1}; PVP-nZVI dosage 0.01 g g^{-1}	The size of papered FVP-nZVI was around 70 nm. The isoelectric point was around 8.5 In terms of TCE removal efficiency of the investigated system, the removal of TCE was around 84.73%.	[64]
nZVI	Cr(VI)	Five ratios were tested (1000/140; 1000/70; 1000/35; 1000/23, 1000/11 mg mg^{-1}); nZVI injection pressure (10,30,100) kPa; Cr(VI) initial concentration 800 mg kg^{-1}	The removal efficiency at (1000/11) mg mg^{-1} ratio was 98% whereas at (1000/140) mg mg^{-1} was 18%. As the pressure increased, the contaminant leaching increased, thus the pressure of 30 kPa was more efficient.	[65]
nZVI and Cu modified nZVI	Nitrate	Initial nitrate concentration (45 mg L^{-1}); NO$_3$-N nZVI and Cu modified (10 g per layer); Flow rate (5 mL min $^{-1}$); Residence time (99 min)	One layer nZVI removed more than 97% of nitrate, whereas two-layer Cu-modified nZVI achieved complete removal.	[66]
Rhamnolipid modified nZVI	Cd(II), Pb(II)	Sediment wight 0.5, 2 g; R-nZVI concentration 2.5, 10 mL with 0.01, 0.03, 0.05,0.1, 0.2 wt%; Reaction time 0, 7, 14, 21, 28, 35, 42 days	R-nZVI was effective for heavy metal immobilization on river sediment.	[67]
nZVI	DDT	nZVI dosage (5% w/w); reaction time (60 min)	DDT removal by nZVI and μZVI was 85% and 11%, respectively, thus revealing the superiority of using nZVI over μZVI on DDT removal from contaminated soil.	[19]
Multi-walled carbon nanotubes (MWCNTs)	Total petroleum hydrocarbons (TPH)	TPH initial concentration 11 000 mg kg^{-1}; microwave treatment 15, 30, 60 s; CTNs concentration 1, 2.5, 5 wt%	Microwave irradiation time and CTNs concentration significantly affect the removal of TPH from contaminated soil.	[68]
MWCNTs	Cr(VI)	Cr(VI) initial concentration 5–60 mg L^{-1}; Citric acid concentration 25–250 mg L^{-1}; MWCNTs concentration 1.25 mg L^{-1}	At pH 5.0, Cr(VI) adsorption capacity could reach 8.09 and 7.85 mg g^{-1} by MWCNT-COOH and MWCNT-OH, respectively. All these data suggest that the catalysis function of CNTs on the reduction of Cr(VI) was decreased by increasing pH	[69]
Modified carbon black nanoparticle (MCBN)	Heavy metal and petroleum	Cd initial concentration 10 mg kg^{-1}; Ni initial concentration (100 mg kg^{-1}); Petroleum initial concentration (2000 mg kg^{-1}); Remediation time 60 days; MCBN dosage (1% w/w)	The result showed that the availability of heavy metals could significantly decrease by using MNCB in Cd and Ni contaminated soil and enhance the growth of the plant	[70]
Single-walled carbon nanotubes (SWCNTs) and MWCNTs	dichlorodiphenyltri-chloroethane (DDT) and hexachlorocyclohexane (HCH)	SWCNTs and MWCNTs dosage 0.058, 0.145, 0.29 wt%; DDT initial concentration 3 mg kg^{-1}; HCH initial concentration l mg kg^{-1}	CNTs could effectively treat DDTs and HCHs, and the optimum condition for the SWCNTs was 0.29 wt% dosage for 4 months. The results suggest that the efficiency of CNTs remediation was highly dependent on dose, type, and sediment–sorbent contact time	[71]
MWCNTs	crude oil	MWNTs concentration 0.1, 0.5, 1 wt%; Crude oil concentration 0.5, 1, 2.5, 5 wt%; Remedation time 30 days	The results showed that using MWCNTs can enhance the degradation of hydrocarbons by increasing the total microbial population	[72]

Table 1. *Cont.*

Nanomaterial	Contaminant	Experimental Condition	Important Results	Ref.
MWCNTs and AC	DDT, HCH	DDT initial concentration 14 ng g^{-1}; HCH initial concentration (5.5 ng g^{-1}); AC dosage (1, 2) wt% MWCNTs dosage (1, 2) wt% Reaction time (30, 45, 150) days	The results suggest that the Ac was more effective than MWCNTS due to its great specific surface area. These findings revealed the promising of using carbon materials as in situ soil remediation	[73].
Al$_2$O$_3$, SiO$_2$, TiO$_2$	Zn, Ni, Cd	Metal NPs (Al$_2$O$_3$, SiO$_2$, TiO$_2$) dosage 1, 3 wt%; Remediation time 30 days	The results showed that SiO$_2$ NPs were feasible sorbent for reduction of the three metals (Zn, CD, Ni) in calcareous soils, whereas Al$_2$O$_3$ NPs were effective for immobilizing Cd and Zn in non-calcareous soil	[74]
Biochar-supported iron phosphate nanoparticle	Cd	Remediation time 28 days; Cd initial concentration (4.25–132.23) mg kg^{-1}	The results indicated that after 25 days, 81.3% of Cd was reduced.	[75]
Goethite nanospheres (nGoethite) and nZVI	As	nZVI dosage (0.5, 2, 5,10) wt%; nGoethite dosage 0.2, 1, 2.5 wt%; As initial concentration 85 mg kg^{-1}	Both nGoethite and nZVI are promising nanomaterials for As immobilization from contaminated soil. Moreover, remediation by a small dosage of nGoethite seems a promising nanoremediation for effective reduction of As in soil.	[76].
MNPs	Cd, Pb	Cd initial concentration 10.91 mg kg^{-1}; Pb initial concentration 190 mg kg^{-1}	The results showed that the organic content of the soil negatively affected the removal of the residual heavy metals, whereas the use of MNPs did not change the chemical composition of the soil.	[77]
Fe$_3$O$_4$@C-COOH MNPs	Pb	Pb initial concentration (737.34) mg kg^{-1}; Fe$_3$O$_4$@C-COOH MNPs dosage (0.6, 1.3, 2.0, 2.6, 3.3, 4.0) wt%; Remediation time 10 days	The migration of Pb was highly reduced, achieving a high degree of remediation of Pb-contaminated soil. The results suggest that Fe$_3$O$_4$@C-COOH MNPs was a promising remediations technology for lead-contaminated soil	[52]
MNPs	As, PAH, TPH	MNPs dosage 0.2, 1, 2, 5 wt%; As initial concentration 1305 mg kg^{-1}; PAHs initial concentration 6777 µg kg^{-1}; TPH initial concentration 384 mg kg^{-1}	MNPs dosage of 1% could immobilize 42% of As, whereas 92.3% was immobilized at 5% dosage. In terms of organic pollutants, at the lowest MNPs dosage, the reductions of PAHs and TPH were 89% and 49%, respectively	[78]

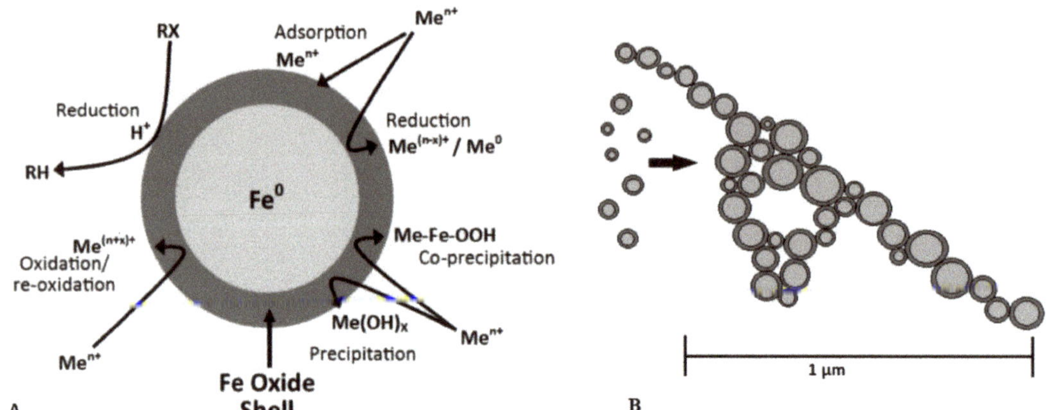

Figure 1. Metal ions removal using nZVI (**A**) is the mechanism and (**B**) is the particles aggregations. Reprinted with permission from [19] (2021, Elsevier).

4.1.2. Carbon Nanotubes

Since the beginning of their application in the water treatment industry, carbon nanotubes (CNTs) have received significant attention from many researchers due to their superior properties, especially their adsorption properties, since CNTs have a strong ability to be attached to the functional groups of pollutants [79]. CNTs can be classified as single-walled carbon tubes and multi-walled carbon nanotubes.

Many studies have been carried out to investigate the performance of CNTs in terms of soil and groundwater remediation. One such attempt has been conducted by Apul et al. [68], who evaluated the performance of microwave-assisted CNTs for removing total petroleum hydrocarbons (TPH) from the soil [68]. Results showed that after using microwave treatment for 60 s, an 82% removal efficiency of TPH was achieved. Zhang et al. [69] assessed the remediation of Cr(VI) contaminated soil using carboxylate or hydroxylated multi-walled carbon nanotubes (MWCNT-COOH or MWCNT-OH) [69]. In addition, the effect of their catalytic activity on the reduction of Cr(VI) by citric acid was evaluated. The results showed that at pH 5, Cr(VI) adsorption capacity was 8.09 and 7.85 mg g^{-1} by MWCNTs-COOH and MWCNT-OH, respectively. In a subsequent study, Cheng et al. [70] studied the efficiency of modified carbon black NPs (MCBN) for petroleum biodegradation and heavy metal immobilization in contaminated soil remediated by plant–microbe combined remediation [70]. The result showed that 65% of petroleum degradation increased in petroleum-Ni co-contaminated soil, whereas in petroleum-Cd co-contaminated soil, the increase in petroleum degradation was 50%. Moreover, the result showed that heavy metals' availability could significantly decrease by using MNCB in Cd- and Ni-contaminated soil, leading to enhancing the plant's growth.

In another study, Gong et al. [80] investigated the performance of single-walled carbon nanotubes (SWCNTs) and multi-walled carbon nanotubes (MWCNTs) for the reduction of dichlorobiphenyls- chloroethane (DDT) and hexachlorocyclohexane (HCH) [80]. The authors used different concentrations of SWCNTs and MWCNTs as well as different remediation times. The result showed that CNTs could effectively treat DDTs and HCHs. Optimum conditions for the SWCNTs were 0.29 wt% dosages for 4 months. In addition, the results suggest that the efficiency of CNTs remediation was highly dependent on dosage and sediment–sorbent contact time. Abbasian et al. [72] used MWCNTs to enhance the bioremediation of crude oil-contaminated soil [72]. They mixed different concentrations of crude oil with MWCNTs for 30 days, and then the microbial diversity of these samples was identified by fluoxetine (FLX) pyrosequencing. The results revealed that using MWCNTs can enhance the degradation of hydrocarbons by increasing the total microbial population. MWCNTs were also used to examine the performance of carbon materials to remediate DDTs and HCHs from sediment. The results showed that sediment remediated with 2 wt% activated carbon (AC) and MWCNTs showed 93% and 59% decrease for DDTs, respectively, and 97% and 75% for HCHs in aqueous equilibrium [73]. The results suggest that the AC was more effective than MWCNTs due to its great specific surface area. These findings revealed the promising of using carbon materials as in situ soil remediation.

4.1.3. Metal and Magnetic Nanoparticles

The use of metal NPs to remediate and immobilize the contaminant from soil and groundwater has attracted much attention recently [74]. Many recent studies examined the performance of using metal NPs for soil remediation. One such attempt has been presented by Peikam and Jalali [74], who studied the remediation of Zn, Ni, and cadmium (Cd) from two contaminated non-calcareous and calcareous soils by SiO_2 NPs [74]. The result showed that the reduction of Cd was maximum with 3% SiO_2 (56.1%) and 1% Al_2O_3 (38.3%) for the calcareous and non-calcareous soils, respectively. In terms of Zi, the highest reduction in calcareous and non-calcareous soil was 57.1% for 3% TiO_2 and 28.8% for 3.0% Al_2O_3. In an earlier study, Qiao et al. [75] examined the performance of biochar-supported iron phosphate NPs stabilized by a sodium carboxymethyl cellulose composite for Cd remediation from contaminated soil [75]. A batch experiment with composite (soil-to-solution ratio 1 g: 10 mL) was used. The results indicated that after 25 days, 81.3% of Cd was reduced. The results suggest that the investigated composite could enhance the immobilization of Cd in soil by reducing bioaccessibilty and leachability. In a recent study, Baragano et al. [76] compared the performance of goethite nanospheres (nGoethite) and nZVI for contaminated soil remediation [76]. The result showed that for 2% nZVI dosage, the decrease was 89.5%. The soil phytotoxicity was reduced in general, and the soil parameters were not negatively affected by using nZVI to remediate the contaminated soil. On the other hand, the use of nGoethite showed an excellent result as a small dosage of nGoethite (0.2%) could decrease the As by 82.5%. However, at high dosage, the soil phytotoxicity increased as the electrical conductivity of the soil increased due to using high dosage. The results suggest that both nGoethite and nZVI are promising nanomaterials for As immobilization from contaminated soil.

Environmental remediation using magnetic NPs (MNPs) has received attention recently because of their facile separation using a magnet and special metal-ion adsorption. Several studies investigated the performance of MNPs for soil and groundwater remediation. Fan et al. (2016) examined new MNPs (core–shell $Fe_3O_4@SiO_2$ NPs coated with iminodiacetic acid chelators) for contaminated soil remediation by non-magnetic heavy metals [77]. The mechanism of removal was chelation and separation by magnetic force. The results indicated that the removal rates of Cd and Pb were 84.9% and 72.2%, respectively. In addition, the results demonstrated that the organic content of the soil negatively affected the removal of the residual heavy metals, whereas the use of MNPs did not change the chemical composition of the soil.

4.2. Groundwater Nanoremediation

The use of NPs in water treatment started in the 1990s and is therefore considered one of the newer technologies. Gillham and Hannesin (1964) were the first researchers to use the idea of using NPs on decontamination of contaminated water. They used nZVI for remediation of the halogenated group [81]. Nevertheless, Wang and Zhang (1997) conducted the first study that used NPs to remediate organo-chlorines from contaminated groundwater. They observed complete and rapid removal of several aromatic chlorinated using bimetallic NPs [82].

4.2.1. Nanoscale Zero-Valent Iron

Since 1997, various nanomaterials have been used for groundwater remediation. However, the application of nZVI for groundwater remediation has received more focus due to their low cost of production and low toxicity [2,63,83]. Figure 2 represents the remediation process by injection of nZVI for DNAPLs. It is reported that nZVI could be used for chlorinated organic compound remediation. Lin et al. (2018) studied the performance of polyethyleneimine (PEI)-coated nZVI (PEI-nZVI) to remediate three DNAPLs (perchloroethene (PCE), trichloroethylene, and 1,2-dichloroethene (1,2-DCE)) by direct injection in the field test [84]. The result showed that after one day of injection, a remarkable reduction in the DNAPLs was recorded. The result showed complete removal (>99%) of the three DNAPLs after one day from the (PEI-nZVI) injection [84]. In a recent study, Chen et al. (2020) investigated the performance of sulfide-modified nZVI (S-nZVI) supported on biochar (BC) for TCE removal from groundwater. In addition, the effect of many factors such as the mass ratio of S-nZVI to BC, pyrolysis temperature of biochar, and initial pH on the TCE removal were examined [20]. The results indicated that the mass ratio of S-nZVI to BC could satisfy the amount of degradation and adsorption of TCE. The pyrolysis temperatures could influence the TCE degradation and adsorption by changing the physicochemical properties of BC. The initial pH had no significant effect on the total TCE removal, whereas the degradation was enhanced at high pH. Moreover, the result showed that within 2 h reaction time, 100% of the TCE was removed at S-nZVI@BC500, where at S-nZVI@BC300 and S-nZVI@BC700, the removal efficiencies of TCE were 79.97% and 86.4%, respectively [20]. In a similar study, Chen et al. [20] studied the effects of Fe/S molar ratio, dissolved oxygen, initial pH, and particle aging on TCE remediations by S-nZVI [20]. The result indicated that Fe/S molar ratio and initial pH remarkably affected the TCE removal, where a higher TCE removal was obtained at Fe/S molar ratio of 60 at pH above 7. A slight decrease in TCE decolorization was observed when S-nZVI was aged up to 20 days, whereas a remarkable decrease was observed at an aging time of 30 days. Finally, dissolved oxygen had a small effect on TCE removal S-nZVI [85]. In another study, Zhu et al. [86] used green technology to synthesize nZVI/Cu from green tea for Cr(VI)-contaminated groundwater remediation [86]. The result showed that the removal efficiency of Cr(VI) was enhanced by decreasing the initial Cr(VI) concentration and initial pH and increasing the temperature, while the presence of humic acids in groundwater decreased the activity of nZVI/Cu. In addition, the result indicated that at optimum conditions (pH = 5, temperature 303 K), the Cr(VI) removal efficiency was 94.7%. Finally, the results suggest that nZVI/Cu is a promising green technology for contaminated groundwater by Cr(VI) [86]. Díaz et al. [21] evaluated the performance of two dosages of commercial nZVI (1 and 5%) for Cu and/or Ni immobilization from water and acidic soil. The results showed that the presence of Cu affected the immobilization of Ni, whereases the presence of Ni did not affect the immobilization of Cu. The efficiency of nZVI was better in water than in soil. The use of 5% dosage completely removed Cu and Ni from water samples, where in soil samples, 5% dosage achieved 54% and 21% embolization for Ni and Cu, respectively [21].

4.2.2. Carbon Nanotubes

In recent years, the use of CNTs for water and groundwater remediation has been increasingly attractive due to their high adsorption affinity. Many recent studies investigated the performance of CNTs for contaminated groundwater remediation. Mpouras et al. [22] investigated Cr(VI) removal from groundwater by MWCNTs. In addition, the effect of operating conditions such as MWCNTs and Cr(VI) concentration, pH, and contact time were examined [22]. The results showed that pH has a significant effect on the adsorption efficiency of MWCNTs; for pH higher than 7, the adsorption process remarkably increased. The adsorption process increased by increasing the MWCNTs concentration. At pH 8, the adsorption percentage increased from 85% to 100% as the concentration of MWCNTs increased from 10 to 50 g L^{-1} [22]. In another study, Lico et al. [87] examined the performance of MWCNTS for the removal of unleaded gasoline from water [87]. They used a lab-scale experiment by adding 20 mL of unleaded gasoline to 30 mL of water and adding different MWCNTs. The results indicated that a small amount of MWCNTs (0.7 g) could remove within 5 min a high percentage of unleaded gasoline. In another study, Liang et al. [88] investigated the efficiency of using alumina-decorated MWCNTs (Al$_2$O$_3$/MWCNTs) for simultaneous remediations of cadmium (Cd(II)) and TCE from groundwater [88]. They conducted a batch experiment for a wide range of conditions. The result showed that the maximum adsorption capacities achieved by Al$_2$O$_3$/MWCNTs were 19.84 mg g^{-1} for Cd(II) and 27.21 mg g^{-1} for TCE. The results suggest that Al$_2$O$_3$/MWCNTs could be a promising technology for Cd(II) and TCE-contaminated groundwater remediations [88]. Table 2 summarizes the recent works conducted in water and groundwater remediation by nanoremediation technologies.

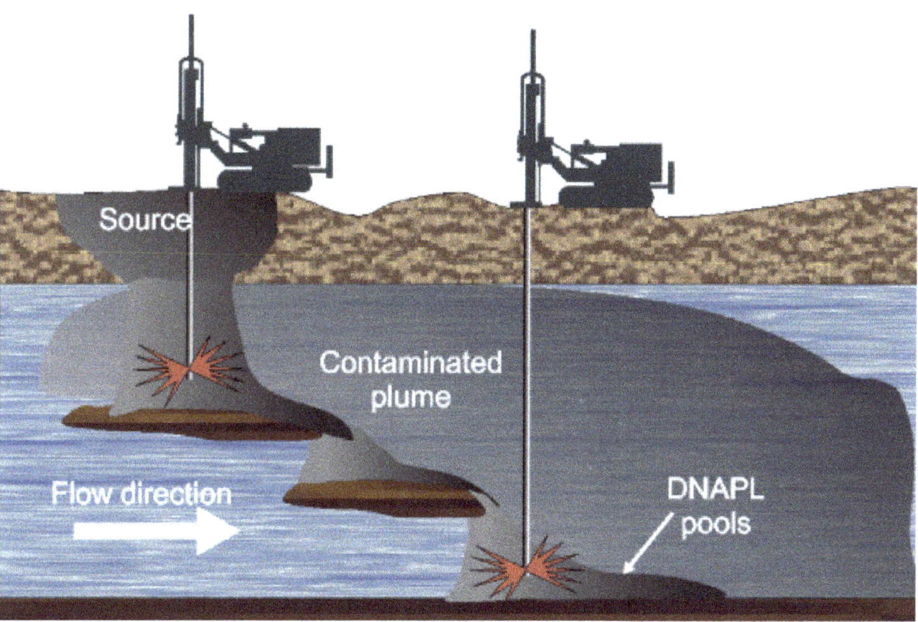

Figure 2. Remediation of an aquifer contaminated by DNAPLs injecting nZVI suspensions directly at the sources of contamination. Reprinted with permission from [83] (2014, Elsevier).

4.2.3. Metal and Magnetic Nanoparticle

The use of metal and magnetic NPs in water and groundwater remediation has received significant attention due to their unique properties. Ou et al. [90] studied the performance of iron-coated aluminum (Fe/Al) BNPs and aluminum-coated iron (Al/Fe) BNPs for the remediations of Cr(VI) from contaminated groundwater [90]. The results indicated that the Cr(VI) removal rate depended on reactive sites and the saturation concentration when (Fe/Al) was used. Moreover, the results showed that the investigated NPs could decrease Cr(VI) to Cr(III). The removal efficiency was 1.47 g g^{-1} when (Fe/Al) BNPs were used and 0.07 g g^{-1} when (Al/Fe) BNPs were used [90]. In a subsequent study, Wang et al. [91] examined the removal of Cr(VI) from contaminated groundwater using iron sulfide NPs (FeS NPs) [91]. The batch test results indicated that a high removal efficiency (1046.1 mg Cr(VI) per gram FeS NPs) was achieved when FeS NPS was used. This high removal efficiency could be attributed to three mechanisms: reduction, adsorption, and co-precipitation. In addition, they found that the pH significantly affected the Cr(VI) removal using FeS NPs. The results suggest that the synthesized Fe NPs could be a promising remediation technology for in situ remediations of Cr(VI) contaminated soil and groundwater [91]. In another study, Xie et al. [92] investigated the immobilization of selenite (Se(IV)) in soil and groundwater using Fe-Me binary oxide NPs [92]. The results showed that high Se(IV) uptake was noticed at a pH range of 5–8, the typical groundwater range. According to Langmuir's maximum capacity, the adsorption capacity was 109 mg Se(IV) per g Fe-Me NPs [92]. In another study, Dong et al. [93] examined the effect of the aging time of Fe/Ni BNPs on particle activity [93]. Moreover, they investigated the reactivity of aged Fe/Ni BNPs by examining their performance in removing tetracycline (TC). The results showed that the aged time plays a significant role in TC removal. The removal efficiency of TC was in the range of 82.3–92.5%. As the aged time increased to 5–15 days, the removal efficiency of TC decreased by 20–50% to reach around 50%, due to oxidation and aggregation of the particles. Finally, the removal efficiency of TC by 90 days using aged Fe/Ni BNPs was around 30% [93].

Groundwater remediations using MNPs have received attention recently because of their facile separation using a magnet and unique metal-ion adsorption. Many studies recently investigated the performance of MNPs for groundwater restoration. Gong et al. (2017) investigated the performance of FeS-coated iron (Fe/FeS) magnetic NPs (MNPs) for the remediation of Cr(VI)-contaminated groundwater (Figure 3) [80]. The results showed that the molar ratio of S/F has a significant effect on the Fe/FeS MNPs. Increasing the S/Fe molar ratio to 0.138 decreased Cr(VI) removal by 42.8%. However, a further increase of 0.207 increased the removal efficiency by 63% within 72 h.

Moreover, the results indicated that the adsorption process of Cr(VI) by Fe/FeS at S/F molar ration of 0.207 fitted with a pseudo-second-order kinetic model and the sorption capacity was 69.7 mg g^{-1}, which was simulated by the Langmuir isotherm model [80]. Huang and Keller [94] developed a regenerable magnetic ligand nanoparticle (Mag-ligand) to rapidly remove Cd and Pb from contaminated water [94]. The results showed high performance of mega-ligand as Cd and Pb were removed from contaminated water quickly, and Cd was removed in less than 2 h where Pb in less than 15 min. The performance of mega-legend in terms of Cd and Pb was not affected by pH (3–10). In addition, the whole regeneration process can be achieved by washed Mega-legend easily by 1% HCl. The results suggest that modified mega-legend is a feasible nanoparticle for efficient, rapid, and convenient removal of Cd and Pb from the contaminated aquatic system [94]. In a recent study, Alani et al. [96] successfully synthesized zero-valent Cu NPs and examined their performance for dye removal from water [96]. The results showed that the removal time was between 5 and 13 min and over 90% removal efficiency was achieved, indicating that the synthesized zero-valent Cu nanoparticle has a great catalytic ability [96]. In another study, Li et al. [95] examined the performance of magnetic mesoporous silica NPs (MMSNPs) for the remediation of uranium (U(VI)) from high and low pH [95]. The result showed that MMSNPs were efficient for U(VI) removal in the pH range of (3.5–9.6) for artificial groundwater. They found that MMSNPs adsorption capacity can reach 133 g U(VI) per g MMSNPs; these results indicate that MMSNPs are a promising solution for treating U(VI) contaminated groundwater at extreme pH [95]. In a recent study, Ari et al. [97] successfully synthesized α-Fe$_2$O$_3$ NPs via a biosynthesis method using leaf extracts of *Azadirachta indica* (neem) and a non-toxic precursor salt (FeCl$_3$·6H$_2$O). In addition, they investigated the potential of using α-Fe$_2$O$_3$ NPs as heterogenous catalyst for tetracycline degradation. The result showed that α-Fe$_2$O$_3$ NPs demonstrated excellent performance as a heterogenous catalyst for degradation of tetracycline aqueous solution by the synergistic effect of the UV/Fenton system [97].

Table 2. Recent studies that employed nanoremediation methods for water and groundwater remediation.

Nanomaterial	Contaminant	Experimental Condition	Important Results	Ref.
(PEI-nZVI)	Perchloroethene (PCE), trichloroethylene (TCE) and 1,2-dichloroethene (1,2-DCE)	TCE initial concentration 94,000 μg L^{-1}; 1,2-DCE initial concentration 2800 μg L^{-1}	The result showed full removal of the three DNAPLs after one day from the (PEI-nZVI) injection.	[84,89]
Sulfide-modified nanoscale zero-valent iron (S-nZVI)	Trichloroethylene (TCE)	TCE initial concentration 20 mg L^{-1}; S-nZVI to BC mass ratio 1:1, 1:3, 3:3; Initial pH 3, 5, 7, 9; pyrolysis temperatures 300, 500, 700 °C.	The results indicated that the mass ratio of S-nZVI to BC could satisfy the amount of degradation and adsorption of TCE. The pyrolysis temperatures could influence the TCE degradation and adsorption by changing the physicochemical properties of BC. The initial pH has no significant effect on the total TCE removal, whereas at high pH, the degradation was enhanced.	[20]
S-nZVI	Trichloroethylene (TCE)	Initial pH 5.57, 7.10, 8.02; TCE initial concentration 30 mg L^{-1}; Aging time 10, 20, and 30 days	The result indicated that Fe/S molar ratio and initial pH remarkably affected the TCE removal, where a higher TCE removal was obtained at Fe/S molar ratio of 60 at pH above than 7.	[85]
nZVI/Cu	Cr(VI)	Cr(VI) initial concentration 10, 10, 15 mg L^{-1}; nZVI/Cu concentration 0.4 g L^{-1}; pH 5, 7, 9; temperature 293, 303, 313 K	At optimum condition (pH = 5, temperature 303 K, the Cr(VI) removal efficiency was 94.7%.	[86]
nZVI	Cu(II), Ni(II)	nZVI dosage 0, 1, 5 wt%; Cu initial concentration 1000 mg kg^{-1}; Ni initial concentration 2000 mg L^{-1}	the presence of Cu affected the immobilization of Ni, whereases the presence of Ni did not affect the immobilization of Cu. The use of 5% dosage completely removed Cu and Ni from water samples, where in soil samples, 5% dosage achieved 54% and 21% immobilization for Ni and Cu, respectively.	[21]
MWCNTs	Cr(VI)	MWCNTs dosage 10–50 g L^{-1}; Cr(VI) initial concentration 11 mg L^{-1} and 250 μg L^{-1}; Contact time 24 h; pH 3–9	The adsorption process was increased by increasing the MWCNTs concentration. At pH 8, the adsorption percentage increased from 85% to 100% as the concentration of MWCNTs increased from 10 g L^{-1} to 50 g L^{-1}.	[22]
MWCNTs	Unleaded gasoline	MWCNTs dosage 0.2–0.8 g; Reaction time 5–120 min	The result indicated that a small amount of MWCNTs (0.7 g) within 5 min could remove a high percentage of unleaded gasoline.	[87]
Al$_2$O$_3$/MWCNTs	Cd(II), TCE	Al$_2$O$_3$/MWCNTs dosage 1 g L^{-1}; Cd(II)/TCE initial concentration 1 mg L^{-1}; pH (4–10)	The result showed that the maximum adsorption capacities achieved by Al$_2$O$_3$/MWCNTs were 19.84 mg g^{-1} for Cd(II) and 27.21 mg g^{-1} for TCE. The results suggest that Al$_2$O$_3$/MWCNTs could be a promising technology for Cd(II)- and TCE-contaminated groundwater remediations.	[88]
Fe/Al BNPs	Cr(VI)	Cr(VI) initial concentration 4–200 mg L^{-1}; (Fe/Al) BNPs and (Al/Fe) BNPs dosage (2.5) g L^{-1}	Removal efficiency was 1.47 g g^{-1} when (Fe/Al) BNPs were used and 0.07 g g^{-1} when(Al/Fe) BNPs was used.	[90]

Table 2. *Cont.*

Nanomaterial	Contaminant	Experimental Condition	Important Results	Ref.
Iron sulfide NPs (FeS NPs)	Cr(VI)	Cr(VI) initial concentration 204.84, 3464 mg kg^{-1}; FeS NPs dosage (28.2, 42.3, 67.7, 84.6 m 117.0) mg L^{-1}	The batch test results indicated that a high removal efficiency (1046.1 mg Cr(VI) per gram FeS NPs) was achieved when FeS NPS was used. This high removal efficiency could be attributed to three mechanisms: reduction, adsorption, and co-precipitation. In addition, they found that the pH significantly affected the Cr(VI) removal using FeS NPs.	[91]
Fe-Me binary oxide NPs.	Selenite (Se(IV))	Se(IV) initial concentration 0–10 mg L^{-1}; Fe-Mn NPs dosage 0.05 g L^{-1}; pH 7	The results showed that high Se(IV) uptake was noticed at a pH range of 5–8, a typical groundwater range. The adsorption capacity was determined according to Langmuir maximum capacity, where it was 109 mg Se(IV) per g Fe-Me NPs.	[92]
Fe/Ni BNPs	Tetracycline	Ageing time 5–90 days; Fe/Ni BNPs dosage 1 g L^{-1}; Ni content (1,3,5) wt%; TC initial concentration 100 mg L^{-1}	The results showed that the aging time plays significant roles in TC removal. The reactivity of Fe/Ni BNPs stayed the same up to 2 days as the removal efficiency of TC was in the range of (82.3–92.5)%. As the aging time increased to 5–15 days, the removal efficiency of TC decreased by 20–50%, to reach around 50%, due to oxidation and aggregation of particles. Finally, the removal efficacy of TC by 90 days aged Fe/Ni BNPs was around 30%.	[93]
FeS-coated iron (Fe/FeS) MNPs	Cr(VI)	S/Fe molar ratio 0, 0.070, 0.138, 0.207; pH 3.5, 5, 7.1, 9; Cr(VI) initial concentration 10, 15, 25, 35, 50, 80 mg L^{-1}	Increasing the S/Fe molar ratio to 0.138 decreased Cr(VI) removal by 42.8%. However, a further increase to 0.207 increased the removal efficiency by 63% within 72 h. Moreover, the results indicated that the adsorption process of Cr(VI) by Fe/FeS at S/F molar ratio of 0.207 was fitted a pseudo-second-order kinetic model, and the sorption capacity was 69.7 mg g^{-1}, which was simulated by the Langmuir isotherm model	[80]
Magnetic ligand nanoparticle	Cd(II), Pb(II)	Cd, Pb initial concentration 1–100 mg L^{-1}; pH 4–10; Mag-Ligand dosage 0.2 g L^{-1}	The results showed high performance of Mega-ligand as Cd and Pb were removed from contaminated water quickly, Cd was removed in less than 2 h, and Pb in less than 15 min. The performance of Mega-legend in terms of Cd and Pb was not affected by pH (3–10). In addition, the full regeneration process could be achieved by washed Mega-legend easily by 1% HCl. The results suggest that modified Mega-legend is a feasible nanoparticle for efficient, rapid, and convenient removal of Cd and Pb from the contaminated aquatic system.	[94]
MMSNPs	U(VI)	U(VI) concentration 2.5×10^{-5} M; MMSNPs dosage 0.075 g in 7.5 mL artificial groundwater	The result showed that MMSNPs were efficient for U(VI) removal in the pH range of (5.5–9.6) for artificial groundwater. They found that MMSNPs adsorption capacity can reach 133 g U(VI) per g MMSNPs	[95]

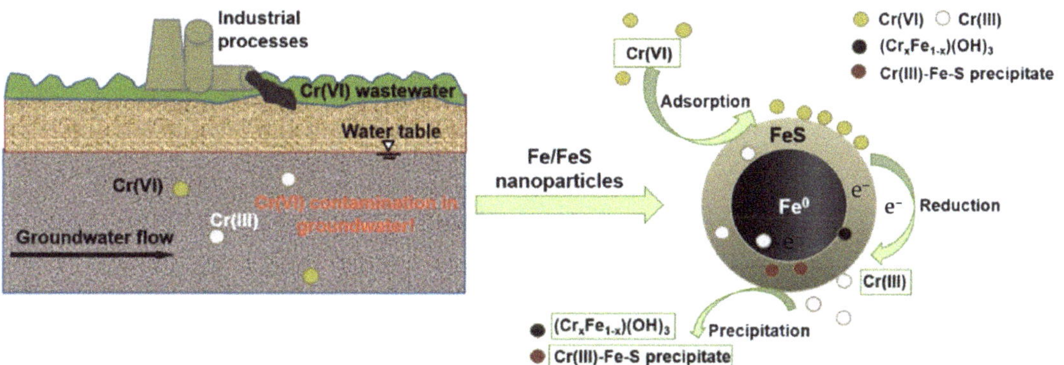

Figure 3. Groundwater remediation using Fe/FeS nanoparticles. Reprinted with permission from [80] (2017, Elsevier).

5. Environmental Risk and Ecotoxicology

Although nanomaterials have been used effectively for soil and groundwater remediation, exposure to nanomaterials may have deleterious effects on humans and environments. Toxicological risk assessments need data on both uptake and exposure of nanomaterials and the immediate effects of NPs when they enter the human system. However, to form a conclusion and recommendations, there are limited data in this domain. The process and factors influencing ecotoxicity are complex. Thus, many factors may determine the impact of synthesized NPs on organisms, such as dissolution potential, particle surface properties, aggregation potential, exposure environment properties, and the physiological, biological, and organism behavior when exposed to NPs [14].

Many studies highlighted the impact of nanomaterials on both humans and environments. For example, iron oxide NPs has a mutagenic impact as it may damage organisms' ability to develop or reproduce [98]. Results indicated exposure to subinhibitory concentrations of amoxicillin-bound iron oxide NPs, in the presence of humic acid, and increased bacterial growth in pseudomonas aeruginosa and Staphylococcus aureus [99]. The joint effects of NPs and other contaminants on terrestrial plants are increasingly investigated but still limited. To provide a sound basis for risk assessment, more research should evaluate the joint effects under realistic conditions [100]. The size and shape of NPs ultimately determine the degree of toxicology. Therefore, not only is monitoring of NPs in soil–plant systems is not essential, but more information is needed on their size allocation and physical properties [101]. Most of the reviewed nano-risk assessment approaches are designed to serve as preliminary risk screening and/or research prioritization tools and are not intended to support regulatory decision making. Although the conventional risk assessment framework is a valuable approach, it may fail to adequately estimate the health and environmental risks from engineering nanomaterials in the near term due to overwhelming methodological limitations and epistemic uncertainties [102]. In this section, an overview of the recent studies about concerns related to the environmental risk of using nanomaterials for soil and groundwater remediation is presented.

Gómez-Sagasti et al. [103] conducted a 3 months experiment to investigate the influence of nZVI concentration (ranging from 1 to 20 mg L^{-1}) on soil microbial properties in two types of soil: sandy-loam and clay-loam soils [103]. The results presented evidence that soil type may affect the degree of potential toxic effects on soil microbial communities by nZVI. The results showed that the accentuated inhibitory impact of nZVI on soil microorganisms in sandy-loam soil was more obvious than clay-loam soil. This can be attributed to the high organic content in clay-loam soil, which acts as a protective agent when nZVI was added to the soil by rendering nZVI inactive, thus prohibiting interaction with soil microorganism cells. Bacterial biomass and arylsulphatase activity, diversity, and richness were negatively influenced by remediation of sandy-loam soil by nZVI. In terms of concentration, they found no obvious concentration–response effect on the soil by nZVI application. The study suggests that many investigations are required using a wide range of soil types and soil proprieties to have clear insight into soil properties' effect and type on the impact of nZVI on soil bacteria communities [103]. In another study, Dong et al. [104] investigated the effects of carboxymethyl cellulose (CMC) surface coating on the cytotoxicity and colloidal stability of nZVI towards Escherichia coli (*E. coli*) and studied the interrelation between cytotoxicity and particle stability [104]. In addition, they examined the influence of CMC ionic strength (Ca^{2+}), concentration, and aging treatment on particle cytotoxicity. The results indicated that nZVI without coating harms *E. coli* and time- and concentration-dependent.

On the other hand, the cytotoxicity of nZVI decreased when the nZVI particles were coated with CMC. This can be attributed to the cell membrane that kept intact in CMC-modified nZVI, whereas cell membrane disruption could be observed when bare nZVI contact with *E. coli*. The aged nZVI and CMC-nZVI did not affect *E. coli* due to the Fe^0 transformation to less toxic iron oxide. The toxicity of nZVI and CMC-nZVI related to the existence of Ca^{2+} was concentration-dependent as it can either decrease or increase. The presence of Ca^{2+} could decrease the toxicity of nZVI by causing aggregation and settling of nZVI.

However, the presence of Ca^{2+} could also increase the toxicity of nZVI by facilitating the adhesion of NPs onto the bacteria surface, forming a more toxic effect [104]. In another study, Chaithawiwat et al. [105] studied the effect of nZVI on the bacterial growth phases on four bacterial strains [105]. The results showed that lag and stationary phases for all bacterial strains were resistant to nZVI, whereas the bacterial strains in exponential and decline phases showed less resistance than lag and stationary phrases. In addition, the results indicated that increasing the nZVI concentration increased bacterial inactivation. The results suggest that it is necessary to consider the bacterial growth phase and nZVI concentration when studying the influence of nZVI on the bacteria [105]. In a subsequent study, Cheng et al. [106] examine the toxicity of S-nZVI to *E. coli* in an aqueous solution [106]. The result indicated that the toxicity of nZVI could be reduced by sulfidation as S/nZVI showed less toxicity at a lower F/S molar ratio, coming out from the higher iron oxide and sulfate and lower Fe^0 content. The results suggest that the typical groundwater contents (i.e., Ca^{2+}, HCO^{3-}, $SO_4{}^{2-}$, and humic acid) could drop the toxicity of nZVI. In addition, in the presence of groundwater mix components, the S/nZVI toxicity was negligible. The results suggest that the implication of S/nZVI could present a low toxicity risk in the ecosystem [106]. In a recent study, Li et al. [107] conducted a long-term study to examine the effect of zeolite-supported nZVI (Z-nZVI) on farmland soils on bacterial communities during the remediation of metals (Cd, As, Pb) [107]. The result indicated that temporary shifts in pH-sensitive, iron resistance/sensitivity, metal resistance, and denitrifying bacteria after adding Z-nZVI were eliminated due to the soil characteristics that drove the re-establishment of the indigenous bacterial community Z-nZVI and restored the bacterial DNA replication and denitrification activity in the soil. The results suggest that Z-nZVI is a promising nanoremediation technology for long-term metal-contaminated soil remediation without ecotoxicity effects [107].

The toxicity of using CNTs in soil and groundwater remediation has been studied by many researchers [108]. However, there are insufficient data related to the effect of CNTs on both humans and the environment. Song et al. [109] studied the effects of MWCNTs different dosages (0.5, 1.0, 2.0, wt%) on bacterial communities, especially the metabolic function, in phenanthrene contaminated sediment [109]. The results indicated that the metabolic function of microbial communities could be significantly changed by the application of high dosage (0.5–2.0, wt%). This can be attributed to the utilization of carbon sources on Biolog ECO microplate. Remotion of phenanthrene-contaminated sediment with 0.5% MWCNTs presented the best microbial activity and Shannon–Wiener diversity index [109].

6. Combined Nanoremediation with Other Remediation Technology

The combination of nanoremediation technologies with other mitigation methods has attracted significant research in recent years. Synergetic studies can be characterized as combining multiple nanoremediation methods simultaneously or combined with soil flushing or with biotreatment. In this section, an overview of the recent work in this domain is presented.

Several studies combined many nanoremediation methods at the same time. Vilardi et al. [110] examined the combination of nZVI and CNTs for the remediation of Cr(VI), selenium (Se), and cobalt (Co) from aqueous solutions by conducted a batch experiment [110]. The result indicated that for Cr(VI), the main removal mechanism the reduction, whereas adsorption was the predominant mechanism for other metals. The results showed that the Cr(VI) removal efficiency was 100% when nZVI was used alone without pH change, whereas it decreased to around 90% when CNTs-nZVI nanocomposite was used. On the other hand, using CNTs-nZVI showed high removal efficiency for Se and Co at 90% and 80%, respectively. The results suggest that the CNTs–nZVI nanocomposite showed high adsorption efficiency for remediation of heavy metals-contaminated water [110]. In another study, Zhang et al. [111] studied the performance of carboxymethyl cellulose (CMC)-stabilized nZVI composited with BC (CMC-nZVI/BC) for remediation of Cr(VI)-contaminated soil [111]. The results indicated that, after 21 days, the immobilization efficiency of Cr(VI) was 19.7, 33.3, and 100% when the dosage of CMC-nZVI/BC was 11, 27.5, and 55 g Kg^{-1}, respectively. The results suggest that the addition of BC to CMC-nZVI could decrease the Cr(VI) transformation slightly, as a small part of CMC-nZVI could be adsorbed to biochar. The Cr$_{total}$ removal efficiency was high because the reduction reaction continued to remediation [69]. In a recent study, Qian et al. [112], for the first time, investigated the performance of biochar-nZVI for the remediation of chlorinated hydrocarbon in the field [112]. They used direct-push and water pressure-driven packer techniques. The field study results demonstrated a sharp reduction of chlorinated solvents in the 24 h after the first injection of nZVI, but within the next two weeks, a rebound of the concentrations in groundwater was observed. However, the implementation of biochar-nZVI highly improved the removal of the chlorinated solvent from groundwater for 42 days (Figure 4). The results suggest that biochar-nZVI is a promising combined technology for chlorinated solvent contaminated groundwater remediation [112].

Galdames et al. [29] developed a new approach combining nanoremediation with bioremediation for hydrocarbon and heavy metals remediation from contaminated soil [29]. Specifically, the method uses a combination of nZVI and compost from organic waste. The results indicated that the combination of nZVI and compost could decrease the aliphatic hydrocarbons concentration up to 60% even under uncontrolled conditions. In addition, they observed a remarkable decrease in ecotoxicity in the bio-pile of soil [29]. In another study, Alabresm et al. [113] studied the combination of PVP-coated magnetite NPs with oil-degrading bacteria for crude oil remediation at the lab scale [113]. The result indicated that NPs alone removed around 70% of high oil concentration after 1 h. However, the removal efficiency did not increase due to the saturation of NPs. On the other hand, bioremediation by oil-degrading bacteria removed 90% of oil after 48 h. Finally, the combination of NPs and oil-degrading bacteria could completely remove the oil within 48 h. This was attributed to the sorption of oil components to NPs and following degradation by bacteria. Further investigation is needed to understand the oil removal mechanism when combining NPs with oil-degrading bacteria are used for oil remediation [113].

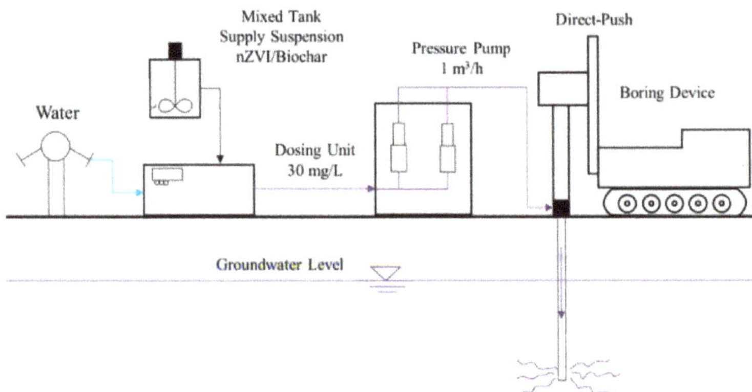

Figure 4. Step of injection procedure. Reprinted with permission from [112] (2020, Elsevier).

Recently, Czinnerova et al. [114] conducted a long-term field study that investigated the degradation of chlorinated ethenes (CEs) by using nZVI supported by electrokinetic (EK) treatment (nZVI-EK) [114]. EK may enhance the nZVI impact on soil bacteria and increased the migration and longevity of nZVI. The results indicated a rapid decrease in cis-1,2-dichloroethene (cDCE) at around 70%, followed by setting new geochemical conditions as a degradation product of CE (ethene, ethane, and methane) was observed. These new conditions enhanced the growth of soil and ground bacteria, such as organohalide-respiring bacteria. The results suggest that nZVI-EK remediation technology is a promising method for CE remediation from soil and groundwater and enhanced bacteria availability in soil and groundwater [114]. In another study, Sierra et al. [115] studied a combination of soil washing and nZVI for the removal and recovery of toxic elements (As, Cu, Hg, Pb, Sb) from polluted soil (Figure 5) [115]. The results showed that a high recovery yield was obtained for Pb, Cu, and Sb in the magnetically separated fraction, whereas Hg was concentrated in a non-magnetic fraction. Taking everything into account, the soil washing efficiency was enhanced by adding nZVI, especially for a larger fraction. The results suggest that the investigated methodology opens the door for NPs' use in soil-washing remediation [115].

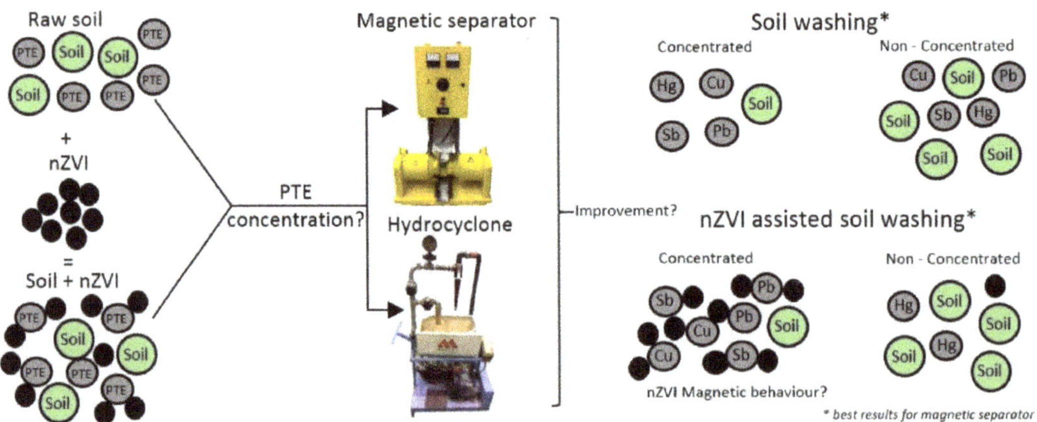

Figure 5. Soil washing assisted nZVI nanoremediation. Reprinted with permission from [115] (2018, Elsevier).

Qu et al. [116] studied the implication of an activated carbon fiber (ACF)-supported nZVI (ACF-nZVI) composite for Cr(VI) remediation from groundwater [116]. In addition, they examined the effect of the operation condition such as nZVI amount on activated carbon fiber, initial Cr(VI) concentration, and pH value on the Cr(VI) removal by conducting a batch experiment. The results indicated that the aggregation of nZVI could be inhabited by ACF, which increases the nZVI reactivity and Cr(VI) removal efficiency. The removal efficiency of Cr(VI) decreased with increasing Cr(VI) initial concentration, whereas, in an acidic environment, complete removal (100%) of Cr(VI) was observed in 1 h reaction time. The proposed removal mechanism consisted of two steps: the first step was the physical adsorption of Cr(VI) on the ACF-nZVI surface area or inner layer, where the second step was a reduction of Cr(VI) to Cr(III) by nZVI [116]. In another study, Huang et al. [117] studied the activation of persulfate (PS) by using a zeolite-supported nZVI composites (PS-Z/nZVI) system and examined its efficiency for TCE degradation. The results indicated that Z/nZVI showed high ability towards PS activation (1.5 mM), and high removal efficiency (98.8%) of TCE was observed at pH 7 within 2 h. Moreover, the PS-Z/nZVI system showed high efficiency in terms of TCE for a wide range of pH (4–7) [117]. Table 3 summarizes the recent works conducted in soil and groundwater remediation by combining nanoremediation technologies with other remediation methods.

Table 3. Recent studies employed the combination of nanoremediation technologies with other remediation methods.

Nanomaterial	Contaminant	Experimental Conditions	Important Results	Ref.
CNTs-nZVI	Cr(VI), Se, Co	Dosage 3 g L^{-1}; Initial concentration 1–10 mg L^{-1}; pH 6–8	The result indicated that for Cr(VI), the main removal mechanism was reduction, whereas adsorption was the predominant mechanism for the metals. The results showed that the Cr(VI) removal efficiency was 100% when nZVI was used alone without the effect of pH change, whereas it decreased to around 90% when CNTs–nZVI nanocomposite was used. On the other hand, using CNTs-nZVI showed high removal efficiency for Se and Co at 90% and 80%, respectively.	[110]
CMC-nZVI/BC	Cr(VI)	CMC-nZVI/BC dosage 11, 27.5, 55 kg g^{-1}; Cr(VI) initial concentration 800 mg kg^{-1}	The results indicate that, after 21 days, the immobilization efficiency of Cr(VI) was 19.7, 33.3, and 100% when the dosage of CMC-nZVI/BC was 11, 27.5, and 55 g Kg^{-1}, respectively.	[69]
Biochar-nZVI	Chlorinated solvents	nZVI dosage (30) g L^{-1}; injection depth (3.5, 4.5, 5.5) m	The field study results demonstrated a sharp reduction of chlorinated solvents in the 24 h after the first injection of nZVI, but within the next two weeks, the re-bond of the concentrations in groundwater was observed. However, the implementation of biochar-nZVI highly improved the removal of the chlorinated solvent from groundwater for 42 days. The results suggest that biochar-nZVI is a promising combined technology for chlorinated-solvent-contaminated groundwater remediation.	[112]
nZVI combined with compost from organic waste	Hydrocarbons (TPH, PAHs) and heavy metals	TPH initial concentration (104.3) mg kg^{-1}; PAHs initial concentration (2.25) mg kg^{-1}	The results indicated that the combination of nZVI and compost could decrease the aliphatic hydrocarbons concentration by up to 60% even under uncontrolled conditions. In addition, they observed a remarkable decrease in ecotoxicity in the biopile of the soil.	[29].
PVP-coated magnetite NPs with oil-degrading bacteria	Crude oil	Oil initial concentration (375) mg L^{-1}; NPs dosage (18) mg L^{-1}	The result indicated that NPs alone removed around 70% of high oil concentration after 1 h. However, the removal efficiency did not increase due to the saturation of NPs. Bioremediation by oil-degrading bacteria removed 90% of oil after 48 h. Finally, the combination of NPs and oil-degrading bacteria could completely remove the oil within 48 h	[113]

Table 3. Cont.

Nanomaterial	Contaminant	Experimental Conditions	Important Results	Ref.
nZVI-EK	Chlorinated ethenes (CEs)	nZVI dosage 3 g L^{-1}; DC voltage 24 V	The results indicated a rapid decrease in cis-1,2-dichloroethene (cDCE) by around 70%, followed by new geochemical conditions as a degradation product of CE (ethene, ethane, and methane) was observed. These new conditions enhanced the growth of soil and ground bacteria such as organohalide-respiring bacteria. The results suggest that nZVI-EK remediation technology not only is a promising method for CE remediation from soil and groundwater but also enhanced the bacteria availability in soil and groundwater	[114]
Soil washing assessed nZVI	As, Cu, Hg, Pb, Sb	nZVI dosage (16) wt%	The results showed that a high recovery yield was obtained for Pb, Cu, and Sb in the magnetically separated fraction, whereas Hg concentrated in the non-magnetic fraction. Taking everything into account, the soil washing efficiency was enhanced by adding nZVI, especially for a larger fraction. The results suggest that the investigated methodology open the door for the use of NPs in soil washing remediation.	[115]
ACF-nZVI	Cr(VI)	Cr(VI) initial concentration (5, 10) mg L^{-1}	The results indicated that the aggregation of nZVI could be inhabited by the presence of ACF, which increases the nZVI reactivity and Cr(VI) removal efficiency. The removal efficiency of Cr(VI) decreased with increasing Cr(VI) initial concentration, whereas, in an acidic environment, full removal (100%) of Cr(VI) was observed in 1 h reaction time. The proposed removal mechanism consisted of two steps: the first step was the physical adsorption of Cr(VI) on the ACF-nZVI surface area or inner layer, while the second step was a reduction of Cr(VI) to Cr(III) by nZVI	[116]
PS-Z/nZVI	TCE	TCE initial concentration (0.15) mM; Z/nZVI dosage (84) mg L^{-1}; PS concentration (1.5) mM	The results indicated that Z/nZVI showed high ability for PS activation (1.5 mM), and high removal efficiency (98.8%) of TCE was observed at pH 7 within 2 h. Moreover, the PS-Z/nZVI system showed high efficiency in terms of TCE for a wide range of pH (4–7).	[117]

7. Conclusions

This review aims to present the latest advances in nanoremediation of contaminated soil and groundwater. The main advantages of using nanomaterials in soil remediation are reduction in cleanup time and overall costs, decreased pollutants to nearly zero in the site, and no need to dispose of polluted soil. The wide use of nZVI nanomaterials in environmental cleanup is due to their high reactivity and high ability to immobilize heavy metals such as Cd, Ni, and Pb. Modifying and/or coating nZVI may decrease the toxicity effects on soil microorganisms. The high adsorption capacity of CNTs is from the large surface area, which makes CNTs a great nanomaterial for organic and inorganic remediation. More studies are needed to investigate the effect of CNTs on the environment.

Soil and groundwater remediation using metal and MNPs is a promising technology due to the unique separation mechanism. Full-scale application of nanoremediation needs further evaluation, particularly in terms of efficiency and potential adverse environmental impacts. Combining nanoremediation with other remediation technology appears to be the future of soil remediation as the combination process increases the sustainable remediation practice towards green environmental protection practice.

8. Recommendation and Future Prospective

This review provides readers with a general overview of using nanoremediation for environmental cleanups, particularly soil and groundwater remediation. More work is needed to developing smarter nanomaterials for soil remediation. For instance, more advanced development could produce NPs with a high ability to work with several functions, such as interacting with hydrophilic and hydrophobic materials or catalyzing many pollutant reactions on the same particle. In addition, further research is needed to design and synthesize NPs that can remediate a wide range of contaminants; and enhance the injecting systems.

Most existing research on nanoremediation is confined to laboratory studies and modeling. Transferring these studies to in situ conditions is a challenge. Thus, more investigations are required in order to develop standard protocols and doses for the application of nanomaterials at the field level. Moreover, efforts should also focus on the application of nanoremediation in the field to understand nanoparticle's fate and transport behavior in soil, water, and sediments and how they affect the environmental variables.

Nanoremediation has been developed and evaluated over the last 20 years. There is, however, concern about its effects on both humans and the environment. With the rapid advancement of nanoremediation techniques, proper evaluation needs to be done to prevent or mitigate any potential environmental or ecological hazards.

In addition, the need for a more thorough understanding of the contaminants' removal processes and the nanomaterials behavior in nature has led to experimentation where no contaminant is present. Many researchers have examined the impacts of nanoremediation on the soil and groundwater bacteria, yet a clear insight into the interaction between nanoremediation materials such as nZVI and microbial activity is still unclear.

Author Contributions: Conceptualization and Funding, M.Y.D.A.; Writing—Original Draft Preparation, A.A. and M.Y.D.A.; Writing—Review and Editing, G.A.M.A.; N.K.C.; S.S.A.A.; M.F.M.A.; and T.A.M.; Validation, M.J.K.B. All authors have read and agreed to the published version of the manuscript.

Funding: The research leading to these results has received funding from Ministry of Higher Education, Research, and Innovation (MoHERI) of the Sultanate of Oman under the Block Funding Program, MoHERI Block Funding Agreement No. MoHERI/BFP/ASU/01/2020.

Institutional Review Board Statement: Not applicable.

Informed Consent Statement: Not applicable.

Data Availability Statement: Data is contained within the article.

Conflicts of Interest: The authors declare no conflict of interest.

References

1. Cheng, M.; Zeng, G.; Huang, D.; Lai, C.; Xu, P.; Zhang, C.; Liu, Y. Hydroxyl radicals based advanced oxidation processes (AOPs) for remediation of soils contaminated with organic compounds: A review. *Chem. Eng. J.* **2016**, *284*, 582–598. [CrossRef]
2. Thomé, A.; Reddy, K.R.; Reginatto, C.; Cecchin, I. Review of Nanotechnology for Soil and Groundwater Remediation: Brazilian Perspectives. *Water Air Soil Pollut.* **2015**, *226*, 121. [CrossRef]
3. Mateas, D.J.; Tick, G.R.; Carroll, K.C. In situ stabilization of NAPL contaminant source-zones as a remediation technique to reduce mass discharge and flux to groundwater. *J. Contam. Hydrol.* **2017**, *204*, 40–56. [CrossRef]
4. Rajan, C.S.R. Nanotechnology in Groundwater Remediation. *Int. J. Environ. Sci. Dev.* **2011**, *2*, 182–187. [CrossRef]
5. Tsai, T.T.; Kao, C.M.; Yeh, T.Y.; Lee, M.S. Chemical Oxidation of Chlorinated Solvents in Contaminated Groundwater: Review. *Pract. Period. Hazard. Toxic Radioact. Waste Manag.* **2008**, *12*, 116–126. [CrossRef]
6. Honetschlägerová, L.; Martinec, M.; Škarohlíd, R. Coupling in situ chemical oxidation with bioremediation of chloroethenes: A review. *Rev. Environ. Sci. Bio/Technol.* **2019**, *18*, 699–714. [CrossRef]
7. Chaney, R.L.; Reeves, P.G.; Ryan, J.A.; Simmons, R.W.; Welch, R.M.; Scott Angle, J. An improved understanding of soil Cd risk to humans and low cost methods to phytoextract Cd from contaminated soils to prevent soil Cd risks. *Biometals* **2004**, *17*, 549–553. [CrossRef]
8. Liu, J.-W.; Wei, K.-H.; Xu, S.-W.; Cui, J.; Ma, J.; Xiao, X.-L.; Xi, B.-D.; He, X.-S. Surfactant-enhanced remediation of oil-contaminated soil and groundwater: A review. *Sci. Total Environ.* **2021**, *756*, 144142. [CrossRef]
9. Yang, C.; Offiong, N.-A.; Zhang, C.; Liu, F.; Dong, J. Mechanisms of irreversible density modification using colloidal biliquid aphron for dense nonaqueous phase liquids in contaminated aquifer remediation. *J. Hazard. Mater.* **2021**, *415*, 125667. [CrossRef] [PubMed]

10. Gupta, V.K.; Agarwal, S.; Sadegh, H.; Ali, G.A.M.; Bharti, A.K.; Hamdy, A.S. Facile route synthesis of novel graphene oxide-β-cyclodextrin nanocomposite and its application as adsorbent for removal of toxic bisphenol A from the aqueous phase. *J. Mol. Liq.* **2017**, *237*, 466–472. [CrossRef]

11. Sadegh, H.; Ali, G.A.M.; Makhlouf, A.S.H.; Chong, K.F.; Alharbi, N.S.; Agarwal, S.; Gupta, V.K. MWCNTs-Fe$_3$O$_4$ nanocomposite for Hg(II) high adsorption efficiency. *J. Mol. Liq.* **2018**, *258*, 345–353. [CrossRef]

12. Lee, S.P.; Ali, G.A.M.; Algarni, H.; Chong, K.F. Flake size-dependent adsorption of graphene oxide aerogel. *J. Mol. Liq.* **2019**, *277*, 175–180. [CrossRef]

13. Arabi, S.M.S.; Lalehloo, R.S.; Olyai, M.R.T.B.; Ali, G.A.M.; Sadegh, H. Removal of congo red azo dye from aqueous solution by ZnO nanoparticles loaded on multiwall carbon nanotubes. *Phys. E Low-Dimens. Syst. Nanostruct.* **2019**, *106*, 150–155. [CrossRef]

14. Karn, B.; Kuiken, T.; Otto, M. Nanotechnology and in Situ Remediation: A Review of the Benefits and Potential Risks. *Environ. Health Perspect.* **2009**, *117*, 1813–1831. [CrossRef]

15. Ganie, A.S.; Bano, S.; Khan, N.; Sultana, S.; Rehman, Z.; Rahman, M.M.; Sabir, S.; Coulon, F.; Khan, M.Z. Nanoremediation technologies for sustainable remediation of contaminated environments: Recent advances and challenges. *Chemosphere* **2021**, *275*, 130065. [CrossRef]

16. Akharame, M.O.; Fatoki, O.S.; Opeolu, B.O. *Regeneration and Reuse of Polymeric Nanocomposites in Wastewater Remediation: The Future of Economic Water Management*; Springer: Berlin/Heidelberg, Germany, 2019; Volume 76, pp. 647–681.

17. Chaturvedi, S.; Dave, P.N. Water purification using nanotechnology an emerging opportunities. *Chem. Methodol.* **2019**, *3*, 115–144.

18. Kumar, A.; Joshi, H.; Kumar, A. Remediation of Arsenic by Metal/Metal Oxide Based Nanocomposites/Nanohybrids: Contamination Scenario in Groundwater, Practical Challenges, and Future Perspectives. *Sep. Purif. Rev.* **2020**, *50*, 283–314. [CrossRef]

19. Blundell, S.P.; Owens, G. Evaluation of enhancement techniques for the dechlorination of DDT by nanoscale zero-valent iron. *Chemosphere* **2021**, *264*, 128324. [CrossRef] [PubMed]

20. Chen, J.; Dong, H.; Tian, R.; Li, R.; Xie, Q. Remediation of Trichloroethylene-Contaminated Groundwater by Sulfide-Modified Nanoscale Zero-Valent Iron Supported on Biochar: Investigation of Critical Factors. *Water Air Soil Pollut.* **2020**, *231*, 432. [CrossRef]

21. Gil-Díaz, M.; Álvarez, M.A.; Alonso, J.; Lobo, M.C. Effectiveness of nanoscale zero-valent iron for the immobilization of Cu and/or Ni in water and soil samples. *Sci. Rep.* **2020**, *10*, 15927. [CrossRef]

22. Mpouras, T.; Polydera, A.; Dermatas, D.; Verdone, N.; Vilardi, G. Multi wall carbon nanotubes application for treatment of Cr(VI)-contaminated groundwater; Modeling of batch & column experiments. *Chemosphere* **2021**, *269*, 128749.

23. Galdames, A.; Ruiz-Rubio, L.; Orueta, M.; Sánchez-Arzalluz, M.; Vilas-Vilela, J.L. Zero-Valent Iron Nanoparticles for Soil and Groundwater Remediation. *Int. J. Environ. Res. Public Health* **2020**, *17*, 5817. [CrossRef]

24. Araújo, R.; Castro, A.C.M.; Fiúza, A. The Use of Nanoparticles in Soil and Water Remediation Processes. *Mater. Today Proc.* **2015**, *2*, 315–320. [CrossRef]

25. Mukhopadhyay, R.; Sarkar, B.; Khan, E.; Alessi, D.S.; Biswas, J.K.; Manjaiah, K.M.; Eguchi, M.; Wu, K.C.W.; Yamauchi, Y.; Ok, Y.S. Nanomaterials for sustainable remediation of chemical contaminants in water and soil. *Crit. Rev. Environ. Sci. Technol.* **2021**, 1–50. [CrossRef]

26. Savolainen, K.; Alenius, H.; Norppa, H.; Pylkkänen, L.; Tuomi, T.; Kasper, G. Risk assessment of engineered nanomaterials and nanotechnologies—A review. *Toxicology* **2010**, *269*, 92–104. [CrossRef]

27. Patil, S.S.; Shedbalkar, U.U.; Truskewycz, A.; Chopade, B.A.; Ball, A.S. Nanoparticles for environmental clean-up: A review of potential risks and emerging solutions. *Environ. Technol. Innov.* **2016**, *5*, 10–21. [CrossRef]

28. Pak, T.; de Lima Luz, L.F.; Tosco, T.; Costa, G.S.R.; Rosa, P.R.R.; Archilha, N.L. Pore-scale investigation of the use of reactive nanoparticles for in situ remediation of contaminated groundwater source. *Proc. Natl. Acad. Sci. USA* **2020**, *117*, 13366–13373. [CrossRef] [PubMed]

29. Galdames, A.; Mendoza, A.; Orueta, M.; de Soto García, I.S.; Sánchez, M.; Virto, I.; Vilas, J.L. Development of new remediation technologies for contaminated soils based on the application of zero-valent iron nanoparticles and bioremediation with compost. *Resour. Effic. Technol.* **2017**, *3*, 166–176. [CrossRef]

30. Thavamani, P.; Smith, E.; Kavitha, R.; Mathieson, G.; Megharaj, M.; Srivastava, P.; Naidu, R. Risk based land management requires focus beyond the target contaminants—A case study involving weathered hydrocarbon contaminated soils. *Environ. Technol. Innov.* **2015**, *4*, 98–109. [CrossRef]

31. Anyika, C.; Abdul Majid, Z.; Ibrahim, Z.; Zakaria, M.P.; Yahya, A. The impact of biochars on sorption and biodegradation of polycyclic aromatic hydrocarbons in soils—A review. *Environ. Sci. Pollut. Res.* **2015**, *22*, 3314–3341. [CrossRef]

32. Shayegan, H.; Ali, G.A.M.; Safarifard, V. Recent Progress in the Removal of Heavy Metal Ions from Water Using Metal-Organic Frameworks. *ChemistrySelect* **2020**, *5*, 124–146. [CrossRef]

33. Moreno-Sader, K.; García-Padilla, A.; Realpe, A.; Acevedo-Morantes, M.; Soares, J.B.P. Removal of Heavy Metal Water Pollutants (Co^{2+} and Ni^{2+}) Using Polyacrylamide/Sodium Montmorillonite (PAM/Na-MMT) Nanocomposites. *ACS Omega* **2019**, *4*, 10834–10844. [CrossRef] [PubMed]

34. Cameselle, C.; Chirakkara, R.A.; Reddy, K.R. Electrokinetic-enhanced phytoremediation of soils: Status and opportunities. *Chemosphere* **2013**, *93*, 626–636. [CrossRef]

35. Rodrigo, M.A.; Oturan, N.; Oturan, M.A. Electrochemically Assisted Remediation of Pesticides in Soils and Water: A Review. *Chem. Rev.* **2014**, *114*, 8720–8745. [CrossRef] [PubMed]

36. Chang, K.-S.; Lo, W.-H.; Lin, W.-M.; Wen, J.-X.; Yang, S.-C.; Huang, C.-J.; Hsieh, H.-Y. Microwave-Assisted Thermal Remediation of Diesel Contaminated Soil. *Eng. J.* **2016**, *20*, 93–100. [CrossRef]
37. Camenzuli, D.; Freidman, B.L.; Statham, T.M.; Mumford, K.A.; Gore, D.B. On-site and in situ remediation technologies applicable to metal-contaminated sites in Antarctica and the Arctic: A review. *Polar Res.* **2013**, *32*, 21522. [CrossRef]
38. Ossai, I.C.; Ahmed, A.; Hassan, A.; Hamid, F.S. Remediation of soil and water contaminated with petroleum hydrocarbon: A review. *Environ. Technol. Innov.* **2020**, *17*, 100526. [CrossRef]
39. Guerra, F.; Attia, M.; Whitehead, D.; Alexis, F. Nanotechnology for Environmental Remediation: Materials and Applications. *Molecules* **2018**, *23*, 1760. [CrossRef]
40. Huang, D.; Qin, X.; Peng, Z.; Liu, Y.; Gong, X.; Zeng, G.; Huang, C.; Cheng, M.; Xue, W.; Wang, X.; et al. Nanoscale zero-valent iron assisted phytoremediation of Pb in sediment: Impacts on metal accumulation and antioxidative system of Lolium perenne. *Ecotoxicol. Environ. Saf.* **2018**, *153*, 229–237. [CrossRef] [PubMed]
41. Peluffo, M.; Rosso, J.A.; Morelli, I.S.; Mora, V.C. Strategies for oxidation of PAHs in aged contaminated soil by batch reactors. *Ecotoxicol. Environ. Saf.* **2018**, *151*, 76–82. [CrossRef] [PubMed]
42. Fouad, O.A.; Makhlouf, S.A.; Ali, G.A.M.; El-Sayed, A.Y. Cobalt/silica nanocomposite via thermal calcination-reduction of gel precursors. *Mater. Chem. Phys.* **2011**, *128*, 70–76. [CrossRef]
43. Fouad, O.A.; Ali, G.A.M.; El-Erian, M.A.I.; Makhlouf, S.A. Humidity sensing properties of cobalt oxide/silica nanocomposites prepared via sol-gel and related routes. *Nano* **2012**, *7*, 1250038. [CrossRef]
44. Ali, G.A.M.; Fouad, O.A.; Makhlouf, S.A. Structural, optical and electrical properties of sol-gel prepared mesoporous Co_3O_4/SiO_2 nanocomposites. *J. Alloys Compd.* **2013**, *579*, 606–611. [CrossRef]
45. Ali, G.A.M.; Fouad, O.A.; Makhlouf, S.A.; Yusoff, M.M.; Chong, K.F. Co_3O_4/SiO_2 nanocomposites for supercapacitor application. *J. Solid State Electrochem.* **2014**, *18*, 2505–2512. [CrossRef]
46. Sadegh, H.; Ali, G.A.M.; Gupta, V.K.; Makhlouf, A.S.H.; Shahryari-ghoshekandi, R.; Nadagouda, M.N.; Sillanpää, M.; Megiel, E. The role of nanomaterials as effective adsorbents and their applications in wastewater treatment. *J. Nanostruct. Chem.* **2017**, *7*, 1–14. [CrossRef]
47. Abdel Ghafar, H.H.; Ali, G.A.M.; Fouad, O.A.; Makhlouf, S.A. Enhancement of adsorption efficiency of methylene blue on Co_3O_4/SiO_2 nanocomposite. *Desalin. Water Treat.* **2015**, *53*, 2980–2989. [CrossRef]
48. Otto, M.; Floyd, M.; Bajpai, S. Nanotechnology for site remediation. *Remediat. J.* **2008**, *19*, 99–108. [CrossRef]
49. Shayegan, H.; Ali, G.A.M.; Safarifard, V. Amide-Functionalized Metal–Organic Framework for High Efficiency and Fast Removal of Pb(II) from Aqueous Solution. *J. Inorg. Organomet. Polym. Mater.* **2020**, *30*, 3170–3178. [CrossRef]
50. Ali, G.A.M.; Makhlouf, A.S.H. Fundamentals of Waste Recycling for Nanomaterial Manufacturing. In *Waste Recycling Technologies for Nanomaterials Manufacturing*; Makhlouf, A.S.H., Ali, G.A.M., Eds.; Springer International Publishing: Cham, Switzerland, 2021; pp. 3–24. [CrossRef]
51. Niemeyer, C.M. Nanoparticles, Proteins, and Nucleic Acids: Biotechnology Meets Materials Science. *Angew. Chem. Int. Ed.* **2001**, *40*, 4128–4158. [CrossRef]
52. Ma, C.; Liu, F.-Y.; Wei, M.-B.; Zhao, J.-H.; Zhang, H.-Z. Synthesis of Novel Core-Shell Magnetic Fe_3O_4@C Nanoparticles with Carboxyl Function for Use as an Immobilisation Agent to Remediate Lead-Contaminated Soils. *Pol. J. Environ. Stud.* **2020**, *29*, 2273–2283. [CrossRef]
53. Meteku, B.E.; Huang, J.; Zeng, J.; Subhan, F.; Feng, F.; Zhang, Y.; Qiu, Z.; Aslam, S.; Li, G.; Yan, Z. Magnetic metal–organic framework composites for environmental monitoring and remediation. *Coord. Chem. Rev.* **2020**, *413*, 213261. [CrossRef]
54. Nematollahzadeh, A.; Seraj, S.; Mirzayi, B. Catecholamine coated maghemite nanoparticles for the environmental remediation: Hexavalent chromium ions removal. *Chem. Eng. J.* **2015**, *277*, 21–29. [CrossRef]
55. Zeng, H.; Zhai, L.; Qiao, T.; Yu, Y.; Zhang, J.; Li, D. Efficient removal of As(V) from aqueous media by magnetic nanoparticles prepared with Iron-containing water treatment residuals. *Sci. Rep.* **2020**, *10*, 9335. [CrossRef]
56. Zhu, S.; Wu, Y.; Qu, Z.; Zhang, L.; Yu, Y.; Xie, X.; Huo, M.; Yang, J.; Bian, D.; Zhang, H.; et al. Green synthesis of magnetic sodalite sphere by using groundwater treatment sludge for tetracycline adsorption. *J. Clean. Prod.* **2020**, *247*, 119140. [CrossRef]
57. Giahi, M.; Pathania, D.; Agarwal, S.; Ali, G.A.M.; Chong, K.F.; Gupta, V.K. Preparation of Mg-doped TiO_2 nanoparticles for photocatalytic degradation of some organic pollutants. *Stud. Univ. Babes Bolyai Chem.* **2019**, *64*, 7–18. [CrossRef]
58. Li, Q.; Chen, X.; Zhuang, J.; Chen, X. Decontaminating soil organic pollutants with manufactured nanoparticles. *Environ. Sci. Pollut. Res.* **2016**, *23*, 11533–11548. [CrossRef] [PubMed]
59. Yaqoob, A.A.; Parveen, T.; Umar, K.; Mohamad Ibrahim, M.N. Role of Nanomaterials in the Treatment of Wastewater: A Review. *Water* **2020**, *12*, 495. [CrossRef]
60. Baby, R.; Saifullah, B.; Hussein, M.Z. Carbon Nanomaterials for the Treatment of Heavy Metal-Contaminated Water and Environmental Remediation. *Nanoscale Res. Lett.* **2019**, *14*, 341. [CrossRef]
61. Benamar, A.; Mahjoubi, F.Z.; Ali, G.A.M.; Kzaiber, F.; Oussama, A. A chemometric method for contamination sources identification along the Oum Er Rbia river (Morocco). *Bulg. Chem. Commun.* **2020**, *52*, 159–171.
62. Zhang, W.-X. Nanoscale Iron Particles for Environmental Remediation: An Overview. *J. Nanopart. Res.* **2003**, *5*, 323–332. [CrossRef]
63. Mondal, A.; Dubey, B.K.; Arora, M.; Mumford, K. Porous media transport of iron nanoparticles for site remediation application: A review of lab scale column study, transport modelling and field-scale application. *J. Hazard. Mater.* **2021**, *403*, 123443. [CrossRef]

64. Tian, H.; Liang, Y.; Yang, D.; Sun, Y. Characteristics of PVP–stabilised NZVI and application to dechlorination of soil–sorbed TCE with ionic surfactant. *Chemosphere* **2020**, *239*, 124807. [CrossRef]

65. Reginatto, C.; Cecchin, I.; Heineck, K.S.; Thomé, A.; Reddy, K.R. Use of Nanoscale Zero-Valent Iron for Remediation of Clayey Soil Contaminated with Hexavalent Chromium: Batch and Column Tests. *Int. J. Environ. Res. Public Health* **2020**, *17*, 1001. [CrossRef]

66. Shubair, T.; Eljamal, O.; Khalil, A.M.E.; Matsunaga, N. Multilayer system of nanoscale zero valent iron and Nano-Fe/Cu particles for nitrate removal in porous media. *Sep. Purif. Technol.* **2018**, *193*, 242–254. [CrossRef]

67. Xue, W.; Huang, D.; Zeng, G.; Wan, J.; Zhang, C.; Xu, R.; Cheng, M.; Deng, R. Nanoscale zero-valent iron coated with rhamnolipid as an effective stabilizer for immobilization of Cd and Pb in river sediments. *J. Hazard. Mater.* **2018**, *341*, 381–389. [CrossRef] [PubMed]

68. Apul, O.G.; Delgado, A.G.; Kidd, J.; Alam, F.; Dahlen, P.; Westerhoff, P. Carbonaceous nano-additives augment microwave-enabled thermal remediation of soils containing petroleum hydrocarbons. *Environ. Sci. Nano* **2016**, *3*, 997–1002. [CrossRef]

69. Zhang, Y.; Yang, J.; Zhong, L.; Liu, L. Effect of multi-wall carbon nanotubes on Cr(VI) reduction by citric acid: Implications for their use in soil remediation. *Environ. Sci. Pollut. Res.* **2018**, *25*, 23791–23798. [CrossRef]

70. Cheng, J.; Sun, Z.; Yu, Y.; Li, X.; Li, T. Effects of modified carbon black nanoparticles on plant-microbe remediation of petroleum and heavy metal co-contaminated soils. *Int. J. Phytoremediat.* **2019**, *21*, 634–642. [CrossRef] [PubMed]

71. Zhang, J.; Gong, J.-L.; Zeng, G.-M.; Yang, H.-C.; Zhang, P. Carbon nanotube amendment for treating dichlorodiphenyl-trichloroethane and hexachlorocyclohexane remaining in Dong-ting Lake sediment—An implication for in-situ remediation. *Sci. Total Environ.* **2017**, *579*, 283–291. [CrossRef]

72. Abbasian, F.; Lockington, R.; Palanisami, T.; Megharaj, M.; Naidu, R. Multiwall carbon nanotubes increase the microbial community in crude oil contaminated fresh water sediments. *Sci. Total Environ.* **2016**, *539*, 370–380. [CrossRef]

73. Hua, S.; Gong, J.-L.; Zeng, G.-M.; Yao, F.-B.; Guo, M.; Ou, X.-M. Remediation of organochlorine pesticides contaminated lake sediment using activated carbon and carbon nanotubes. *Chemosphere* **2017**, *177*, 65–76. [CrossRef]

74. Naderi Peikam, E.; Jalali, M. Application of three nanoparticles (Al_2O_3, SiO_2 and TiO_2) for metal-contaminated soil remediation (measuring and modeling). *Int. J. Environ. Sci. Technol.* **2018**, *16*, 7207–7220. [CrossRef]

75. Qiao, Y.; Wu, J.; Xu, Y.; Fang, Z.; Zheng, L.; Cheng, W.; Tsang, E.P.; Fang, J.; Zhao, D. Remediation of cadmium in soil by biochar-supported iron phosphate nanoparticles. *Ecol. Eng.* **2017**, *106*, 515–522. [CrossRef]

76. Baragaño, D.; Alonso, J.; Gallego, J.R.; Lobo, M.C.; Gil-Díaz, M. Zero valent iron and goethite nanoparticles as new promising remediation techniques for As-polluted soils. *Chemosphere* **2020**, *238*, 124624. [CrossRef]

77. Fan, L.; Song, J.; Bai, W.; Wang, S.; Zeng, M.; Li, X.; Zhou, Y.; Li, H.; Lu, H. Chelating capture and magnetic removal of non-magnetic heavy metal substances from soil. *Sci. Rep.* **2016**, *6*, 21027. [CrossRef]

78. Baragaño, D.; Alonso, J.; Gallego, J.R.; Lobo, M.C.; Gil-Díaz, M. Magnetite nanoparticles for the remediation of soils co-contaminated with As and PAHs. *Chem. Eng. J.* **2020**, *399*, 125809. [CrossRef]

79. Savage, N.; Diallo, M.S. Nanomaterials and Water Purification: Opportunities and Challenges. *J. Nanopart. Res.* **2005**, *7*, 331–342. [CrossRef]

80. Gong, Y.; Gai, L.; Tang, J.; Fu, J.; Wang, Q.; Zeng, E.Y. Reduction of Cr(VI) in simulated groundwater by FeS-coated iron magnetic nanoparticles. *Sci. Total Environ.* **2017**, *595*, 743–751. [CrossRef]

81. Gillham, R.W.; O'Hannesin, S.F. Enhanced Degradation of Halogenated Aliphatics by Zero-Valent Iron. *Groundwater* **1994**, *32*, 958–967. [CrossRef]

82. Wang, C.-B.; Zhang, W.-X. Synthesizing Nanoscale Iron Particles for Rapid and Complete Dechlorination of TCE and PCBs. *Environ. Sci. Technol.* **1997**, *31*, 2154–2156. [CrossRef]

83. Tosco, T.; Petrangeli Papini, M.; Cruz Viggi, C.; Sethi, R. Nanoscale zerovalent iron particles for groundwater remediation: A review. *J. Clean. Prod.* **2014**, *77*, 10–21. [CrossRef]

84. Lin, K.-S.; Mdlovu, N.V.; Chen, C.-Y.; Chiang, C.-L.; Dehvari, K. Degradation of TCE, PCE, and 1,2–DCE DNAPLs in contaminated groundwater using polyethylenimine-modified zero-valent iron nanoparticles. *J. Clean. Prod.* **2018**, *175*, 456–466. [CrossRef]

85. Dong, H.; Zhang, C.; Deng, J.; Jiang, Z.; Zhang, L.; Cheng, Y.; Hou, K.; Tang, L.; Zeng, G. Factors influencing degradation of trichloroethylene by sulfide-modified nanoscale zero-valent iron in aqueous solution. *Water Res.* **2018**, *135*, 1–10. [CrossRef]

86. Zhu, F.; Ma, S.; Liu, T.; Deng, X. Green synthesis of nano zero-valent iron/Cu by green tea to remove hexavalent chromium from groundwater. *J. Clean. Prod.* **2018**, *174*, 184–190. [CrossRef]

87. Lico, D.; Vuono, D.; Siciliano, C.; Nagy, J.B.; De Luca, P. Removal of unleaded gasoline from water by multi-walled carbon nanotubes. *J. Environ. Manag.* **2019**, *237*, 636–643. [CrossRef] [PubMed]

88. Liang, J.; Liu, J.; Yuan, X.; Dong, H.; Zeng, G.; Wu, H.; Wang, H.; Liu, J.; Hua, S.; Zhang, S.; et al. Facile synthesis of alumina-decorated multi-walled carbon nanotubes for simultaneous adsorption of cadmium ion and trichloroethylene. *Chem. Eng. J.* **2015**, *273*, 101–110. [CrossRef]

89. Mdlovu, N.V.; Lin, K.-S.; Chen, C.-Y.; Mavuso, F.A.; Kunene, S.C.; Carrera Espinoza, M.J. In-situ reductive degradation of chlorinated DNAPLs in contaminated groundwater using polyethyleneimine-modified zero-valent iron nanoparticles. *Chemosphere* **2019**, *224*, 816–826. [CrossRef]

90. Ou, J.-H.; Sheu, Y.-T.; Tsang, D.C.W.; Sun, Y.-J.; Kao, C.-M. Application of iron/aluminum bimetallic nanoparticle system for chromium-contaminated groundwater remediation. *Chemosphere* **2020**, *256*, 127158. [CrossRef] [PubMed]

91. Wang, T.; Liu, Y.; Wang, J.; Wang, X.; Liu, B.; Wang, Y. In-situ remediation of hexavalent chromium contaminated groundwater and saturated soil using stabilized iron sulfide nanoparticles. *J. Environ. Manag.* **2019**, *231*, 679–686. [CrossRef] [PubMed]

92. Xie, W.; Liang, Q.; Qian, T.; Zhao, D. Immobilization of selenite in soil and groundwater using stabilized Fe–Mn binary oxide nanoparticles. *Water Res.* **2015**, *70*, 485–494. [CrossRef] [PubMed]

93. Dong, H.; Jiang, Z.; Deng, J.; Zhang, C.; Cheng, Y.; Hou, K.; Zhang, L.; Tang, L.; Zeng, G. Physicochemical transformation of Fe/Ni bimetallic nanoparticles during aging in simulated groundwater and the consequent effect on contaminant removal. *Water Res.* **2018**, *129*, 51–57. [CrossRef]

94. Huang, Y.; Keller, A.A. EDTA functionalized magnetic nanoparticle sorbents for cadmium and lead contaminated water treatment. *Water Res.* **2015**, *80*, 159–168. [CrossRef]

95. Li, D.; Egodawatte, S.; Kaplan, D.I.; Larsen, S.C.; Serkiz, S.M.; Seaman, J.C. Functionalized magnetic mesoporous silica nanoparticles for U removal from low and high pH groundwater. *J. Hazard. Mater.* **2016**, *317*, 494–502. [CrossRef] [PubMed]

96. Alani, O.A.; Ari, H.A.; Offiong, N.-A.O.; Alani, S.O.; Li, B.; Zeng, Q.-r.; Feng, W. Catalytic Removal of Selected Textile Dyes Using Zero-Valent Copper Nanoparticles Loaded on Filter Paper-Chitosan-Titanium Oxide Heterogeneous Support. *J. Polym. Environ.* **2021**, *29*, 2825–2839. [CrossRef]

97. Ari, H.A.; Alani, O.A.; Zeng, Q.-r.; Ugya, Y.A.; Offiong, N.-A.O.; Feng, W. Enhanced UV-assisted Fenton performance of nanostructured biomimetic α-Fe$_2$O$_3$ on degradation of tetracycline. *J. Nanostruct. Chem.* **2021**. [CrossRef]

98. Dissanayake, N.M.; Current, K.M.; Obare, S.O. Mutagenic Effects of Iron Oxide Nanoparticles on Biological Cells. *Int. J. Mol. Sci.* **2015**, *16*, 23482–23516. [CrossRef] [PubMed]

99. Current, K.M.; Dissanayake, N.M.; Obare, S.O. Effect of Iron Oxide Nanoparticles and Amoxicillin on Bacterial Growth in the Presence of Dissolved Organic Carbon. *Biomedicines* **2017**, *5*, 55. [CrossRef] [PubMed]

100. Du, W.; Xu, Y.; Yin, Y.; Ji, R.; Guo, H. Risk assessment of engineered nanoparticles and other contaminants in terrestrial plants. *Curr. Opin. Environ. Sci. Health* **2018**, *6*, 21–28. [CrossRef]

101. Shrivastava, M.; Srivastav, A.; Gandhi, S.; Rao, S.; Roychoudhury, A.; Kumar, A.; Singhal, R.K.; Jha, S.K.; Singh, S.D. Monitoring of engineered nanoparticles in soil-plant system: A review. *Environ. Nanotechnol. Monit. Manag.* **2019**, *11*, 100218. [CrossRef]

102. Hristozov, D.R.; Gottardo, S.; Critto, A.; Marcomini, A. Risk assessment of engineered nanomaterials: A review of available data and approaches from a regulatory perspective. *Nanotoxicology* **2012**, *6*, 880–898. [CrossRef]

103. Gómez-Sagasti, M.T.; Epelde, L.; Anza, M.; Urra, J.; Alkorta, I.; Garbisu, C. The impact of nanoscale zero-valent iron particles on soil microbial communities is soil dependent. *J. Hazard. Mater.* **2019**, *364*, 591–599. [CrossRef]

104. Dong, H.; Xie, Y.; Zeng, G.; Tang, L.; Liang, J.; He, Q.; Zhao, F.; Zeng, Y.; Wu, Y. The dual effects of carboxymethyl cellulose on the colloidal stability and toxicity of nanoscale zero-valent iron. *Chemosphere* **2016**, *144*, 1682–1689. [CrossRef]

105. Chaithawiwat, K.; Vangnai, A.; McEvoy, J.M.; Pruess, B.; Krajangpan, S.; Khan, E. Impact of nanoscale zero valent iron on bacteria is growth phase dependent. *Chemosphere* **2016**, *144*, 352–359. [CrossRef] [PubMed]

106. Cheng, Y.; Dong, H.; Lu, Y.; Hou, K.; Wang, Y.; Ning, Q.; Li, L.; Wang, B.; Zhang, L.; Zeng, G. Toxicity of sulfide-modified nanoscale zero-valent iron to Escherichia coli in aqueous solutions. *Chemosphere* **2019**, *220*, 523–530. [CrossRef]

107. Li, Z.; Wang, L.; Wu, J.; Xu, Y.; Wang, F.; Tang, X.; Xu, J.; Ok, Y.S.; Meng, J.; Liu, X. Zeolite-supported nanoscale zero-valent iron for immobilization of cadmium, lead, and arsenic in farmland soils: Encapsulation mechanisms and indigenous microbial responses. *Environ. Pollut.* **2020**, *260*, 114098. [CrossRef] [PubMed]

108. Lam, C.W. Pulmonary Toxicity of Single-Wall Carbon Nanotubes in Mice 7 and 90 Days After Intratracheal Instillation. *Toxicol. Sci.* **2003**, *77*, 126–134. [CrossRef] [PubMed]

109. Song, B.; Chen, M.; Ye, S.; Xu, P.; Zeng, G.; Gong, J.; Li, J.; Zhang, P.; Cao, W. Effects of multi-walled carbon nanotubes on metabolic function of the microbial community in riverine sediment contaminated with phenanthrene. *Carbon* **2019**, *144*, 1–7. [CrossRef]

110. Vilardi, G.; Mpouras, T.; Dermatas, D.; Verdone, N.; Polydera, A.; Di Palma, L. Nanomaterials application for heavy metals recovery from polluted water: The combination of nano zero-valent iron and carbon nanotubes. Competitive adsorption non-linear modeling. *Chemosphere* **2018**, *201*, 716–729. [CrossRef]

111. Zhang, R.; Zhang, N.; Fang, Z. In situ remediation of hexavalent chromium contaminated soil by CMC-stabilized nanoscale zero-valent iron composited with biochar. *Water Sci. Technol.* **2018**, *77*, 1622–1631. [CrossRef]

112. Qian, L.; Chen, Y.; Ouyang, D.; Zhang, W.; Han, L.; Yan, J.; Kvapil, P.; Chen, M. Field demonstration of enhanced removal of chlorinated solvents in groundwater using biochar-supported nanoscale zero-valent iron. *Sci. Total Environ.* **2020**, *698*, 134215. [CrossRef]

113. Alabresm, A.; Chen, Y.P.; Decho, A.W.; Lead, J. A novel method for the synergistic remediation of oil-water mixtures using nanoparticles and oil-degrading bacteria. *Sci. Total Environ.* **2018**, *630*, 1292–1297. [CrossRef] [PubMed]

114. Czinnerová, M.; Vološčuková, O.; Marková, K.; Ševců, A.; Černík, M.; Nosek, J. Combining nanoscale zero-valent iron with electrokinetic treatment for remediation of chlorinated ethenes and promoting biodegradation: A long-term field study. *Water Res.* **2020**, *175*, 115692. [CrossRef] [PubMed]

115. Boente, C.; Sierra, C.; Martínez-Blanco, D.; Menéndez-Aguado, J.M.; Gallego, J.R. Nanoscale zero-valent iron-assisted soil washing for the removal of potentially toxic elements. *J. Hazard. Mater.* **2018**, *350*, 55–65. [CrossRef] [PubMed]

116. Qu, G.; Kou, L.; Wang, T.; Liang, D.; Hu, S. Evaluation of activated carbon fiber supported nanoscale zero-valent iron for chromium (VI) removal from groundwater in a permeable reactive column. *J. Environ. Manag.* **2017**, *201*, 378–387. [CrossRef]
117. Huang, J.; Yi, S.; Zheng, C.; Lo, I.M.C. Persulfate activation by natural zeolite supported nanoscale zero-valent iron for trichloroethylene degradation in groundwater. *Sci. Total Environ.* **2019**, *684*, 351–359. [CrossRef] [PubMed]

 *water*

Review

Insight on Extraction and Characterisation of Biopolymers as the Green Coagulants for Microalgae Harvesting

Teik-Hun Ang [1], Kunlanan Kiatkittipong [2,*], Worapon Kiatkittipong [3], Siong-Chin Chua [1], Jun Wei Lim [4], Pau-Loke Show [5], Mohammed J. K. Bashir [6] and Yeek-Chia Ho [1,*]

[1] Department of Civil and Environmental Engineering, Centre of Urban Resource Sustainability, Institute of Self-Sustainable Building, Universiti Teknologi PETRONAS, Seri Iskandar 32610, Perak Darul Ridzuan, Malaysia; teikhun0422@hotmail.com (T.-H.A.); chuasiongchin95@gmail.com (S.-C.C.)

[2] Department of Chemical Engineering, Faculty of Engineering, King Mongkut's Institute of Technology Ladkrabang, Bangkok 10520, Thailand

[3] Department of Chemical Engineering, Faculty of Engineering and Industrial Technology, Silpakorn University, Nakhon Pathom 73000, Thailand; KIATKITTIPONG_W@su.ac.th

[4] Department of Fundamental and Applied Sciences, HICoE-Centre for Biofuel and Biochemical Research, Institute of Self-Sustainable Building, Universiti Teknologi PETRONAS, Seri Iskandar 32610, Perak Darul Ridzuan, Malaysia; junwei.lim@utp.edu.my

[5] Department of Chemical and Environmental Engineering, University of Nottingham Malaysia, Broga Road, Semenyih 43500, Selangor Darul Ehsan, Malaysia; PauLoke.Show@nottingham.edu.my

[6] Department of Environmental Engineering, Faculty of Engineering and Green Technology (FEGT), Universiti Tunku Abdul Rahman, Kampar 31900, Perak Darul Ridzuan, Malaysia; jkbashir@utar.edu.my

* Correspondence: kunlanan.kia@kmitl.ac.th (K.K.); yeekchia.ho@utp.edu.my (Y.-C.H.)

Received: 25 March 2020; Accepted: 3 May 2020; Published: 14 May 2020

Abstract: This review presents the extractions, characterisations, applications and economic analyses of natural coagulant in separating pollutants and microalgae from water medium, known as microalgae harvesting. The promising future of microalgae as a next-generation energy source is reviewed and the significant drawbacks of conventional microalgae harvesting using alum are evaluated. The performances of natural coagulant in microalgae harvesting are studied and proven to exceed the alum. In addition, the details of each processing stage in the extraction of natural coagulant (plant, microbial and animal) are comprehensively discussed with justifications. This information could contribute to future exploration of novel natural coagulants by providing description of optimised extraction steps for a number of natural coagulants. Besides, the characterisations of natural coagulants have garnered a great deal of attention, and the strategies to enhance the flocculating activity based on their characteristics are discussed. Several important characterisations have been tabulated in this review such as physical aspects, including surface morphology and surface charges; chemical aspects, including molecular weight, functional group and elemental properties; and thermal stability parameters including thermogravimetry analysis and differential scanning calorimetry. Furthermore, various applications of natural coagulant in the industries other than microalgae harvesting are revealed. The cost analysis of natural coagulant application in mass harvesting of microalgae is allowed to evaluate its feasibility towards commercialisation in the industrial. Last, the potentially new natural coagulants, which are yet to be exploited and applied, are listed as the additional information for future study.

Keywords: natural coagulant; production; characterisation; application; microalgae harvesting; cost analysis; coagulation and flocculation

---

1. Introduction

The fast-developing countries rely heavily on large-scale industrialisation to improve their global economic competitiveness. Concurrently, the growing amount of waste produced in modern society has become a global issue. In many contexts, developing countries have produced tons of wastes from industrial revolution. As the world is moving towards green technology, natural coagulant, which can be extracted from plant tissues, animals or microorganisms, has been a major point of interest. Notably, studies from past researchers have shown the effectiveness of natural coagulant in wastewater treatment, such as turbidity removal through neutralisation of anionic suspended particles with cationic polymers. To provide a more focused discussion, coagulation is an important process in surface water treatment, for example, *Moringa oleifera* has been used in native communities in treating river water for drinking purpose. On the other hand, natural coagulant could take place in treating commercial wastewater, for instance, *Maerua decumbent* has been used to treat paint wastewater, which marked 99% of turbidity removal by using 1 kg·m^{-3} in dosage [1].

Recently, natural coagulant has also emerged as a promising solution in microalgae harvesting as it will not create by-product, such as suspended alum residual in microalgae biomass, which is needed to be further removed before lipid extraction. Concurrently, natural coagulant requires a lower dosage in mass harvesting of microalgae compared to alum. Natural coagulants are usually used as point-of-use products in less developed countries because they are fairly cost-effective as compared with the alum and could be easily processed in the usable form [2]. Moreover, natural coagulant gains advantages over alum in terms of reduced sludge production, produces treated water with less extreme pH and it is in line with sustainable development. The use of natural plant-derived materials to coagulate and flocculate microalgae biomass produced in mass cultivation system are not a new idea, for instance, natural coagulants have been used to clear turbid water since ancient times, even before the emergence of chemical coagulants [3]. In view of microalgae cultivation, the most common approach is the suspended growing method as it allows the microalgae to distribute evenly in the medium for nutrient intake. In contrast, a non-suspended mode of cultivation allows microalgae to grow on the surface to form a biofilm. It is more commercially feasible because the harvesting process of microalgae biomass will be easier. The typical example of non-suspended microalgae are *Scenedesmus obliquus* sp. and *Botryococcus braunii* sp. [4].

Apparently, natural coagulant is an emerging solution to green and sustainable water treatment. Besides microalgae harvesting, it has been utilised on various new sectors, for example, electro-coagulation in microbial fuel cell system to precipitate heavy metals for self-power system [5] and membrane manufacturing plant wastewater treatment [6]. Moreover, natural coagulant has also been proven to remove 97% of copper ions in 3 h from wastewater of rotating triboelectric Nano generator (R-TENG), to remove lead (II) ion by 88% at pH 5 from simulated wastewater using cashew nut coagulant [7], to remove wastewater by coupling coagulation using anion exchange membranes (AEMs) in electrodialysis [8] and last to be used in bioreactor for aerobic sewage wastewater treatment using natural coagulant (micro-based coagulant) such as *Bacillus* species, *Achromobacter* species and *Comamonas* species. [9].

There is also strong evidence that the use of biopolymer and plant-based materials has been increasing and penetrating into various fields, for example, the technology of reusing cysteine-containing protein materials from keratinous waste to produce tough keratin fibre [10], fabrication of sustainable membrane using bamboo fibre to enhance cross-flow filtration performance [11] and perforated lotus leaf to treat oil spillage [12]. Besides, biopolymer is also widespread in other fields such as utilisation of natural fatty acids for drug releases in the medical field [13], biophenol coatings on nanofiltration membranes to improve its performance on the separation of organic media [14] and glucose-based biopolymer to modify the interlayer of the solar cell, which enables 95% of enhancement in power conversion efficiency [15].

Prior to the application, the extraction of natural coagulant from plant, animal or microbes are needed. However, the current extraction method poses a significant drawback, which is time consuming as it involves several stages of pretreatment. The preparation stages associated with each type of natural coagulant (plant, microbial and animal) are varied as well. In addition, each plant, animal or microbial coagulant has different optimum extraction methods. Sometimes, water extraction of natural coagulants commonly used by native communities could be further incorporated with currently employed techniques such as salt and acid extractions to maximise the extraction efficiency.

Furthermore, the previous papers are mainly focused on the performance of natural coagulant in coagulation and flocculation, for instance, the optimum operating condition of 21 types of plant based coagulants and their barrier to commercialisation [16]; the optimum operating conditions of *Dolichas lablab, Azadirachta indica, Moringa oleifera, Hibiscus rosa sinensis* [17]; and the modification of functional group of natural coagulants in enhancing the flocculating activity [18]. Thus, to provide a more comprehensive discussion, this review will include the technical aspects, such as the extraction processes of natural coagulant with detailed explanations on its necessity, because the extraction stage is as important as the performance stage and should not be neglected. By reviewing the extraction processes step by step, further studies on discovering the new natural coagulant will be easier because the relevant extraction processes could be referred here based on their nature of characteristic. To illustrate, the study of extraction method of new coagulant, Aloe Veragel, could be referred to Aloe Vera as they are from the same genus. Besides, the recent trend of research is mainly on plant-based coagulant; thus, in this review, further explorations on animal- and microbe-based coagulants are carried out and their optimum operating conditions are technical discussed. Characterisation of natural coagulant is explored as well in accordance with Ang [18], covering additional information such as surface morphology, molecular weight, zeta potential, TGA and others. The promising future of microalgae as next generation of energy is presented with the application of natural coagulant in microalgae harvesting, followed by its advantages and disadvantages with respect to alum. In summary, natural coagulants derived from plants, microorganism and animals are reviewed for their extractions, characterisations and applications in microalgae harvesting. Cost analysis of natural coagulant for large scale application in industrial is carried out to provide an appropriate platform for future researchers to intensify on microalgae harvesting by using these natural materials.

2. The Promising Future of Microalgae

2.1. Microalgae as Next Generation of Biofuel

Renewable energy plays a vital role in energy resources, and access to green and cheap energy has become a trend in modern society. Biofuel is a type of renewable energy in a form of liquid and gaseous fuels produced from biomass, namely bio-ethanol, bio-methanol, bio-oil, bio-diesel and bio-gas. Generally, there are four types of biofuel generation and their characterisations are based on the nature of the feedstock. The first generation of biofuel is extracted from food crops, for example, from sugarcane through chemical process such as fermentation. However, a series of problems regarding to fuel vs. food dilemma have been attributed to the production of the second-generation of biofuel, in which its extraction is from non-food crops. Likewise, the second generation of biofuel does experience unexpected demise as the first generation of biofuel. Arnold et al. [19] noted the innovation of second-generation biofuel was relatively constant in the mid-1990s and followed by decline in the following years. The long-term development of the second generation of biofuels is a step in the right direction; however, it has several drawbacks such as cost effectiveness and technological barriers in dealing with biomass [20,21]. To address these concerns, the third generation of biofuel, which is derived from microbes, has been introduced.

Microalgae is touted to be a sustainable energy source of the third generation of biofuel. The Solar Energy Research Institute, USA, has proposed microalgae as an intermediate tool for biofuel production since the 1940s [22]. In comparison with other energy crops, microalgae biofuel has the advantage of quick growth, high lipid, carbohydrate content and excellent biomass yield with the lowest capacity of land used [23]. Previous studies have established that microalgae can be the replacement of fossil fuel due to its high amount of intracellular accumulated oils [20]. Additionally, the cultivation of microalgae using waste is the essence of the research vanguard. In view of sustainability, cultivation of microalgae using waste such as Palm Oil Mill Effluent (POME) is an added point to the environment.

Besides, the strong requirement for clean energy production and conversion technology development at a global scale led to mass researches on microalgae as a feedstock to generate biofuel. Developed nations, for instance, the USA, Australia and Mexico have focused their researches towards the efficient cultivation of microalgae and simultaneous wastewater treatment in the past few years [24]. To further enhance biofuel production, the fourth generation of biofuel, which uses genetically modified microalgae in production, has attracted enormous attention. The improvement in metabolic activity, photosynthesis efficiency, light penetration and reduction of photo-inhibition of genetically modified microalgae lead to enhancement of fourth generation of biofuels in term of quality and quantity [25].

Further, comparative studies evinced that microalgae also help in absorbing carbon from the industrial gases and utilisation of nitrogen and phosphorus from industrial and municipal wastewater [23,26]. At present, microalgae are competitive and are becoming a trend of future energy resources.

2.2. Bioprocess Approach of Microalgae Biofuel

The bioprocess approaches of microalgae biofuel are divided into four phases: (1) microalgae cultivation, (2) harvesting, (3) cell disruption and extraction and (4) fatty acid profiling [27]. In the cultivation stage, the selection of a cultivation medium is relatively important. Cultivation medium is a source of energy, nutrients or growth factors that design to grow certain targeted species. The favourable medium for microalgae growth consists of nutrients such as nitrogen and phosphorus, moderate pH, feasible to light penetration and allow CO_2 circulation. Therefore, several wastewaters, such as POME, rubber mill effluent and landfill leachate, have been studied to be used as cultivation medium, with the condition that nitrogen and phosphorus are present in the composition [20,23,24]. Alternatively, there are many standard solutions available in the market, prepared for the cultivation of microalgae, called standard cultivation medium. Bold's Basal Medium (BBM) is one of the mediums consisting of (1) 10 mL per litre of culture medium with the following chemicals, sodium nitrate (25 $g \cdot L^{-1}$), calcium chloride dihydrate (2.5 $g \cdot L^{-1}$), magnesium sulfate heptahydrate (7.5 $g \cdot L^{-1}$), dipotassium phosphate (7.5 $g \cdot L^{-1}$), monopotassium phosphate (17.5 $g \cdot L^{-1}$) and sodium chloride (2.5 $g \cdot L^{-1}$) [20]. Noteworthy, the starvation phase in the pre-harvesting cultural stage is also proven to trigger the accumulation of lipids after the stage where microalgae growth is maximised [28].

Subsequently, the harvesting of microalgae is the main focus of this review. Indeed, alum is mainly used as a coagulant in microalgae harvesting industrial and the usage of natural coagulant is limited to academic research. However, an obvious drawback has arisen in conventional microalgae harvesting as the alum will result in extreme pH of the treated end product, especially in mass harvesting of microalgae biomass with the addition of a huge amount of alum. Some of the reflections are gathered and stated that using alum with coagulants is ineffective in low temperature, and has high procurement costs and detrimental effects on human health [29]. It has, somewhat, been noted that the mass harvesting process faces drawbacks such as the reduction in lipid due to the addition of alum [30]. Further, alum might be the cause of Alzheimer's disease, which deposition of alum in body has significantly impact to our health [2]. In some aspects, such as the harvesting of EPA/DHA enriched microalgae oil using alum might result in a high aluminium level in microalgae oil. Thus, an alternative solution is sought, and natural coagulants are deemed and anticipated to overcome this problem. Subsequently, optimum microalgae recovery could be carried out for mass production.

Compared to alum, the usage of natural coagulant in the harvesting process is promising due to its non-toxic nature and that it is safe for consumption [31]. Therefore, progressive research on natural coagulant in microalgae harvesting should be done, especially for EPA/DHA dietary microalgae oil. To date, the disadvantages of natural coagulant are mainly due to its feasibility in terms of production time, commercialisation in industrial and quality control. Figure 1 shows the disadvantages of utilising natural coagulant in microalgae harvesting.

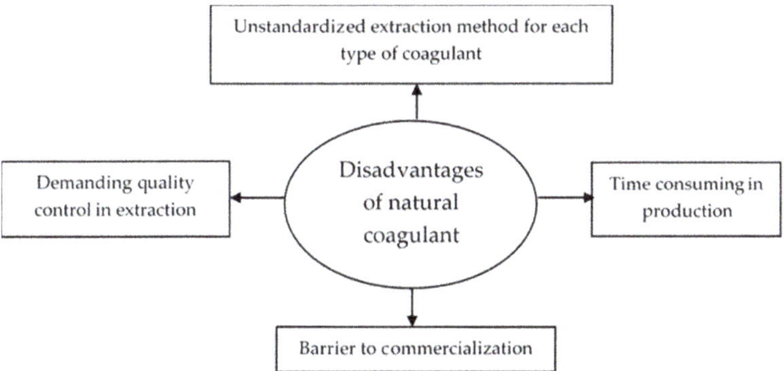

Figure 1. Disadvantages of utilising natural coagulant in microalgae harvesting.

In sum, the extraction process of natural coagulant from plant, animal and microbes is complicated and time-consuming. Different optimum extraction methods for each type of natural coagulant further disrupt the commercialisation of natural coagulant in the industrial.

Up to now, studies have shown progressive optimisation in the extraction process of natural coagulant. Significantly, the extraction process could be enhanced and modified based on domain knowledge. In this review, justifications on the necessity of each sub-steps to be carried out in the extraction process of natural coagulant (plant, animal and microbial) are discussed. Notably, this information will help in understanding of the concept of extraction and also provide references for the extraction of new natural coagulant in the future.

Furthermore, it is also technically proven that flocculating activity could be increased by modifying the characteristic of natural coagulant. Thus, this could be applied to all types of natural coagulant to offset the disadvantages as mentioned. To address this, strategies to enhance the flocculating activity of natural coagulant based on their physical, chemical and thermal characteristic have been espoused in Section 4.

3. Extraction of Natural Coagulants

3.1. Plant Based Polymers

Over the past few years, researches have been conducted on various types of natural coagulants derived from plant wastes and fruit pieces, for instance, Nirmali seeds, *Moringa oleifera*, Surjana seed, Arabic gum, maize seed, tannin, Cactaceae, etc. had demonstrated significant coagulant capacities [32,33]. Among all, the plant-based coagulant recently received the greatest level of attention is the seed of *Moringa oleifera* native to Sudan [2]. Research by Vijayaraghawan et al. [34] shows that the water extracted *M. oleifera* seed has a comparative result with aluminium salt (alum). Moreover, there are standardised and well organised extraction steps in extracting the plant-based coagulant [35]. The general processing steps of the extraction of plant-based coagulants can be categorised into three major stages: primary, secondary and tertiary (Figure 2). Further, there is green extraction technology, for instance, utilisation of salt solvent will increase the extraction efficiency and flocculating activity of peanut seed coagulant as compared to water extraction, which has an improvement of 61% in turbidity removal [36]. Ultimately, it will reduce the cost of extraction and energy used along the harvesting process.

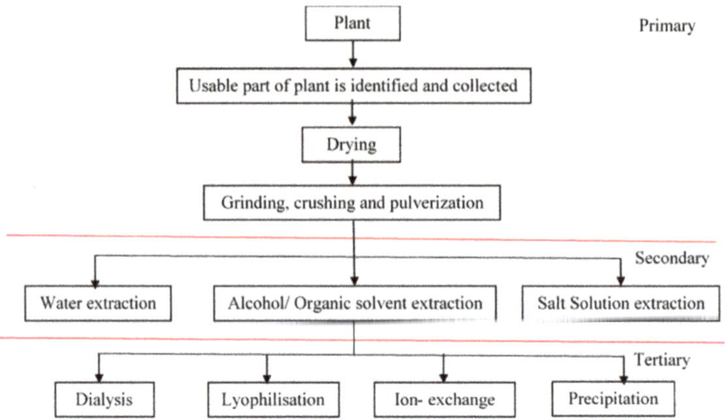

Figure 2. General processing steps in plant based coagulant extraction.

3.1.1. Primary Processing

The primary processing stage involves choosing of usable parts of the plant. The important factor to be considered in the choice of usable part is associated with their respective coagulating properties. In the case of cacti, the usable part is the vascular tissue of the plant, and therefore the skin and spines are eliminated. However, the usable part of aloe species is different, which are the mature leaves and the perimeter spines. At this stage, there must be no sign of contamination, parasites, presence of external insects and organisms on the surface of plants [35]. Furthermore, the selected usable part must be washed with plenty of water to eliminate impurities such as sand stones or grain wastes and to prevent the presence of fungi and yeasts due to bulk handling, fractionation and packaging [35]. In regard to this, the authors of [37] have introduced formaldehyde or known as acid–alkaline wash of plant, which significantly enhanced the pretreatment phase of natural coagulant extraction. The acid helps in removing the minerals on the surface while the alkali acts as neutralising agent to acid. These solvents had been proven to remove organic materials and in the same way, it is traditionally used in the ion-exchanged technology. Therefore, by utilising acid– alkaline wash, it can be assumed that certain degree of organic materials on the surface of plant had been removed prior to the secondary processing and this ultimately reduces the leaching of organic matters inside the usable part. The presence of organic matter will has negative impact on drinking water treatment such as causing colour, odour and taste problem. In the drying stage, the materials are ubiquitously carried out at oven or outdoor to evaporate the water content and reduce moisture level. The presence of water will affect the extraction, while dried plants will reduce the possibility of further enzymatic or metabolic alteration of plant. It is important to be carried out in warm tropical climate with temperature ranges between 20 °C and 35 °C, low humidity between 50% and 70% on average and most importantly, it is highly advised not to dampen by rain or other water sources [35]. Afterwards, the crushing, mechanical grinding and powdering of the dried extract could be carried out with machine followed by passing through a mill to pulverise the material [35]. Ultimately, it is sieved to obtain a very fine powder and stored in airtight containers to avoid hydration prior to the subsequent use in secondary processing of coagulants [35].

3.1.2. Secondary Processing

In the secondary processing stage, the active coagulating agents of each plant could be extracted via different solvents (organic, water or salt solutions). This comes as a surprise at first glance as each type of plant has a unique chemical structure and electrostatic properties providing novelty. Additionally, different solvents could be used in sequence at the secondary processing stage, for example, solvent extraction of valuable and edible oil from *M. oleifera* (MO) seed [38] followed by water extraction of active component for coagulants from *M. oleifera* (MO) seed waste. In the African countries, MO seed residue as a by-product of oil extraction is used to extract natural coagulant for water treatment [2]. Indeed, the oil content in MO seed is not attributed to flocculating activity and the oil content will actually affect the performance of natural coagulant especially in heavy metal removal activities [39]. The oil will reduce the efficiency of coagulation by making re-stabilisation of destabilised particles and ultimately reduce the binding sites for coagulation. Therefore, the oil content in each plant should be processed through a pre-secondary treatment [39]. In many cases, the extraction using water is evidently the most accepted option due to its abundance and cost-effectiveness provided that the plant's active component is water-soluble protein [33].

The application of salt solution extraction is rather recent and more effective compared to the water extraction method. It is found that the coagulation capacity of the MO using salt solution extraction is 7.4 times higher than the water-based extraction in the study of the removal of suspended kaolinite [40]. To illustrate, the delipidation is involved in the salt extraction process and this will lead to least possible of lipid content in the extracted MO active component. Ultimately, the decrease in lipid will result in the increase in coagulation capacity [40]. The previous study by Ndabigengesere, Narasiah and Talbot [34] was first proposed that one of the disadvantages of water-based extraction of MO was the increase in dissolved organic carbon (DOC) residual of the treated water. The DOC is usually due to the presence of organic materials and a precursor of disinfection by-products in drinking water treatment. The presence of DOC could result in an increase in chemical oxygen demand (COD). The increase, however, does not affect the salt solution extraction due to the salting-in mechanism where increasing of ionic strength of a solution will increase the solubility of the solute.

Nonetheless, significant setback emerges because the prepared powder (biopolymer) contains not just the coagulating active agents, but also plant tissues. The latter is rich in plant tissue, thereby increasing the organic loading in the treated water, which may exacerbate the situation further rather than improving the efficiency of treatment after coagulation and flocculation [38]. This problem can be addressed by processing the powder through tertiary (purification) stages.

3.1.3. Tertiary Processing

The tertiary processing is rarely performed and is limited to academic research [34,38] as this increases the overall processing cost. After the secondary processing stage, the active coagulating agents appear as supernatants in the solution. Preliminary studies suggested that dialysis, lyophilisation and ion-exchange were feasible purification methods in tertiary processing stage. A recent review of literature on this topic found that the oldest and simplest coagulant recovery technologies are solid–liquid filtration and settlement to remove just gross solids from the extracted coagulant [41]. All of these are still applied in industrial applications; however, modern technologies do discriminate natural coagulants from contaminants by molecular size and charge. These principles have been applied using membranes and adsorbents. Certainly, there are several studies on tertiary recovery of coagulant using ultrafiltration (UF) at the bench as well as pilot scale [42,43]. In these studies of tertiary recovery, the rationale was to select ultrafiltration pore sizes that allowing trivalent metal to penetrate while retaining natural organic material. Further, membranes with a molecular weight cut-offs of 10 kDa allowed aluminium permeation to exceed 90% and total organic carbon rejections of 50–66% [44]. Although it is not extensively reported in past researches, the main drawback of ultrafiltration was the fouling and quality issues. In other words, the molecular weight, functionality and nature of organic compounds are varied widely and depended heavily on environmental conditions and heavy metals do have similar cationic and molecular weight characteristics as natural coagulants. Due to the overlap in molecular weights of natural coagulants and organic contaminants, researches proposed the coagulant separation technologies using molecular charge as the principal means to differentiate the cationic coagulant from anionic or neutral contaminants. The tertiary recovery process of ionic exchange has been in the form of ion-exchange media such as liquids, resins, and dialysis membranes [41]. Besides, it is recognised, for example, *M. oleifera* is highly biodegradable natural coagulant with a very limited shelf life. Lyophilisation, often known as freeze-drying, is a technique used to retain biological material by freezing the extraction mixture, extract the supernatant (natural coagulant), then drying at quite low temperatures through a vacuum. The relevant study also showed that the freeze-dried *M. oleifera* retained its high coagulation efficiency for up to 11 months regardless of storage temperatures and packaging methods [45,46].

3.2. Microbial Based Polymers

Apart from plant-based coagulants, there are coagulants produced by bacteria and fungi. Particularly, different microorganisms could yield different flocculating coagulant from their respective bacteria strain, i.e., proteoglycan coagulants (98% polysaccharide and 1.6% protein) is yielded from *Bacillus mojavensis* strain 32A with an interesting flocculating activity of 96% recorded at pH 10 [47]. Various factors have to be considered in the selection of bacteria. The predominant step in the preparation of microbial-based coagulant starts with the preliminary screening of bacterium strain based on its mucoid and ropy colony morphology characteristics. It is then followed by the biochemical identification of the strain based on 29 biochemical and enzymatic reaction tests (BBL Crystal Gram-Positive ID System). After the identification of the microbial-based coagulants from bacteria strain, batch cultures are prepared to cultivate bacteria and produce natural coagulant at room temperature. Subsequently, the flocculating activity of each natural coagulant is determined via kaolin assays [47]. It had been observed that variation in cultivation medium of bacteria would affect the growth of microorganisms and its ability in producing the expected exopolysaccharides or natural coagulant [48]. Researches had been performed to identify the bacterium strain that aid in flocculating activity and shown in Table 1. General preparation processes of microbial-based coagulant are summarised as below.

1. Preliminary identification of the natural coagulants-producing bacterium strain based on its mucoid and ropy colony morphology characteristics.
2. Screening of bacteria and fungi to find microbial-based coagulants from bacterium strain.
3. Determining the flocculating activity of microbial-based coagulants (natural coagulants) yielded from each bacterium strain by kaolin clay suspension.
4. Optimising the culture conditions of bacteria to produce a higher amount of natural coagulant.

Table 1. Microbial strains and their respective flocculating activities.

Bacterial Strain	Flocculating Activity in Removal of Kaolin (%)	Reference
Bacillus agaradhaerens C9	81	[49]
Bacillus sp. XF-56	94	[49]
Arthrobacter sp. B4	99	[50]
Bacillus licheniformis X14	98	[47,51]
Bacillus velezensis 40B	>98	[52]
Chryseobacterium daeguense W6	97	[48]
Klebsiella sp. ZZ-3	95	[53]
Streptomyces sp. MBRC-91	96	[54]
Aspergillus flavus (source NI 3)	>90	[55]
Penicillum strain HHE-P7	96	[47,56]
Aspergillus flavus (source NI)	97	[55]
Rhizopus sp. M9 & *Rhizopus* sp. M17	90	[57]
Talaromyces sp.	93	[58]

There are several screening methods that could be applied on testing of a bacterium strain. A colorimetric method is an approach to determine the concentration of chemical compounds with the aid of colour. Further, optimisation of cultivation medium of bacteria and fungi could be conducted using statistical analyses, which discovers the pattern and trend of bacteria growth with equation. Experimental design of various microbes is carried out by cultivating them in different sources of nutrients to produce natural coagulants with different characteristics. These data are collected and useful for interpretation by a statistical linear regression method to find the relationship between each factor and ultimately lead to production of higher amount of natural coagulant.

Besides, the essential nutrients for microbial growth are mainly carbon and nitrogen elements. At the same time, wastewater and sludge are abundant with carbon, nitrogen, phosphorus and micronutrients, which could sustain the microbial growth for natural coagulants production. In this context, studies also showed that agro-industrial wastes, such as sugarcane, starch molasses, corn-steep liquor, soybean juice, etc., which are mainly composed of polysaccharides, could be used for microbial growth for natural coagulants production. To sum up, the optimisation in cultivation medium of each type of bacteria is different and should be studied accordingly through experimental works.

3.3. Animal Based Polymers

The animal-based coagulant is derived mainly from chitin, which is a natural polymer from two marine crustaceans, namely, shrimp and crabs. Chitin, the most common polysaccharide after cellulose, is a non-elastic and nitrogenous natural polymer structured as a linear chain by the 2-acetoamido-2-deoxy-β-D-glucopyranose monomers [59]. Chitosan-based materials are the potentially eco-friendly coagulants and flocculants in harvesting process because of their natural biological characteristics and biodegradability. Generally, the mechanism involved in the harvesting process of chitosan is bridging. Chitosan is commonly used in laboratory for microalgae harvesting, for example, to harvest *Chlorella* sp. from its cultivation medium [60]. Furthermore, its advantages of recyclability and as an excellent chelating agent for arsenic, molybdenum, cadmium, chromium, lead and cobalt ions make it an excellent choice for industrial wastewater treatment [60]. Table 2 shows the flocculation abilities of chitosan at its optimum operating conditions in removing various pollutants or separating microalgae.

Table 2. Flocculation abilities of chitosan at its best conditions to separate various pollutants from the aqueous medium.

Chitosan	Operating Condition	Flocculation Ability	Reference
Chitosan	214 mg·L^{-1}, pH 8 and 131 rpm	92% removal of *Chlorella vulgaris*	[61]
Chitosan (*Plaemon serratus*)	15 mg·L^{-1} at 67 nephelometric turbidity units (NTU) raw water, flocculation time of 20 min	89% removal of sewage wastewater	[62]
Chitosan (shrimp)	4 mg·L^{-1}, pH 6 and flocculation time of 10 min	95% removal of *Thalassiosira pseudonana* microalgae	[63]
Chitosan	30 mg·L^{-1}, pH 7, flocculation time of 20 min	98% removal of *Chlorella vulgaris*	[64]
Chitosan	4 mg·L^{-1}, pH 4 and flocculation time of 10 min	90% removal of *Thalassiosira pseudonana* microalgae	[63]
Chitosan	20 mg·L^{-1}, pH 9.9, flocculation time of 10 min	90% removal of *Thalassiosira pseudonana* microalgae	[63]
Xanthated chitosan	50 mg·L^{-1}, pH 6.0, slow stirring for 10 min and settling for 10 min	>97% removal of Cu^{2+}	[65]

However, chitosan is insoluble in either water or solvent. Thus, diluted acids such as acetic acid and hydrochloric acid are used. When acid is added, the free amino groups are protonated and the biopolymer becomes fully soluble [66]. Most of the preparation techniques of chitosan rely on chemical processes for extracting the protein and removing of inorganic matter. The processes involved extraction by solvent, followed by oxidation of remaining residues [67]. Overall, the extraction of chitosan from raw material includes the following stages; (1) grinding of raw materials (processing), (2) translating the mineral components of raw material into the soluble form (demineralisation), (3) removing the protein fractions (deproteinisation) and (4) deacetylating of chitin in obtaining the chitosan (Figure 3).

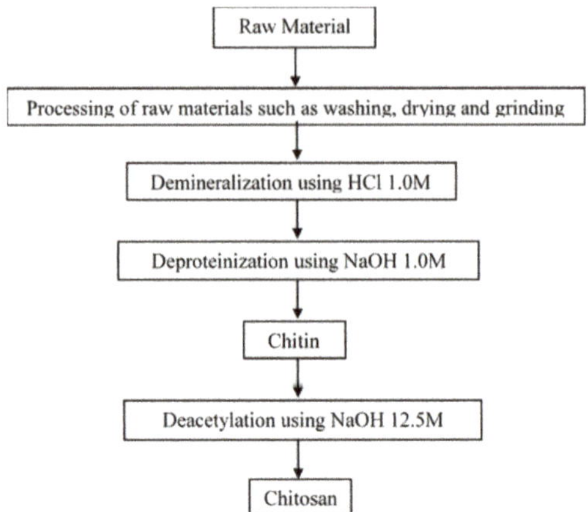

Figure 3. Flow of conventional preparation of chitosan.

The first step in the extraction of chitosan is the processing of raw materials, e.g., crab shell is removed from crab, washed, dried, grinded and filtered before it can proceed to demineralisation. During the demineralisation process, both metal ions and salt anions are removed via ion exchange. In this process, strong acid cation in the form of H^+ converts the dissolved salts into their conjugate acids. Demineralisation involves three sub-steps: (1) reaction of shell powder with hydrochloric acid to release carbon dioxide bubbles, followed by (2) washing using distilled water and (3) oven drying [59]. In addition to the removal of hardness in demineralisation stage, this process removes all dissolved solids such as sodium, silica, alkalinity and the mineral anions. Deproteinisation is carried out right after the demineralisation. By definition, deproteinisation is a process of removing protein and various enzymes in the sample prior to extraction of chitosan. As a cleaning agent, sodium hydroxide saponifies fats and dissolves proteins. Moreover, its hydrolysing power can be further enhanced with the presence of chlorine [68]. The change in colour of sodium hydroxide to clear indicates an index of full deproteinisation [59]. Prior to deacetylation stage, the precipitant must be drained and washed with distilled water repeatedly until its pH is dropped to neutral. Traditionally, deproteinisation and demineralisation steps are repeated twice to aid in higher yield of chitin from the shells. The last step is deacetylation, which refers to the process of removing acetyl groups. In general, alkali could be used to partially deacetylate chitin to produce a mixture of chitin and chitosan. As compared with chitin in terms of chemical structure, chitosan only lacks in acetyl group. Thus, deacetylation is a process of removing acetyl group. Deacetylation started by dissolving the demineralised and deproteinised product (chitin) in high concentration of sodium hydroxide. Heating can be introduced to increase the degree of deacetylation to produce the final product of chitosan. The product can be tested with acetic acid, in which the solubility of the resulting product in acetic acid will indicate a high degree of deacetylation [59].

4. Strategy to Enhance Performance of Natural Coagulants in Microalgae Harvesting

After the extraction processes, the final end product is the natural coagulant (plant, animal or microbes). Prior to application in coagulation and flocculation, the characterisation of natural coagulant is vital. Modification of the characteristic of natural coagulant could help in improving its performance in terms of flocculating activity in microalgae harvesting. Table 3 shows the physical, chemical and thermal characteristics of various natural coagulants. Additionally, the performance of various natural coagulants in different application is tabulated in Table 3. Subsequently, the interpretation of these characteristic in related to flocculating activity and their roles in enhancing the performance of natural coagulant in microalgae harvesting are discussed.

Table 3. Characterisation of natural coagulants and its performances.

Natural Coagulant	Surface Morphology	Surface Charge	Molecular Weight	Functional Group	Elemental Property	Thermogravimetry Analysis	Differential Scanning Calorimetry	Performance	Reference
Banana peel (*Musa acuminate*)	-N/A-	-N/A-	-N/A-	C=O, O-H, N–H	-N/A-	-N/A-	-N/A-	0.4 g·L^{-1} dosage, 67% removal of chemical oxygen demand (COD) from municipal wastewater	[69]
Banana pith	-N/A-	-N/A-	-N/A-	O-H, C-H, C-OO, C-H, COOH	O (44%), C (32%), (36 %), H (4.2%), N (1.5%), S (0.86%)	-N/A-	-N/A-	0.1 kg·m^{-3} dosage, pH 4, 99% removal of COD from river water	[70]
Brachystegia eurycoma extract	Compact structure with dispersed but continuous crack-like openings, absence cf irregular surfaces, randomly formed aggregates and/or loosely bcund cluster	-N/A-	-N/A-	O–H, N–H, O=H, C–N, C≡C, C=C–H and H–C–H	-N/A-	334.44 °C to 361.73 °C	−1.708 mV	5 g·L^{-1} dosage, pH 8, 97% removal of COD from paint wastewater	[71]
Brassica spp. seed protein	Pollen grain surface	−6.8 mV	6.5 kDa	-N/A-	-N/A-	95 °C	-N/A-	-N/A-	[29,72]
Cassava peel starch	Polygonal and spherical starch granules, rough surface	-4.37 mV	1.057×10^5 kDa	O-H, C-H	Ca, K and Na	-N/A-	-N/A-	7.5 mg·L^{-1} dosage, pH 7, 93% removal of total suspended solid (TSS) from dam water 50 mg·L^{-1} dosage, pH 7, 100% removal of *E. coli* from dam water	[73]

Table 3. *Cont.*

Natural Coagulant	Surface Morphology	Surface Charge	Molecular Weight	Functional Group	Elemental Property	Thermogravimetry Analysis	Differential Scanning Calorimetry	Performance	Reference
Cactus leaves	Presence of cracks and cavities	-N/A-	-N/A-	O-H, C=O, COOH	Na, K, Ca, Mg	-N/A-	-N/A-	$10~mg\cdot L^{-1}$ dosage, 90% removal of kaolin	[2,74]
Cassava Peel (periderm and cortex)	Non-porous and heterogeneous characteristics, smooth and globular in shape	-N/A-	-N/A-	O-H, CH, CH_2, C=O, C-O, COOH	K_2O (5.5%), CaO (4.2%), Fe_2O_3 (1.5%), SO_3 and SiO_2 (0.87%), Al_2O_3 (0.74%), C (0.10%),	-N/A-	-N/A-	-N/A-	[75]
Cassia obtusifolia seed gum	Fibrous networks with rough surface and porosity	-N/A-	-N/A-	O-H, C-H, C=O,	-N/A-	289 °C	-N/A-	$2.47~g\cdot L^{-1}$ dosage, 82% removal of TSS, settling time of 35.16 min	[76]
Ceratonia silique seed gums	Rough cuticle on the adaxial and the abaxial surface, stomatal pores	-N/A-	5–8 kDa	O-H	-N/A-	-N/A-	-N/A-	-N/A-	[29,77]
Chitin	Microporous, fish scale shaped nanofibrous surface	+18 mV	-N/A-	N-H, O-H, C-H, C=O	-N/A-	-N/A-	-N/A-	$0.3~g\cdot L^{-1}$ dosage, pH 6, 68% removal of turbidity from surface water	[78–80]
Chitosan extracted from lobster shell (*Thenus unimaculatus*)	Rough surface, irregular block, crystalline with cluster and porosity structure	-N/A-	-N/A-	$R-NH_2$, O-H	Ca, K, Na, Mg and Fe	-N/A-	-N/A-	-N/A-	[81,82]
Citrus *Limettioides* peels	Porous structure	-N/A-	-N/A-	CH, CH_2, CH_3, C=O, COOH, M(RCOO)n,	O, Na, Ca	-N/A-	-N/A-	-N/A-	[75,83]

Table 3. *Cont.*

Natural Coagulant	Surface Morphology	Surface Charge	Molecular Weight	Functional Group	Elemental Property	Thermogravimetry Analysis	Differential Scanning Calorimetry	Performance	Reference
C. obtusifolia seed gum	Rough, fibrous, porous and bulky	+6.41 mV	-N/A-	O-H, C-H, CH_3, CH_2	-N/A-	280–300 °C	-N/A-	19×10^{-3} mol gum, 6×10^{-2} mol of NaOH, 87% removal of TSS and 85% removal of COD from palm oil mill effluent (POME) at 50 °C	[84,85]
Cocos nucifera seed protein	Porous structure, clustered, aggregated shapes	-N/A-	5.6 kDa	O-H, N-H	-N/A-	-N/A-	-N/A-	10 g·L^{-1} dosage, 96% removal of As(III) in 8 h, 80 rpm and 50 °C	[29,86]
Cucumis melo peels	-N/A-	-N/A-	54 kDa	O-H, N-H, CH, CH_2, CH_3, C=O, R–COOH, M(RCOO)n, C-O or –C-N	-N/A-	-N/A-	-N/A-	0.5 g·L^{-1} dosage, pH 7, 91% removal of Mn(II) 0.5 g·L^{-1} dosage, pH 6.5, 91% removal of Pb(II)	[75,87,88]
Cyamopsis tetragonoloba seed gums	Nanoparticles	−6.66 mV	50–800 kDa	O-H	-N/A-	-N/A-	-N/A-	-N/A-	[29,89]
Dolichos lablab seed gums	Aggregated free, rough	-N/A-	-N/A-	N–H, O-H, C–H, C–C, –COOH	C, O	-N/A-	-N/A-	0.6 mL·L^{-1} dosage, pH 11, 99% removal of turbidity	[29,90]
Garden cress (*Lepidium Sativum*)	Flake-shaped structures with non-uniform distribution and emerged as interconnected channels, porous and heterogenous characteristics	−16 mV	-N/A-	O–H, C-H, C=O, OCH_3	-N/A-	-N/A-	-N/A-	15 mg·L^{-1} dosage, pH 5, 99% removal of turbidity from river water	[91]

Table 3. *Cont.*

Natural Coagulant	Surface Morphology	Surface Charge	Molecular Weight	Functional Group	Elemental Property	Thermogravimetry Analysis	Differential Scanning Calorimetry	Performance	Reference
Grafted 2-methacryloylo xyethyl trimethyl ammonium chloride lentil extract	More compact and less porous compared to lentil extract	+15.08 mV	-N/A-	-N/A-	C (62%), O (36%), Cl (2.0%)	-N/A-	-N/A-	5.09 mL·g^{-1} dosage, pH 10, 99% removal of turbidity in surface water and industrial wastewater	[92]
H. esculentus	Compact, cross linkage of molecules	-N/A-	100 kDa	O–H, C–H, C=O	-N/A-	180 °C	36.12 mV	-N/A-	[3,93]
Kenaf crude extract (KCE)	-N/A-	−8.3 mV	-N/A-	-N/A-	-N/A-	-N/A-	-N/A-	100 mg·L^{-1} dosage, 85% removal of kaolin, 40 mg·L^{-1}, 83% removal of turbidity from river water	[94]
Klebsiella pneumoniae	-N/A-	-N/A-	-N/A-	COO$^-$, O-H, N-H	C, N, O	-N/A-	-N/A-	pH 7, 40% removal of Cd	[2,95]
Lens culinaris	Rough surface with pores and obvious surface abrasions	−3.58 mV	-N/A-	O–H, C-H, COOH, C=O, C-O	C (60%), O (40%), K (0.39%)	-N/A-	-N/A-	26.3 mg·L^{-1} dosage, 99% removal of kaolin, 3 min settling time	[96]
Lentil extract	Highly porous surface, scattered pieces of compounds attached	−5.91 mV	-N/A-	O–H, C-H, C=O, N-H, C-O-C	C (59%), O (39%)	280 °C	-N/A-	-N/A-	[92]

Table 3. *Cont.*

Natural Coagulant	Surface Morphology	Surface Charge	Molecular Weight	Functional Group	Elemental Property	Thermogravimetry Analysis	Differential Scanning Calorimetry	Performance	Reference
Maerua decumbent	-N/A-	-N/A-	-N/A-	O-H, C-H, N-H, C=O, C-O, C-N	C (39%), O (42%) H (3.8%), N (1.2%), S (0.31%)	-N/A-	-N/A-	1 kg·m^{-3} dosage, pH 5.56, settling time 52.31 min, 99% removal of turbidity from paint industry wastewater 0.8 kg·m^{-3} dosage, pH 5.11, settling time 53.53 min, 79% removal of COD from paint industry wastewater	[1]
Malva nut gum	A branch-like surface structure	−58.7 mV	2.3×10^5 kDa	-N/A-	-N/A-	-N/A-	-N/A-	0.06 mg·L^{-1} dosage, pH 3.01, 97% removal of kaolin	[97]
Mango peels	Well-pronounced heterogeneous cavities that are well distributed	-N/A-	-N/A-	O-H, N-H, CH, CH$_2$, CH$_3$, C=O, C-O or –C-N	C, H, N, S	-N/A-	-N/A-	-N/A-	[75,98]
Moringa oleifera	Group-like, composed of many small particles	+6 mV	6.5 kDa	O-H, C-H, C=O, N-H, C-OH, S=O	-N/A-	-N/A-	-N/A-	50 mg·L^{-1} dosage, 94% removal of kaolin	[2,94,99,100]
Nirmali seeds	highly porous with reticulated structure	-N/A-	12 kDa	COOH, O-H	-N/A-	-N/A-	-N/A-	1.5 mg·L^{-1} dosage, 96% removal of turbidity from surface water	[2,101]

Table 3. *Cont.*

Natural Coagulant	Surface Morphology	Surface Charge	Molecular Weight	Functional Group	Elemental Property	Thermogravimetry Analysis	Differential Scanning Calorimetry	Performance	Reference
okra	Porous and rough	−8.3 mV	-N/A-	-N/A-	Mg (7.2%), Al (4.1%), Si (3.7%), P (11.8%), S (8.2%), Cl (7.7%), K (22.0%), Ca (7.5%), O (27.8%)	-N/A-	-N/A-	3 g·L^{-1} dosage, 85% removal of fluoride from hydrofluoric acid synthetic wastewater 20 mg·L^{-1} dosage, 94% removal of kaolin, 40 mg·L^{-1} dosage, 98% removal of turbidity from river water	[99]
Prosopis spp. seed gums	Homogenous in size and shape with a flake-like morphology	-N/A-	62 kDa	-N/A-	-N/A-	Ca, Mg, Fe, Zn	-N/A-	-N/A-	[29,94,102]
Sabdariffa crude extract (SCE)	-N/A-	−6.4 mV	-N/A-	-N/A-	-N/A-	-N/A-	-N/A-	60 mg·L^{-1} dosage, 88% removal of kaolin, 40 mg·L^{-1} dosage, 96% removal of turbidity from river water	[94]
Sago	Smooth and solid surface with no pores	-N/A-	-N/A-	N-H, O-H, C=O	-N/A-	-N/A-	-N/A-	0.1 g·L^{-1} dosage, pH 7, 69% removal of turbidity from surface water	[78]
Tannin	-N/A-	−13.6 mV	1250 kDa	O-H, R-NH^2, C=O, COOH	-N/A-	200 °C	-N/A-	14 mg·L^{-1} dosage, 75% removal from kaolin 11 mg·L^{-1} dosage, pH 5 to 7, 97% removal of *Chlorella vulgaris*	[2,29,103]

Table 3. *Cont.*

Natural Coagulant	Surface Morphology	Surface Charge	Molecular Weight	Functional Group	Elemental Property	Thermogravimetry Analysis	Differential Scanning Calorimetry	Performance	Reference
Tamarindus indica seed gums	No fissures, cracks or interruptions	-N/A-	700–880 kDa	-N/A-	-N/A-	97.67 °C	128.40 J/g	15 ppm dosage, 94% removal of turbidity from river water	[29,104,105]
Telfairia occidentalis seed	Coarse fibrous substance largely composed of cellulose and lignin, presence of pores (micro-, macro- and mesopores, compact net structure	-N/A-	-N/A-	O-H, N-H, C=H	-N/A-	-N/A-	-N/A-	247.40 mg·L^{-1} dosage, pH 2, 99% removal of dye in 34.32 mg·L^{-1} concentration with 540 settling time	[106,107]
T. foenum graecum seed gums	-N/A-	-N/A-	32.3 kDa	O-H, C-H, C=O, N-H, C-OH, C-O-C	C,O	295 °C to 430 °C	-N/A-	-N/A-	[29,108]
Vegetable tannin	-N/A-	-N/A-	-N/A-	-N/A-	-N/A-	430 °C	-N/A-	pH 7, removal of color and turbidity from dairy wastewater	[109]
Vigna unguiculata seed proteins	Fairly uniform, hexagonal structure, spiked or rugged surface, rough surface, coarse fibrous	-N/A-	6 kDa	O-H, N-H, C=O, C=C-H, C=CH, C-H	-N/A-	-N/A-	-N/A-	256.09 mg·L^{-1} dosage, pH 2, 99% removal of dye of 16.7 mg·L^{-1} with 540 min settling time	[29,106,107,110]

Note: -N/A- denotes unavailable.

4.1. Physical Characteristics

The most important physical aspects of natural coagulant that could be studied are surface morphology and surface charges. Surface morphology refers to the imaging of an exposed surface of any object under the microscope, which cannot be seen by the naked eye. By analysing the surface morphology, the active groups attributed to flocculation function can be identified, for example, the citral. According to the Essential Oil-Bearing Grasses, the genus Cymbopogon by Akhila, the oil in citral would help in the blood coagulation–fibrinolysis system [111]. Besides, citral is an antimicrobial element that will protect coagulant such as chitosan from microbial damage [112]. Moreover, the presence of pores (micro-, macro- and mesopores) on natural coagulant could be clearly identified via surface morphology analysis, and they are favourable for the attachment of suspended particles through adsorption, intraparticle bridging or electrostatic contacts during coagulation and flocculation. In addition, the previous study by Obiora-Okafo and Onukwuli [107] proved that a compact net structure coagulant showed higher flocculating activity as compared with a branched structure. Furthermore, changes to the surface morphology of coagulants after coagulation and flocculation show proof of interaction between the coagulants and suspended particles. In view of surface morphology as a strategy to enhance the flocculating activity, modification on physical structures such as grafting could be done to create a high density of pores and ultimately more favourable to coagulation. With these, the mass harvesting of microalgae in the industrial scale is applicable.

On the other hand, surface charge, or zeta potential, is one of the factors that will affect the flocculating activity. Theoretically, zeta potential is the measure of the electrical charge of particles that are suspended in liquid [113]. Practically, the higher the negative surface charge of natural coagulant, the greater it's flocculating activity against positive suspended particles and vice versa for the positive surface charge of natural coagulant against negatively suspended particles. Thus, the study of surface charge shows a preliminary estimation of flocculating activity of natural coagulant. Besides, the nature of surface charge (positive or negative) indicates the potential treated group of suspended particles, to illustrate, a negatively charged coagulant is used to remove cation heavy metals or the other way round. Chemically and structurally modified of natural coagulant such as quaternary agent 3-chloro-2-hydroxypropyltrimethylammonium chloride (CHPTAC) grafted on cellulose nanocrystals (CNC) could be applied to enhance the zeta potential to extreme positive or negative [114]. Above all, natural coagulant with positive zeta potential is favourable in microalgae harvesting due to the anionic nature of microalgae.

Moreover, different molecules of the same compound could have different molecular masses because they contain different isotopes with different mass number. The physical aspect of coagulants, molecular weight, could reflect their flocculating mechanism and activity. Yin [2] noted that high molecular weight of natural coagulant played a role in improving aggregation. The higher the molecular weight of natural coagulant, the stronger the bridge formed onto the particle surface than natural coagulant with a lower molecular weight. Thus, the formed flocs were stronger, larger and denser for a larger molecular weight natural coagulant and permitted better settling, also improving the harvesting efficiency [115]. Additionally, the high molecular weight allows natural coagulant's chains to stretch sufficiently far from the particle surfaces; thus, favourable for bridging to form [81]. Another study by Muylaert et al. [116] also showed that the high molecular weight polyelectrolytes (i.e., lignosulfonate) were a better bridging agent. On the other hand, the molecular mass of natural coagulant often reveals its undergoing mechanism in flocculation, for example, the lower molecular weight of natural coagulants, such as polyethyleneamine are usually undergoing flocculation via the charge patch mechanism [116]. It had also been well reported that the high molecular weight of natural coagulant would usually predominant in bridging mechanisms. Yin [2] also suggested that the dimeric cationic proteins with the molecular mass of 12–14 kDa and isoelectric point (pI) between 10 and 11 were predominant in adsorption and charge neutralisation mechanisms. Therefore, by studying the molecular weights of natural coagulants in advance prior to the application, the underlying coagulation mechanism of natural coagulant could be defined and modification could be made based on their respective mechanism. All in all, by knowing the molecular weight, the same compounds can be operated as dispersants (e.g., dextrin, low molecular weight) or coagulants (e.g., starch, high molecular weight). Generally, a dispersant is used to prevent fine particles from aggregating and normally being utilised in a selective flocculation process, in which gangue minerals are dispersed while flocculating valuable or desired minerals [117]. Such approach is suitable to be used in microalgae harvesting.

4.2. Chemical Characteristics

The flocculating activity of natural coagulants also depends on the specific chemical properties of the polymer. One of the key polymer characteristics includes various functional groups. The particular functional groups to be evaluated are COO^- and OH^- as their existence usually contributes to the flocculating activity of natural coagulant. Besides, the increase in positively charged functional groups allows more interactions with the negatively charged suspended particles, and thus improve the binding capabilities of natural coagulants [116]. Modification on functional groups of natural coagulants is also proposed and evinced by researchers in the past studies to increase the flocculating activity. For example, functionalising of cationic starch and TANFLOC, in which, the starch and tannins added with quaternary ammonium groups to increase the flocculating activity and serve as the low-cost as well as more effective alternatives for flocculation process [116]. Additionally, natural coagulants often perform poorly in harvesting marine microalgae [118]. The underlying reason is the high ionic strength of seawater will cause coiling, and this will decrease the effective size of natural coagulants. Therefore, an alternative had been proposed to modify the structure to a more rigid molecule such as tannin-based natural coagulants or functionalised nanoparticles, namely, nanocellulose [116]. Furthermore, in microalgae harvesting, the functional group of natural coagulants can be furthered enhanced with magnetoresponsive Fe_3O_4 nanoparticle to separate the flocculated microalgae from the medium using a magnetic field [119]. To summarise, the modification of functional group begins with characterisation of natural coagulant, which is an important factor that influences the effectiveness of natural coagulant in microalgae harvesting.

Elemental property of natural coagulant affects the flocculating activity. The trivalent cation is the most efficient in flocculating the negatively charged suspended particles. However, trivalent cation is commonly found only in inorganic coagulant such as alum. In plant-based coagulants, divalent cation is predominant instead. Besides, numerous studies have shown that when there are more phenolic groups available in a tannin structure, the coagulation capability could be enhanced [2]. Correspond to this statement, it was reported that the legume-based coagulant was rich in phenolic compounds, and it had also proven to exhibit antibacterial property [3]. These could aid in removing pathogenic bacteria such as *Salmonella parathyphi* that is presented in wastewater due to the leaching of sewage effluents. Thus, phenolic groups provide the –OH group not only for bridging, but to indirectly inactivate the pathogenic bacteria in the wastewater [3]. Therefore, the phenolic group deserves attention in wastewater treatment as well as microalgae harvesting, especially in extraction of DHA microalgae oil. Moreover, there is also one characteristic that has been ubiquitously used as a preselection criterion for new plant-based coagulants, namely, mucilage. Mucilage is a thick, gluey and adhesive substance produced by nearly all plants and some microorganisms. Evidently, the high bridging–coagulation capability of Opuntia with the presence of mucilage will promote the bounding action of particulates to mucilage without directly contact of particulate and has been widely used in water treatment in North America [2,120]. Besides, the recent study on biopolymer coagulant showed that 73% (from 320.0 to 88.0 mg·L^{-1}) of Fe^{3+} reduction and ~36% of COD removal with an addition of 3.20 mg·L^{-1} of okra mucilage during the harvesting process [120]. The presence of galacturonic acid in mucilage will act as an active coagulating agent and provide a bridge for particles adsorption. Further, the partial deprotonation of carboxylic functional group of mucilage in aqueous solution has given rise to the chemisorption between charged particle with COO^- and OH^- [2]. Therefore, it will aid in flocculating activity. To conclude, the selection of natural coagulant for microalgae harvesting should be focused on mucilage as its primary concern.

4.3. Thermal Characteristics

The thermal stability of natural coagulants is also a crucial parameter to be studied in enhancing the flocculating activity. Indeed, an optimum temperature will increase the flocculating activity. However, the temperature higher than 80 °C will usually destroy the chemical composition of natural coagulants [121]. Moreover, the temperature has direct effects on floc formation, breakage and reformation. To illustrate, floc formation is slower at a lower temperature, whereas breakage of floc is greater at higher temperatures. On the other hand, thermogravimetry analysis determines the minimum temperature causing decomposition of organic components in natural coagulant and differential scanning calorimetry allows study relating to the heat flow required to decompose the natural coagulant. In general, the thermal characteristics reveal the thermal stability of natural coagulant and it has no direct impact on microalgae harvesting because coagulation will not occur in extreme temperature.

5. Application of Natural Coagulant in Microalgae Harvesting

In the previous section, the extraction and characteristic of natural coagulant, as well as the strategies to enhance its flocculating activity, are reviewed. In this section, the application of natural coagulant in microalgae harvesting will be the focal point. To recall, alum always appears to be the first option in industrial applications when comes to the selection of coagulant for microalgae harvesting. The reason being, it is widely available, it promotes coagulation by neutralisation and most importantly it is ready to be dissolved with water.

However, the emerging usage of plant-based coagulant has achieved higher harvesting efficiency compared with chemical coagulant and there are reviews on their effectiveness and relevant coagulating mechanisms for the treatment of wastewater and microalgae harvesting [120,122,123]. To illustrate, the plant-based coagulant could be applied on microalgae harvesting at relatively low cost [124]. Compared to alum, the natural coagulant is deemed to be environmentally friendly because it is extracted from plants, animal or microbial and usually existed in non-toxic form [125]. The water soluble active compound in natural coagulant will be removed after several cycle of kidney filtration, leaving less possibility of producing toxicity in the body [126]. In view of sludge production after the harvesting process, natural coagulant does not produce suspended alum residual and indeed produces less organic residual due to its biodegradability. In contrast, alum requires chemical reaction to break down and will not decompose naturally. In a specific type of microalgae harvesting, for instance, extraction of DHA rich microalgae oil as a dietary supplement, natural coagulant appears to be the best option as it harvests a higher amount of microalgae biomass compared to alum and at the same time, it is safe for consumption. Thus, it will not pose any health concern even there is residual remained in algae biomass. The natural coagulant is proven to achieve higher flocculating activity in comparison to alum and their performance is shown in Tables 3 and 4. In addition, by utilising the natural coagulants, it reduces the alum dependency and ultimately achieves sustainability in the microalgae-based biofuel production industry as well as various fields, including wastewater treatment and medical to name a few. Figure 4 shows the advantages of natural coagulant in microalgae harvesting.

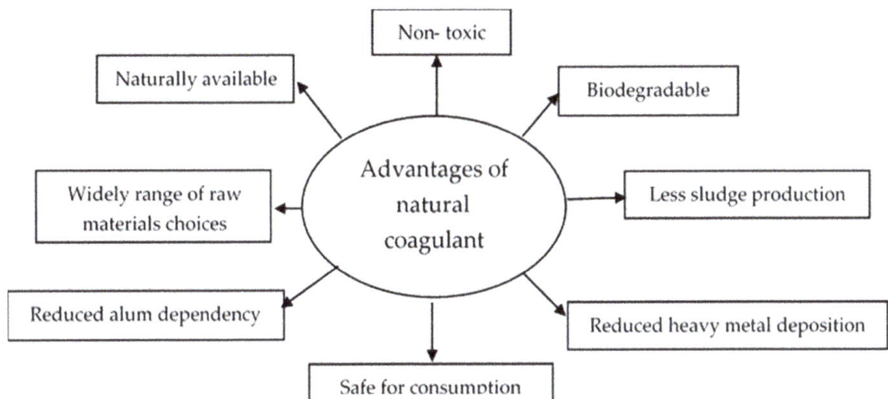

Figure 4. Advantages of utilising natural coagulant in microalgae harvesting.

Furthermore, natural coagulants have also been proven by other researchers as an effective way to harvest microalgae. It was found that the usage of bio-coagulants for harvesting microalgae could eliminate the toxicity contamination on harvested microalgae biomass [127]. The study carried out by Tran et al. [128] to harvest *Chlorella vulgaris* with alkyl-grafted chiton Fe_3O_4–SiO_2 showed 90% of biomass removal by merely employing $0.013 g \cdot L^{-1}$ dosage. On the other hand, a plant-based coagulant, *M. oleifera*, showed a 76% of harvesting efficiency on *Chlorella* sp. biomass after 100 min with $8 \ mg \cdot L^{-1}$ dosage and 96% of harvesting efficiency in 20 min when combining *M. oleifera* with chitosan [129]. Furthermore, 60% of microalgae removal efficiency was achieved with $12 \ mg \cdot mL^{-1}$ of *F. indica* extract after 120 min of settling time [130]. To sum up, the utilisation of natural coagulants in microalgae harvesting is a trend of research in the past few years. Unfortunately, it was set up and investigated merely at a laboratory scale. Table 4 shows the application of natural coagulants on microalgae harvesting.

Table 4. Application of microalgae harvesting using natural coagulants.

Natural Coagulant	Operating Condition	Performance	Reference
Alkyl-grafted chiton Fe_3O_4–SiO_2	$0.013 \ g \cdot L^{-1}$ dosage	90% removal of *Chlorella vulgaris*	[128]
M. oleifera	$8 \ mg \cdot L^{-1}$ dosage	76% removal of *Chlorella vulgaris*	[129]
M.oleifera with chitosan	$8 \ mg \cdot L^{-1}$ dosage	96% removal of *Chlorella vulgaris*	[129]
F. indica	$12 \ mg \cdot mL^{-1}$ dosage	60% removal of microalgae	[130]
Pleurotus ostreatus strain HEI-8	pH 3, glucose content $20 \ g \cdot L^{-1}$, fungi pelletisation time 7 days, 100 rpm	65% removal of *Chlorella* sp.	[131]
Citrobacter freundii (No. W4) and *Mucor circinelloides*	pH 7, glucose concentration $1.47 g \cdot L^{-1}$	97% removal of *Chlorella pyrenoidosa*	[132]
Tannin	$11 \ mg \cdot L^{-1}$ dosage, pH 5 to 7	97% removal of *Chlorella vulgaris*	[133]
Tannin	$5 \ mg \cdot L^{-1}$ dosage, pH 7	80% removal of *Oocystis* microalgae	[134]
Eucalyptus globulus	$20 \ mg \cdot L^{-1}$ dosage	95% removal of Scenedesmus sp.	[135]
Cassia gum	$80 \ mg \cdot L^{-1}$ dosage	93% removal of *Chlamydomonas* sp.	[136]
Cassia gum	$35 \ mg \cdot L^{-1}$ dosage	92% removal of *Chlorella* sp.	[136]

As an additional point, statistical modelling approaches could be studied to identify the optimum operating condition of natural coagulant. After several trials in the coagulation process, a statistical approach such as linear regression method is feasible in extracting the optimum parameters of natural coagulant in coagulation with collected data and equations.

6. Cost analysis of Natural Coagulants in Microalgae Harvesting

In general, the natural coagulants can be utilised for various applications as demonstrated in Figure 5.

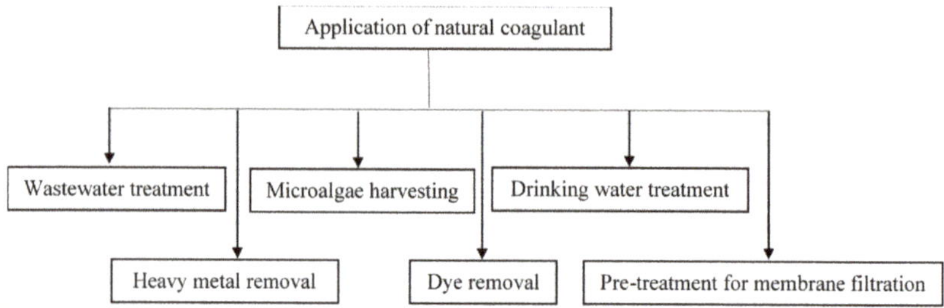

Figure 5. Potential applications of natural coagulants.

For specific instances, natural coagulants reduce suspended solids, traps *E. coli*, reduces turbidity, removes COD, adsorbs heavy metals, harvests microalgae, decolorises dye and others. With regards to the preparation stages, the natural coagulants derived either from plant, animal or microbial feedstock can be facilely produced as opposed to chemical-based coagulant, namely, alum [2]. Moreover, natural coagulants are also more sustainable off late, thus research should be intensified on the exploration of new natural coagulants to substitute the conventional alum. Nonetheless, it is postulated that an abundance of new natural coagulants is yet to be discovered.

On another note, the main drawback of utilising natural coagulant in industries is their low availability for large scale employment [2] as compared with alum. It had been reported by Mubarak et al. [127] on the suitability for large-scale application of natural coagulant and it is limited by the cost of preparation. This has directly led to the necessity of cost assessment on life cycle and cost analysis of different natural coagulants and to compare with alum as depicted in Table 5 [127,129].

Table 5. Comparison of cost analyses between natural coagulants and alum to harvest microalgae [127].

Coagulant	Energy Consumption (Mega Joule per Metric Tons, MJ/MT of Microalgae)	Greenhouse Gas (GHG) Emission (kg CO_2 eqv/MT of Microalgae)	Cost Analysis ($/MT)
Chitosan	300	70	9.02
Alum	200	50	0.28
Plant-based coagulant	175	40	0.037

Although the cost analysis by Behera and Balasubramanian [129] showed that the plant-based natural coagulant was relatively cheaper as compared with alum and chitosan in harvesting with a basis of a unit MT of microalgae, it only covered the cost of the harvesting process. As a matter of fact, the extraction process is generally time-extensive. A good illustration has been presented in Figure 2, in which the various processes are involved during the extraction of natural coagulants such as the addition of acid, the aid of equipment as well as the refining tertiary stage. Besides, the extraction is largely confined in the laboratory scale, which may not be feasible in terms of process scalability for industrial applications. An evaluation and approval from the local governing bodies are also part of concerns to commercialise the natural coagulants in the industry. Moreover, the overall costing must take into account of the stringent screenings and documentations that are needed to ensure product compliance to the respective standards [16]. Though the presented costing values in Table 5 are exclusively limited to only the harvesting process, the commercialisation and regulatory authorities are new inputs and cost-effective extraction techniques are vital to scale up the application of natural coagulants in the future. Further, researchers should pay close attention to the costing of natural coagulants from the primary stage, which involves the plant, animal and microbe selection to the final end product, i.e., plant-based, animal-based and microbe-based coagulant. Exploration in further research should be focused on economical extraction technology of natural coagulant to replace the alum in the near future.

7. Potentially New Natural Coagulant Yet to Be Exploited and Applied

To summarise, our study provides an additional list of potential new natural coagulant to be studied in the future. To recall, mucilage is a criterion of selection for new natural coagulant and it attributed to COO^- and OH^- functional groups, which are mainly associated with the flocculating activity. Besides, it has been espoused in Section 4.2 that galacturonic acid in mucilage is the active component that aids the coagulation–flocculation. Therefore, the most reliable method to predetermine the potentially new natural coagulant is to study the chemical composition, galacturonic acid, in each natural coagulant. A previous study [137] noted that pectic acid (polygalacturonic acid) extracted from sugar beet pectin comprises approximately 68 percent of the galacturonic acid. Moreover, pectic acid from flax pectin was found out to be made up of 61 percent of galacturonic acid, and the pectic acid from orange peelings was composed of 73.7 percent of galacturonic acid [137]. To sum up, natural coagulant often been extracted from materials with galacturonic acid as its chemical composition. Thus, the chemical components, galacturonic acid in mucilage is a point of interest in the selection of potentially new natural coagulant. Table 6 shows the potentially new natural coagulant.

Table 6. Potentially new natural coagulant.

Possible Natural Coagulant	Scientific Name	Reference
Cowpea	*Vigna unguiculata*	[138]
Chia seeds	*Salvia hispanica L.*	[139]
Rockcress	*Arabidopsis thaliana*	[140]
Quince seed	*Cydonia oblonga*	[141]
Jujube	*Ziziphus mauritiana Lam*	[142]
Seashore mallow	*Kosteletzkya virginica*	[143]
Watershield	*Brasenia schreberi*	[144]
Beet root	*Alyssum homolocarpum*	[145]
Levant wormseed	*Artemisia sphaerocephala*	[146]
Fenugreek seed	*Trigonella foenum-graecum L.*	[147]
Cress seed	*Lepidium sativum*	[148]

8. Conclusions

The usage of natural coagulants derived from plant, animal and microbial sources in industry is a trend of sustainable environmental development in the 21st century. Particularly, natural coagulant should be given priority in microalgae harvesting as it is highly effectual in flocculating activity and will not leave a negative impact to the end product due to its biodegradability.

On the other hand, the extraction processes differ for each type of natural coagulant. A comprehensive review is conducted to explain and justify the necessity to carry out each sub-step in the extraction process of natural coagulant. This information is useful in the exploration of new natural coagulant in the future.

Furthermore, the characterisation of natural coagulants is vital in enhancing their flocculating activity. The modification such as grafting could be used to increase or decrease the zeta potential and to provide more functional groups for attachment, which fundamentally enhancing the flocculating activity of natural coagulant. Moreover, molecular weight determines the coagulation mechanism of natural coagulant, for example, high molecular weight of natural coagulant (when they are more than 1×10^6 kDa) would usually predominant in bridging mechanisms. The functional group of natural coagulant identifies the effective groups, which help in the coagulation–flocculation process, typically O-H and C-H groups.

The applications of natural coagulant in the industry are summarized in this review, for instance, wastewater and drinking water treatment, heavy metal and dye removal and pretreatment for membrane filtration and microalgae harvesting. In view of the current studies, there is no doubt that the application of natural coagulant in microalgae harvesting will play a significant role in upscaling for mass production. To illustrate this, chitosan requires only 0.013 $g \cdot g^{-1}$ algae dosage to remove 90% of *C. vulgaris* as compared to 0.101 $g \cdot g^{-1}$ of polyaluminium chloride to remove 93% of algae [128] or 30 $mg \cdot L^{-1}$ of alum (aluminium sulphate) to remove 95% of *C. vulgaris* [149]. Importantly, using chitosan as coagulant does not inhibit the downstream process of transesterification biodiesel production on both the enzyme and chemical catalysed while other coagulants do [128]. Furthermore, tannin requires 11 $mg \cdot L^{-1}$ dosage to remove 97% of *C. vulgaris* [133], and *M. oleifera* with chitosan requires 8 $mg \cdot L^{-1}$ dosage to remove 96% removal of *C. vulgaris* [129]. These results show that natural coagulant is efficient in harvesting algae with a relatively lower dosage than alum. Additionally, natural coagulant deserves an attention on the harvesting of microalgae that produced DHA oil due to its non-toxic and non-chemical nature.

As noted in Section 2, the extractions of plant-based coagulants require specialised knowledge in identifying the potential plants as a coagulant and require performing detailed extraction stages, which are time-consuming. Common problems are also encountered in the preparation of animal- and microbial-based coagulants. At this point, the mass production of natural coagulants is still economically infeasible due to its complexity in bulk processing, low-volume in market demands and lack of supportive regulation that stipulates the quality of the natural coagulant extracts [2]. In view of this, the natural coagulant is currently restricted to small-scale usage and academic research, but it has the potential, especially for bulk microalgae harvesting in industries. Moreover, optimisation of natural coagulant based on their respective characteristic will further enhance its efficiency in coagulation and the result will be significant regarding mass harvesting of microalgae. The key effort of this paper includes the production of economical and sustainable natural–organic coagulants in the future.

Author Contributions: Conceptualisation, T.-H.A., J.W.L. and Y.-C.H.; literature review, T.H.A., M.J.K.B., Y.-C.H. and S.-C.C.; writing—original draft, T.H.A., K.K., J.W.L. and Y.-C.H.; writing—review and editing, W.K., P.-L.S. and S.-C.C.; funding acquisition, K.K. All authors have read and agreed to the published version of the manuscript.

Funding: Kunlanan Kiatkittipong wishes give thanks for the financial support received from the King Mongkut's Institute of Technology Ladkrabang, KMITL with the Grant no. KREF046209.

Acknowledgments: The authors would like to extend sincere thanks to all participants in the different rounds of consultations. The authors would like to express deepest gratitude to Mdm. Norhayama Bt Ramli for technical assistance and Universiti Teknologi PETRONAS for providing laboratory facilities. The authors would also like to thank the reviewers for all their comments. Moreover, the financial support received from the King Mongkut's Institute of Technology Ladkrabang, KMITL with the Grant no. KREF046209 is gratefully acknowledged.

Conflicts of Interest: The authors declare no conflict of interest.

References

1. Kakoi, B.; Kaluli, J.W.; Ndiba, P.; Thiong'o, G. Optimization of Maerua Decumbent bio-coagulant in paint industry wastewater treatment with response surface methodology. *J. Clean. Prod.* **2017**, *164*, 1124–1134. [CrossRef]
2. Yin, C.-Y. Emerging usage of plant-based coagulants for water and wastewater treatment. *Process Biochem.* **2010**, *45*, 1437–1444. [CrossRef]
3. Choy, S.Y.; Prasad, K.M.N.; Wu, T.Y.; Ramanan, R.N. A review on common vegetables and legumes as promising plant-based natural coagulants in water clarification. *Int. J. Environ. Sci. Technol.* **2013**, *12*, 367–390. [CrossRef]
4. Katarzyna, L.; Sai, G.; Singh, O.A. Non-enclosure methods for non-suspended microalgae cultivation: Literature review and research needs. *Renew. Sustain. Energy Rev.* **2015**, *42*, 1418–1427. [CrossRef]

5. Gajda, I.; Stinchcombe, A.; Greenman, J.; Melhuish, C.; Ieropoulos, I. Microbial fuel cell—A novel self-powered wastewater electrolyser for electrocoagulation of heavy metals. *Int. J. Hydrog. Energy* **2017**, *42*, 1813–1819. [CrossRef]

6. Razali, M.; Kim, J.; Attfield, M.; Budd, P.; Drioli, E.; Lee, Y.M.; Szekely, G. Sustainable wastewater treatment and recycle in membrane manufacturing. *Green Chem.* **2015**, *17*, 5196–5205. [CrossRef]

7. Senthil Kumar, P. Adsorption of lead(II) ions from simulated wastewater using natural waste: A kinetic, thermodynamic and equilibrium study. *Environ. Prog. Sustain. Energy* **2014**, *33*, 55–64. [CrossRef]

8. Cseri, L.; Baugh, J.; Alabi, A.; AlHajaj, A.; Zou, L.; Dryfe, R.A.; Budd, P.M.; Szekely, G. Graphene oxide–polybenzimidazolium nanocomposite anion exchange membranes for electrodialysis. *J. Mater. Chem. A* **2018**, *6*, 24728–24739. [CrossRef]

9. Mehrotra, T.; Srivastava, A.; Rao, P.R.; Singh, R. A Novel Immobilized Bacterial Consortium Bioaugmented in a Bioreactor For Sustainable Wastewater Treatment. *J. Pure Appl. Microbiol.* **2019**, *13*, 371–383. [CrossRef]

10. Mu, B.; Hassan, F.; Yang, Y. Controlled assembly of secondary keratin structures for continuous and scalable production of tough fibers from chicken feathers. *Green Chem.* **2020**, *22*, 1726–1734. [CrossRef]

11. Le Phuong, H.A.; Izzati Ayob, N.A.; Blanford, C.F.; Mohammad Rawi, N.F.; Szekely, G. Nonwoven Membrane Supports from Renewable Resources: Bamboo Fiber Reinforced Poly(Lactic Acid) Composites. *ACS Sustain. Chem. Eng.* **2019**, *7*, 11885–11893. [CrossRef]

12. Zhang, C.; Zhang, Y.; Xiao, X.; Liu, G.; Xu, Z.; Wang, B.; Yu, C.; Ras, R.H.; Jiang, L. Efficient separation of immiscible oil/water mixtures using a perforated lotus leaf. *Green Chem.* **2019**, *21*, 6579–6584. [CrossRef]

13. Zhu, C.; Huo, D.; Chen, Q.; Xue, J.; Shen, S.; Xia, Y. A Eutectic Mixture of Natural Fatty Acids Can Serve as the Gating Material for Near-Infrared-Triggered Drug Release. *Adv. Mater.* **2017**, *29*, 1703702. [CrossRef] [PubMed]

14. Fei, F.; Le Phuong, H.A.; Blanford, C.F.; Szekely, G. Tailoring the Performance of Organic Solvent Nanofiltration Membranes with Biophenol Coatings. *ACS Appl. Polym. Mater.* **2019**, *1*, 452–460. [CrossRef]

15. Lin, P.-C.; Wong, Y.-T.; Su, Y.-A.; Chen, W.-C.; Chueh, C.-C. Interlayer Modification Using Eco-friendly Glucose-Based Natural Polymers in Polymer Solar Cells. *ACS Sustain. Chem. Eng.* **2018**, *6*, 14621–14630. [CrossRef]

16. Choy, S.Y.; Choy, S.Y.; Choy, S.Y.; Raghunandan, M.E.; Ramanan, R.N. Utilization of plant-based natural coagulants as future alternatives towards sustainable water clarification. *J. Environ. Sci.* **2014**, *26*, 2178–2189. [CrossRef]

17. Saravanan, J.; Priyadharshini, D.; Soundammal, A.; Sudha, G.; Suriyakala, K. Wastewater Treatment using Natural Coagulants. *Int. J. Civ. Eng.* **2017**, *4*, 40–42. [CrossRef]

18. Ang, W.L.; Mohammad, A. State of the art and sustainability of natural coagulants in water and wastewater treatment. *J. Clean. Prod.* **2020**, *262*, 121267. [CrossRef]

19. Arnold, M.; Tainter, J.A.; Strumsky, D. Productivity of innovation in biofuel technologies. *Energy Policy* **2019**, *124*, 54–62. [CrossRef]

20. Lam, M.K.; Lee, K.T. Potential of using organic fertilizer to cultivate *Chlorella vulgaris* for biodiesel production. *Appl. Energy* **2012**, *94*, 303–308. [CrossRef]

21. Pandey, A.; Pathak, V.V.; Kothari, R.; Black, P.N.; Tyagi, V.V. Experimental studies on zeta potential of flocculants for harvesting of algae. *J. Environ. Manag.* **2019**, *231*, 562–569. [CrossRef]

22. Singh, A.; Olsen, S.I. A critical review of biochemical conversion, sustainability and life cycle assessment of algal biofuels. *Appl. Energy* **2011**, *88*, 3548–3555. [CrossRef]

23. Jayakumar, S.; Yusoff, M.M.; Rahim, M.H.A.; Maniam, G.P.; Govindan, N.J.R.; Reviews, S.E. The prospect of microalgal biodiesel using agro-industrial and industrial wastes in Malaysia. *Renew. Sustain. Energy Rev.* **2017**, *72*, 33–47. [CrossRef]

24. Salama, E.-S.; Kurade, M.B.; Abou-Shanab, R.A.; El-Dalatony, M.M.; Yang, I.-S.; Min, B.; Jeon, B.-H. Recent progress in microalgal biomass production coupled with wastewater treatment for biofuel generation. *Renew. Sustain. Energy Rev.* **2017**, *79*, 1189–1211. [CrossRef]

25. Abdullah, B.; Muhammad, S.A.F.a.S.; Shokravi, Z.; Ismail, S.B.; Kassim, K.A.; Mahmood, N.A.B.N.; Aziz, M.M.A. Fourth generation biofuel: A review on risks and mitigation strategies. *Renew. Sustain. Energy Rev.* **2019**, *107*, 37–50. [CrossRef]

26. Thomas, D.M.; Mechery, J.; Paulose, S.V. Carbon dioxide capture strategies from flue gas using microalgae: A review. *Environ. Sci. Pollut. Res.* **2016**, *23*, 16926–16940. [CrossRef]

27. Gerde, J.A.; Yao, L.; Lio, J.; Wen, Z.; Wang, T. Microalgae flocculation: Impact of flocculant type, algae species and cell concentration. *Algal Res.* **2014**, *3*, 30–35. [CrossRef]

28. Wong, Y.K. Growth Medium Screening for Chlorella vulgaris Growth and Lipid Production. *J. Aquac. Mar. Biol.* **2017**, 6. [CrossRef]

29. Saleem, M.; Bachmann, R.T. A contemporary review on plant-based coagulants for applications in water treatment. *J. Ind. Eng. Chem.* **2019**, *72*, 281–297. [CrossRef]

30. Zhu, L.; Li, Z.; Hiltunen, E. Microalgae Chlorella vulgaris biomass harvesting by natural flocculant: Effects on biomass sedimentation, spent medium recycling and lipid extraction. *Biotechnol. Biofuels* **2018**, *11*, 183. [CrossRef]

31. Ho, Y.-C.; Chua, S.-C.; Chong, F.-K. Coagulation-Flocculation Technology in Water and Wastewater Treatment. In *Handbook of Research on Resource Management for Pollution and Waste Treatment*; IGI Global: Hershey, PA, USA, 2020; pp. 432–457.

32. Chethana, M.; Sorokhaibam, L.G.; Bhandari, V.M.; Raja, S.; Ranade, V.V. Green Approach to Dye Wastewater Treatment Using Biocoagulants. *ACS Sustain. Chem. Eng.* **2016**, *4*, 2495–2507. [CrossRef]

33. Vijayaraghavan, G.; Sivakumar, T.; Adichakkravarthy, V. Application of plant based coagulants for waste water treatment. *Int. J. Adv. Eng. Res. Stud.* **2011**, *1*, 88–92.

34. Ndabigengesere, A.; Narasiah, K.S.; Talbot, B.G. Active agents and mechanism of coagulation of turbid waters using Moringa oleifera. *Water Res.* **1995**, *29*, 703–710. [CrossRef]

35. Epalza, J.; Jaramillo, J.; Guarín, O. Extraction and Use of Plant Biopolymers for Water Treatment. In *Desalination and Water Treatment*; IntechOpen: London, UK, 2018. [CrossRef]

36. Birima, A.H.; Hammad, H.A.; Desa, M.N.M.; Muda, Z.C. Extraction of natural coagulant from peanut seeds for treatment of turbid water. *IOP Conf. Ser. Earth Environ. Sci.* **2013**, *16*, 012065. [CrossRef]

37. Sowmeyan, R.; Santhosh, J.; Latha, R. Effectiveness of herbs in community water treatment. *Int. Res. J. Biochem. Bioinform.* **2011**, *1*, 297–303.

38. Ghebremichael, K.A.; Gunaratna, K.R.; Henriksson, H.; Brumer, H.; Dalhammar, G. A simple purification and activity assay of the coagulant protein from Moringa oleifera seed. *Water Res.* **2005**, *39*, 2338–2344. [CrossRef]

39. Shan, T.C.; Matar, M.A.; Makky, E.A.; Ali, E.N. The use of Moringa oleifera seed as a natural coagulant for wastewater treatment and heavy metals removal. *Appl. Water Sci.* **2016**, *7*, 1369–1376. [CrossRef]

40. Okuda, T.; Baes, A.; Nishijima, W.; Okada, M. Isolation and characterization of coagulant extracted from Moringa oleifera seed by salt solution. *Water Res.* **2001**, *35*, 405–410. [CrossRef]

41. Keeley, J.; Jarvis, P.; Judd, S.J. Coagulant Recovery from Water Treatment Residuals: A Review of Applicable Technologies. *Crit. Rev. Environ. Sci. Technol.* **2014**, *44*, 2675–2719. [CrossRef]

42. Meng, S.; Zhang, M.; Yao, M.; Qiu, Z.; Hong, Y.; Lan, W.; Xia, H.; Jin, X. Membrane Fouling and Performance of Flat Ceramic Membranes in the Application of Drinking Water Purification. *Water* **2019**, *11*, 2606. [CrossRef]

43. Ulmert, H.D. Method for Treatment of Sludge from Waterworks and Wastewater Treatment Plants. Google Patents, U.S. Patent No. 7,713,419, 11 May 2010.

44. Shukla, A.K.; Alam, J.; Alhoshan, M.; Dass, L.A.; Muthumareeswaran, M.R. Development of a nanocomposite ultrafiltration membrane based on polyphenylsulfone blended with graphene oxide. *Sci Rep.* **2017**, *7*, 41976. [CrossRef]

45. Katayon, S.; Ng, S.C.; Johari, M.M.N.M.; Ghani, L.A.A. Preservation of coagulation efficiency ofMoringa oleifera, a natural coagulant. *Biotechnol. Bioprocess Eng.* **2006**, *11*, 489–495. [CrossRef]

46. Kansal, S.K.; Kumari, A. Potential of M. oleifera for the treatment of water and wastewater. *Chem. Rev.* **2014**, *114*, 4993–5010. [CrossRef]

47. Ben Rebah, F.; Mnif, W.; Siddeeg, M.S. Microbial Flocculants as an Alternative to Synthetic Polymers for Wastewater Treatment: A Review. *Symmetry* **2018**, *10*, 556. [CrossRef]

48. Adebami, G.; Adebayo-Tayo, B. Comparative effect of medium composition on bioflocculant production by microorganisms isolated from wastewater samples. *Rep. Opin.* **2013**, *5*, 46–53.

49. Liu, W.; Cong, L.; Yuan, H.; Yang, J. The mechanism of kaolin clay flocculation by a cation-independent bioflocculant produced by *Chryseobacterium daeguense* W6. *AIMS Environ. Sci.* **2015**, *2*, 169–179. [CrossRef]

50. Li, Y.; Li, Q.; Hao, D.; Hu, Z.; Song, D.; Yang, M. Characterization and flocculation mechanism of an alkali-activated polysaccharide flocculant from *Arthrobacter* sp. B4. *Bioresour. Technol.* **2014**, *170*, 574–577. [CrossRef]

51. Li, Z.; Zhong, S.; Lei, H.-y.; Chen, R.-w.; Yu, Q.; Li, H.-L. Production of a novel bioflocculant by Bacillus licheniformis X14 and its application to low temperature drinking water treatment. *Bioresour. Technol.* **2009**, *100*, 3650–3656. [CrossRef]

52. Zaki, S.A.; Elkady, M.F.; Farag, S.; Abd-El-Haleem, D. Characterization and flocculation properties of a carbohydrate bioflocculant from a newly isolated Bacillus velezensis 40B. *J. Environ. Biol.* **2013**, *34*, 51–58.

53. Li, Y.; Xu, Y.; Song, R.; Tian, C.; Liu, L.; Zheng, T.; Wang, H. Flocculation characteristics of a bioflocculant produced by the actinomycete Streptomyces sp. hsn06 on microalgae biomass. *BMC Biotechnol.* **2018**, *18*, 58. [CrossRef]

54. Manivasagan, P.; Kang, K.H.; Kim, D.G.; Kim, S.K. Production of polysaccharide-based bioflocculant for the synthesis of silver nanoparticles by *Streptomyces* sp. *Int. J. Biol. Macromol.* **2015**, *77*, 159–167. [CrossRef]

55. Aljuboori, A.H.; Idris, A.; Abdullah, N.; Mohamad, R. Production and characterization of a bioflocculant produced by Aspergillus flavus. *Bioresour. Technol.* **2013**, *127*, 489–493. [CrossRef]

56. Li-Fan, L.; Cheng, W. Characteristics and culture conditions of a bioflocculant produced by *Penicillium* sp. *Biomed. Environ. Sci.* **2010**, *23*, 213–218.

57. Pu, S.Y.; Qin, L.L.; Che, J.P.; Zhang, B.R.; Xu, M. Preparation and application of a novel bioflocculant by two strains of Rhizopus sp. using potato starch wastewater as nutrilite. *Bioresour. Technol.* **2014**, *162*, 184–191. [CrossRef]

58. Fang, D.; Shi, C. Characterization and flocculability of a novel proteoglycan produced by Talaromyces trachyspermus OU5. *J. Biosci. Bioeng.* **2016**, *121*, 52–56. [CrossRef]

59. Rasti, H.; Parivar, K.; Baharara, J.; Iranshahi, M.; Namvar, F. Chitin from the Mollusc Chiton: Extraction, Characterization and Chitosan Preparation. *Iran. J. Pharm. Res. IJPR* **2017**, *16*, 366–379.

60. Ummalyma, S.B.; Mathew, A.K.; Pandey, A.; Sukumaran, R.K. Harvesting of microalgal biomass: Efficient method for flocculation through pH modulation. *Bioresour. Technol.* **2016**, *213*, 216–221. [CrossRef]

61. Riaño, B.; Molinuevo, B.; García-González, M.C. Optimization of chitosan flocculation for microalgal-bacterial biomass harvesting via response surface methodology. *Ecol. Eng.* **2012**, *38*, 110–113. [CrossRef]

62. Tavera, M.J. Extraction and Efficiency of Chitosan from Shrimp Exoskeletons as Coagulant for Lentic Water Bodies. *Int. J. Appl. Eng. Res.* **2018**, *13*, 1060–1067.

63. Han, N.T.; Trung, T.S.; Khanh Huye, N.T.; Minh, N.C.; Trang, T.T.L. Optimization of Harvesting of Microalgal Thalassiosira pseudonana Biomass Using Chitosan Prepared from Shrimp Shell Waste. *Asian J. Agric. Res.* **2016**, *10*, 162–174. [CrossRef]

64. Mohd Yunos, F.H.; Nasir, N.M.; Wan Jusoh, H.H.; Khatoon, H.; Lam, S.S.; Jusoh, A. Harvesting of microalgae (*Chlorella* sp.) from aquaculture biofloc using an environmental-friendly chitosan-based bio-coagulant. *Int. Biodeterior. Biodegrad.* **2017**, *124*, 243–249. [CrossRef]

65. Yang, K.; Wang, G.; Chen, X.; Wang, X.; Liu, F. Treatment of wastewater containing Cu2+ using a novel macromolecular heavy metal chelating flocculant xanthated chitosan. *Colloids Surf. A Phys. Eng. Asp.* **2018**, *558*, 384–391. [CrossRef]

66. Pontius, F.W. Chitosan as a Drinking Water Treatment Coagulant. *Am. J. Civ. Eng.* **2016**, *4*, 205–215. [CrossRef]

67. De Queiroz Antonino, R.; Lia Fook, B.R.P.; de Oliveira Lima, V.A.; de Farias Rached, R.I.; Lima, E.P.N.; da Silva Lima, R.J.; Peniche Covas, C.A.; Lia Fook, M.V. Preparation and Characterization of Chitosan Obtained from Shells of Shrimp (Litopenaeus vannamei Boone). *Mar. Drugs* **2017**, *15*, 141. [CrossRef]

68. Hyde, A.M.; Zultanski, S.L.; Waldman, J.H.; Zhong, Y.-L.; Shevlin, M.; Peng, F. Development. General principles and strategies for salting-out informed by the Hofmeister series. *Org. Process Res. Dev.* **2017**, *21*, 1355–1370. [CrossRef]

69. Maurya, S.; Daverey, A. Evaluation of plant-based natural coagulants for municipal wastewater treatment. *3 Biotech* **2018**, *8*, 77. [CrossRef]

70. Kakoi, B.; Kaluli, J.W.; Ndiba, P.; Thiong'o, G. Banana pith as a natural coagulant for polluted river water. *Ecol. Eng.* **2016**, *95*, 699–705. [CrossRef]

71. Menkiti, M.C.; Okoani, A.O.; Ejimofor, M.I. Adsorptive study of coagulation treatment of paint wastewater using novel Brachystegia eurycoma extract. *Appl. Water Sci.* **2018**, *8*, 189. [CrossRef]

72. Koen, J.; Slabbert, M.M.; Booyse, M.; Bester, C. Honeybush (Cyclopia spp.) pollen viability and surface morphology. *S. Afr. J. Bot.* **2020**, *128*, 167–173. [CrossRef]

73. Asharuddin, S.M.; Othman, N.; Zin, N.S.M.; Tajarudin, H.A.; Din, M.F.M. Flocculation and antibacterial performance of dual coagulant system of modified cassava peel starch and alum. *J. Water Process Eng.* **2019**, *31*, 100888. [CrossRef]

74. Wahab, M.A.; Boubakri, H.; Jellali, S.; Jedidi, N. Characterization of ammonium retention processes onto cactus leaves fibers using FTIR, EDX and SEM analysis. *J. Hazard. Mater.* **2012**, *241–242*, 101–109. [CrossRef]

75. Zainorizuan, M.J.; Mohd-Asharuddin, S.; Othman, N.; Mohd Zin, N.S.; Tajarudin, H.A.; Yee Yong, L.; Alvin John Meng Siang, L.; Mohamad Hanifi, O.; Siti Nazahiyah, R.; Mohd Shalahuddin, A. A Chemical and Morphological Study of Cassava Peel: A Potential Waste as Coagulant Aid. *MATEC Web Conf.* **2017**, *103*, 06012. [CrossRef]

76. Shak, K.P.Y.; Wu, T.Y. Optimized use of alum together with unmodified Cassia obtusifolia seed gum as a coagulant aid in treatment of palm oil mill effluent under natural pH of wastewater. *Ind. Crops Prod.* **2015**, *76*, 1169–1178. [CrossRef]

77. Kolyva, F.; Stratakis, E.; Rhizopoulou, S.; Chimona, C.; Fotakis, C. Leaf surface characteristics and wetting in Ceratonia siliqua L. *Flora Morphol. Distrib. Funct. Ecol. Plants* **2012**, *207*, 551–556. [CrossRef]

78. Saritha, V.; Karnena, M.K.; Dwarapureddi, B.K. "Exploring natural coagulants as impending alternatives towards sustainable water clarification"—A comparative studies of natural coagulants with alum. *J. Water Process Eng.* **2019**, 32. [CrossRef]

79. Kaya, M.; Sofi, K.; Sargin, I.; Mujtaba, M. Changes in physicochemical properties of chitin at developmental stages (larvae, pupa and adult) of Vespa crabro (wasp). *Carbohydr. Polym.* **2016**, *145*, 64–70. [CrossRef]

80. Peng, N.; Ai, Z.; Fang, Z.; Wang, Y.; Xia, Z.; Zhong, Z.; Fan, X.; Ye, Q. Homogeneous synthesis of quaternized chitin in NaOH/urea aqueous solution as a potential gene vector. *Carbohydr. Polym.* **2016**, *150*, 180–186. [CrossRef]

81. Lee, C.S.; Robinson, J.; Chong, M.F. A review on application of flocculants in wastewater treatment. *Process. Saf. Environ. Prot.* **2014**, *92*, 489–508. [CrossRef]

82. Arasukumar, B.; Prabakaran, G.; Gunalan, B.; Moovendhan, M. Chemical composition, structural features, surface morphology and bioactivities of chitosan derivatives from lobster (Thenus unimaculatus) shells. *Int. J. Biol. Macromol* **2019**, *135*, 1237–1245. [CrossRef]

83. Sudha, R.; Srinivasan, K.; Premkumar, P. Removal of nickel(II) from aqueous solution using Citrus Limettioides peel and seed carbon. *Ecotoxicol. Environ. Saf.* **2015**, *117*, 115–123. [CrossRef]

84. Shak, K.P.Y.; Wu, T.Y. Synthesis and characterization of a plant-based seed gum via etherification for effective treatment of high-strength agro-industrial wastewater. *Chem. Eng. J.* **2017**, *307*, 928–938. [CrossRef]

85. Subramonian, W.; Wu, T.Y.; Chai, S.-P. An application of response surface methodology for optimizing coagulation process of raw industrial effluent using Cassia obtusifolia seed gum together with alum. *Ind. Crops Prod.* **2015**, *70*, 107–115. [CrossRef]

86. Nashine, A.L.; Tembhurkar, A.R. Equilibrium, kinetic and thermodynamic studies for adsorption of As(III) on coconut (Cocos nucifera L.) fiber. *J. Environ. Chem. Eng.* **2016**, *4*, 3267–3273. [CrossRef]

87. Mohd Asharuddin, S. Optimization of Biosorption Process Using Cucumis Melo Rind for the Removal of Fe, Mn and Pb Ions from Groundwater. Master's Thesis, Universiti Tun Hussein Onn Malaysia, Parit Raja, Malaysia, 2015.

88. Devi, B. Isolation, Partial Purification And Characterization Of Alkaline Serine Protease From Seeds of Cucumis Melo Var Agrestis. *Int. J. Res. Eng. Technol.* **2014**, *3*, 88–97. [CrossRef]

89. Raghavendra, C.K.; Srinivasan, K. Influence of dietary tender cluster beans (Cyamopsis tetragonoloba) on biliary proteins, bile acid synthesis and cholesterol crystal growth in rat bile. *Steroids* **2015**, *94*, 21–30. [CrossRef]

90. Kahsay, M.H.; RamaDevi, D.; Kumar, Y.P.; Mohan, B.S.; Tadesse, A.; Battu, G.; Basavaiah, K. Synthesis of silver nanoparticles using aqueous extract of Dolichos lablab for reduction of 4-Nitrophenol, antimicrobial and anticancer activities. *OpenNano* **2018**, *3*, 28–37. [CrossRef]

91. Lim, B.-C.; Lim, J.-W.; Ho, Y.-C. Garden cress mucilage as a potential emerging biopolymer for improving turbidity removal in water treatment. *Process Saf. Environ. Prot.* **2018**, *119*, 233–241. [CrossRef]

92. Chua, S.C.; Chong, F.K.; Ul Mustafa, M.R.; Mohamed Kutty, S.R.; Sujarwo, W.; Abdul Malek, M.; Show, P.L.; Ho, Y.C. Microwave radiation-induced grafting of 2-methacryloyloxyethyl trimethyl ammonium chloride onto lentil extract (LE-g-DMC) as an emerging high-performance plant-based grafted coagulant. *Sci. Rep.* **2020**, *10*, 3959. [CrossRef]

93. Zaharuddin, N.D.; Noordin, M.I.; Kadivar, A. The use of Hibiscus esculentus (Okra) gum in sustaining the release of propranolol hydrochloride in a solid oral dosage form. *Biomed. Res. Int* **2014**, *2014*, 735891. [CrossRef]

94. Ndahi Jones, A. The Zeta Potential In Crude Extracts Of Some Hibiscus Plants For Water Treatment. *Arid Zone J. Eng. Technol. Environ.* **2018**, *14*, 85–93.

95. Khan, Z.; Hussain, S.Z.; Rehman, A.; Zulfiqar, S.; Shakoori, A.J.P.J.o.Z. Evaluation of cadmium resistant bacterium, Klebsiella pneumoniae, isolated from industrial wastewater for its potential use to bioremediate environmental cadmium. *Pak. J. Zool.* **2015**, *47*, 1533–1543.

96. Chua, S.C.; Malek, M.A.; Chong, F.K.; Sujarwo, W.; Ho, Y.C. Red Lentil (Lens culinaris) Extract as a Novel Natural Coagulant for Turbidity Reduction: An Evaluation, Characterization and Performance Optimization Study. *Water* **2019**, *11*. [CrossRef]

97. Ho, Y.C.; Norli, I.; Alkarkhi, A.F.; Morad, N. Extraction, characterization and application of malva nut gum in water treatment. *J. Water Health* **2015**, *13*, 489–499. [CrossRef]

98. Fauzi Abdullah, M.; Jawad, A.H.; Hanani Mamat, N.F.; Ismail, K. Adsorption of methylene blue onto acid-treated mango peels: Kinetic, equilibrium and thermodynamic. *Desalin. Water Treat.* **2016**, *59*, 210–219. [CrossRef]

99. Lim, W.L.K.; Chung, E.C.Y.; Chong, C.H.; Ong, N.T.K.; Hew, W.S.; Kahar, N.b.; Goh, Z.J. Removal of fluoride and aluminium using plant-based coagulants wrapped with fibrous thin film. *Process Saf. Environ. Prot.* **2018**, *117*, 704–710. [CrossRef]

100. Chen, C.; Zhang, B.; Huang, Q.; Fu, X.; Liu, R.H. Microwave-assisted extraction of polysaccharides from Moringa oleifera Lam. leaves: Characterization and hypoglycemic activity. *Ind. Crops Prod.* **2017**, *100*, 1–11. [CrossRef]

101. Kagithoju, S.; Godishala, V.; Nanna, R.S. Eco-friendly and green synthesis of silver nanoparticles using leaf extract of Strychnos potatorum Linn.F. and their bactericidal activities. *3 Biotech* **2015**, *5*, 709–714. [CrossRef]

102. Sennu, P.; Choi, H.-J.; Baek, S.-G.; Aravindan, V.; Lee, Y.-S. Tube-like carbon for Li-ion capacitors derived from the environmentally undesirable plant: Prosopis juliflora. *Carbon* **2016**, *98*, 58–66. [CrossRef]

103. Ibrahim, A.; Yaser, A.Z. Colour removal from biologically treated landfill leachate with tannin-based coagulant. *J. Environ. Chem. Eng.* **2019**, *7*, 103483. [CrossRef]

104. Rahman, M.M.; Sarker, P.; Saha, B.; Jakarin, N.; Shammi, M.; Uddin, M.K.; Sikder, M.T. Removal of turbidity from the river water using tamarindus indica and litchi chinensis seeds as natural coagulant. *Int. J. Environ. Prot. Policy* **2015**, *3*, 19. [CrossRef]

105. Alpizar-Reyes, E.; Carrillo-Navas, H.; Gallardo-Rivera, R.; Varela-Guerrero, V.; Alvarez-Ramirez, J.; Pérez-Alonso, C. Functional properties and physicochemical characteristics of tamarind (*Tamarindus indica* L.) seed mucilage powder as a novel hydrocolloid. *J. Food Eng.* **2017**, *209*, 68–75. [CrossRef]

106. Obiora-Okafo, I. Characterization and removal of colour from aqueous solution using bio-coagulants: Response surface methodological approach. *J. Chem. Technol. Metall.* **2019**, *54*, 1.

107. Obiora-Okafo, I.; Onukwuli, O. Characterization and optimization of spectrophotometric colour removal from dye containing wastewater by Coagulation-Flocculation. *Pol. J. Chem. Technol.* **2018**, *20*, 49–59. [CrossRef]

108. Radini, I.A.; Hasan, N.; Malik, M.A.; Khan, Z. Biosynthesis of iron nanoparticles using Trigonella foenum-graecum seed extract for photocatalytic methyl orange dye degradation and antibacterial applications. *J. Photochem. Photobiol. B* **2018**, *183*, 154–163. [CrossRef]

109. Justina, M.D.; Muniz, B.R.B.; Bröring, M.M.; Costa, V.J.; Skoronski, E. Using vegetable tannin and polyaluminium chloride as coagulants for dairy wastewater treatment: A comparative study. *J. Water Process Eng.* **2018**, *25*, 173–181. [CrossRef]

110. Kanneganti, A.; Talasila, M. MoO 3 Nanoparticles: Synthesis, Characterization and Its Hindering Effect on Germination of Vigna Unguiculata Seeds. *Int. J. Eng. Res. Appl.* **2014**, *4*, 116–120.

111. Akhila, A. *Essential oil-Bearing Grasses: The Genus Cymbopogon*; CRC press: New York, NY, USA, 2009.

112. Arnon-Rips, H.; Porat, R.; Poverenov, E. Enhancement of agricultural produce quality and storability using citral-based edible coatings; the valuable effect of nano-emulsification in a solid-state delivery on fresh-cut melons model. *Food Chem.* **2019**, *277*, 205–212. [CrossRef]

113. Chua, S.-C.; Chong, F.-K.; Yen, C.-H.; Ho, Y.-C. Valorization of conventional rice starch in drinking water treatment and optimization using response surface methodology (RSM). *Chem. Eng. Commun.* **2019**, 1–11. [CrossRef]

114. Morantes, D.; Muñoz, E.; Kam, D.; Shoseyov, O.J.N. Highly charged cellulose nanocrystals applied as a water treatment flocculant. *Nanomaterials* **2019**, *9*, 272. [CrossRef]

115. The, C.Y.; Budiman, P.M.; Shak, K.P.Y.; Wu, T.Y. Recent Advancement of Coagulation–Flocculation and Its Application in Wastewater Treatment. *Ind. Eng. Chem. Res.* **2016**, *55*, 4363–4389. [CrossRef]

116. Muylaert, K.; Bastiaens, L.; Vandamme, D.; Gouveia, L. Harvesting of microalgae: Overview of process options and their strengths and drawbacks. In *Microalgae-Based Biofuels and Bioproducts*; Gonzalez-Fernandez, C., Muñoz, R., Eds.; Woodhead Publishing: Cambridge, UK, 2017; pp. 113–132. [CrossRef]

117. Su, T.; Chen, T.; Zhang, Y.; Hu, P. Selective Flocculation Enhanced Magnetic Separation of Ultrafine Disseminated Magnetite Ores. *Minerals* **2016**, *6*, 86. [CrossRef]

118. Chatsungnoen, T.; Chisti, Y. Flocculation and Electroflocculation for Algal Biomass Recovery. In *Biomass, Biofuels and Biochemicals: Biofuels from Algae*; Elsevier: Amsterdam, The Netherlands, 2019; pp. 257–286. [CrossRef]

119. Branyikova, I.; Prochazkova, G.; Potocar, T.; Jezkova, Z.; Branyik, T. Harvesting of Microalgae by Flocculation. *Fermentation* **2018**, *4*, 93. [CrossRef]

120. Freitas, T.K.F.S.; Oliveira, V.M.; de Souza, M.T.F.; Geraldino, H.C.L.; Almeida, V.C.; Fávaro, S.L.; Garcia, J.C. Optimization of coagulation-flocculation process for treatment of industrial textile wastewater using okra (A. esculentus) mucilage as natural coagulant. *Ind. Crops Prod.* **2015**, *76*, 538–544. [CrossRef]

121. Singh, A.; Barman, R.; Anantha Singh, T.S. Effect of thermal Energy on Artificial Coagulation for the Treatment of Wastewater. *J. Energy Res. Environ. Technol.* **2017**, *4*, 117–119.

122. Guo, H.; Hong, C.; Zhang, C.; Zheng, B.; Jiang, D.; Qin, W.J.B.t. Bioflocculants' production from a cellulase-free xylanase-producing Pseudomonas boreopolis G22 by degrading biomass and its application in cost-effective harvest of microalgae. *Bioresour. Technol.* **2018**, *255*, 171–179. [CrossRef]

123. Jindal, M.; Kumar, V.; Rana, V.; Tiwary, A.K. Aegle marmelos fruit pectin for food and pharmaceuticals: Physico-chemical, rheological and functional performance. *Carbohydr. Polym.* **2013**, *93*, 386–394. [CrossRef]

124. Nagappan, S.; Devendran, S.; Tsai, P.-C.; Dinakaran, S.; Dahms, H.-U.; Ponnusamy, V.K.J.F. Passive cell disruption lipid extraction methods of microalgae for biofuel production—A review. *Fuel* **2019**, *252*, 699–709. [CrossRef]

125. Amran, A.; Zaidi, N.S.; Muda, K.; Liew, W.L. Effectiveness of Natural Coagulant in Coagulation Process: A Review. *Int. J. Eng. Technol.* **2018**, *7*, 34–37. [CrossRef]

126. Anku, W.W.; Mamo, M.A.; Govender, P. Phenolic compounds in water: Sources, reactivity, toxicity and treatment methods. In *Phenolic Compounds-Natural Sources, Importance and Applications*; InTechOpen: Rijeka, Croatia, 2017.

127. Mubarak, M.; Shaija, A.; Suchithra, T.V. Flocculation: An effective way to harvest microalgae for biodiesel production. *J. Environ. Chem. Eng.* **2019**, *7*, 103221. [CrossRef]

128. Tran, D.T.; Le, B.H.; Lee, D.J.; Chen, C.L.; Wang, H.Y.; Chang, J.S. Microalgae harvesting and subsequent biodiesel conversion. *Bioresour. Technol.* **2013**, *140*, 179–186. [CrossRef]

129. Behera, B.; Balasubramanian, P. Natural plant extracts as an economical and ecofriendly alternative for harvesting microalgae. *Bioresour. Technol.* **2019**, *283*, 45–52. [CrossRef] [PubMed]

130. Zainorizuan, M.J.; Kumar, V.; Othman, N.; Asharuddin, S.; Yee Yong, L.; Alvin John Meng Siang, L.; Mohamad Hanifi, O.; Siti Nazahiyah, R.; Mohd Shalahuddin, A. Applications of Natural Coagulants to Treat Wastewater—A Review. *MATEC Web Conf.* **2017**, *103*. [CrossRef]

131. Luo, S.; Wu, X.; Jiang, H.; Yu, M.; Liu, Y.; Min, A.; Li, W.; Ruan, R. Edible fungi-assisted harvesting system for efficient microalgae bio-flocculation. *Bioresour. Technol.* **2019**, *282*, 325–330. [CrossRef] [PubMed]

132. Jiang, J.; Jin, W.; Tu, R.; Han, S.; Ji, Y.; Zhou, X. Harvesting of Microalgae Chlorella pyrenoidosa by Bio-flocculation with Bacteria and Filamentous Fungi. *Waste Biomass Valorization* **2020**. [CrossRef]

133. Mezzari, M.P.; da Silva, M.L.B.; Pirolli, M.; Perazzoli, S.; Steinmetz, R.L.R.; Nunes, E.O.; Soares, H.M. Assessment of a tannin-based organic polymer to harvest Chlorella vulgaris biomass from swine wastewater digestate phycoremediation. *Water Sci. Technol.* **2014**, *70*, 888–894. [CrossRef] [PubMed]

134. Barrado-Moreno, M.M.; Beltrán-Heredia, J.; Martín-Gallardo, J. Removal of Oocystis algae from freshwater by means of tannin-based coagulant. *J. Appl. Phycol.* **2016**, *28*, 1589–1595. [CrossRef]

135. Cancela, Á.; Sánchez, Á.; Álvarez, X.; Jiménez, A.; Ortiz, L.; Valero, E.; Varela, P. Pellets valorization of waste biomass harvested by coagulation of freshwater algae. *Bioresour. Technol.* **2016**, *204*, 152–156. [CrossRef]

136. Banerjee, C.; Ghosh, S.; Sen, G.; Mishra, S.; Shukla, P.; Bandopadhyay, R. Study of algal biomass harvesting through cationic cassia gum, a natural plant based biopolymer. *Bioresour. Technol.* **2014**, *151*, 6–11. [CrossRef]

137. Link, K.P.; Dickson, A.D. The preparation of d-galacturonic acid from lemon pectic acid. *J. Biol. Chem.* **1930**, *86*, 491–497.

138. Horst, W.J.; Wagner, A.; Marschner, H. Mucilage protects root meristems from aluminium injury. *Zeitschrift für Pflanzenphysiologie* **1982**, *105*, 435–444. [CrossRef]

139. Muñoz, L.A.; Cobos, A.; Diaz, O.; Aguilera, J.M. Chia seeds: Microstructure, mucilage extraction and hydration. *J. Food Eng.* **2012**, *108*, 216–224. [CrossRef]

140. Voiniciuc, C.; Schmidt, M.H.-W.; Berger, A.; Yang, B.; Ebert, B.; Scheller, H.V.; North, H.M.; Usadel, B.; Günl, M. MUCILAGE-RELATED10 produces galactoglucomannan that maintains pectin and cellulose architecture in Arabidopsis seed mucilage. *Plant Physiol.* **2015**, *169*, 403–420. [CrossRef] [PubMed]

141. Jouki, M.; Mortazavi, S.A.; Yazdi, F.T.; Koocheki, A. Characterization of antioxidant–antibacterial quince seed mucilage films containing thyme essential oil. *Carbohydr. Polym.* **2014**, *99*, 537–546. [CrossRef] [PubMed]

142. Thanatcha, R.; Pranee, A.J.I.F.R.J. Extraction and characterization of mucilage in Ziziphus mauritiana Lam. *Int. Food Res. J.* **2011**, *18*, 201–212.

143. Ghanem, M.E.; Han, R.-M.; Classen, B.; Quetin-Leclerq, J.; Mahy, G.; Ruan, C.-J.; Qin, P.; Perez-Alfocea, F.; Lutts, S. Mucilage and polysaccharides in the halophyte plant species Kosteletzkya virginica: Localization and composition in relation to salt stress. *J. Plant. Physiol.* **2010**, *167*, 382–392. [CrossRef]

144. Li, J.; Liu, Y.; Luo, J.; Liu, P.; Zhang, C. Excellent lubricating behavior of Brasenia schreberi mucilage. *Langmuir* **2012**, *28*, 7797–7802. [CrossRef]

145. Koocheki, A.; Mortazavi, S.A.; Shahidi, F.; Razavi, S.M.A.; Kadkhodaee, R.; Milani, J. Optimization of mucilage extraction from Qodume shirazi seed (*Alyssum homolocarpum*) using response surface methodology. *J. Food Process Eng.* **2010**, *33*, 861–882. [CrossRef]

146. Yang, X.; Dong, M.; Huang, Z. Role of mucilage in the germination of *Artemisia sphaerocephala* (Asteraceae) achenes exposed to osmotic stress and salinity. *Plant Physiol. Biochem.* **2010**, *48*, 131–135. [CrossRef]

147. Nayak, A.K.; Pal, D.; Pradhan, J.; Hasnain, M.S. Fenugreek seed mucilage-alginate mucoadhesive beads of metformin HCl: Design, optimization and evaluation. *Int. J. Biol. Macromol.* **2013**, *54*, 144–154. [CrossRef]

148. Behrouzian, F.; Razavi, S.M.A.; Phillips, G.O. Cress seed (*Lepidium sativum*) mucilage, an overview. *Bioact. Carbohydr. Diet. Fibre* **2014**, *3*, 17–28. [CrossRef]

149. Gani, P.; Mohamed Sunar, N.; Matias-Peralta, H.; Abdul Latiff, A.A. Effect of pH and alum dosage on the efficiency of microalgae harvesting via flocculation technique. *Int. J. Green Energy* **2017**, *14*, 395–399. [CrossRef]

MDPI

St. Alban-Anlage 66

4052 Basel

Switzerland

Tel. +41 61 683 77 34

Fax +41 61 302 89 18

www.mdpi.com

Water Editorial Office

E-mail: water@mdpi.com

www.mdpi.com/journal/water

* 9 7 8 3 0 3 6 5 3 2 7 4 5 *